International Conference on e-Health 2018, International Conference on ICT, Society, and Human Beings 2018 and International Conference on Web Based Communities and Social Media 2018

Held at the Multi-Conference on Computer Science and Information Systems (MCCSIS 2018)

Madrid, Spain
17 – 20 July 2018

Editors:

Mario Macedo
Piet Kommers

ISBN: 978-1-5108-7014-7

Printed from e-media with permission by:

Curran Associates, Inc.
57 Morehouse Lane
Red Hook, NY 12571

Some format issues inherent in the e-media version may also appear in this print version.

Printed by Curran Associates, Inc. (2018)

For permission requests, please contact International Association for Development of the Information Society (IADIS) at the address below.

International Association for Development of the Information Society (IADIS)
Rua Sao Sebastiao Da Pedreira
No. 100, 30, 1050-209
Lisbon, Portugal

Phone: 351 21 3151373
Fax: 351 21 3151244

secretariat@iadis.org

Additional copies of this publication are available from:

Curran Associates, Inc.
57 Morehouse Lane
Red Hook, NY 12571 USA
Phone: 845-758-0400
Fax: 845-758-2633
Email: curran@proceedings.com
Web: www.proceedings.com

INTERNATIONAL CONFERENCES ON

E-HEALTH 2018

ICT, SOCIETY AND HUMAN BEINGS 2018

and

WEB BASED COMMUNITIES AND SOCIAL MEDIA 2018

part of the

MULTI CONFERENCE ON COMPUTER SCIENCE AND INFORMATION SYSTEMS 2018

PROCEEDINGS OF THE INTERNATIONAL CONFERENCES

E-HEALTH 2018

ICT, SOCIETY AND HUMAN BEINGS 2018

and

WEB BASED COMMUNITIES AND SOCIAL MEDIA 2018

Madrid, Spain
JULY 17 - 19, 2018

Organised by
IADIS
International Association for Development of the Information Society

Volume Editors:
Mário Macedo and Piet Kommers

Computer Science and Information Systems Series Editors:
Piet Kommers and Pedro Isaías

Associate Editor: Luís Rodrigues

TABLE OF CONTENTS

FULL PAPERS

E-HEALTH

SHORT PAPERS

E-HEALTH

ICT, SOCIETY, AND HUMAN BEINGS

REFLECTION PAPERS

POSTERS

E-HEALTH

WEB BASED COMMUNITIES AND SOCIAL MEDIA

DOCTORAL CONSORTIUM

E-HEALTH

FOREWORD

These proceedings contain the papers and poster of the 10th International Conference on e-Health 2018, the 11th International Conference on ICT, Society, and Human Beings 2018 and of the 15th International Conference Web Based Communities and Social Media 2018, which was organised by the International Association for Development of the Information Society, from 17 - 19 July, 2018. These conferences are part of the Multi Conference on Computer Science and Information Systems 2018, 17 - 20 July, which had a total of 617 submissions.

The use of ICTs (Information and Communication Technologies) in Healthcare Services is the main mechanism to improve efficiency and effectiveness. The e-Health (EH) 2018 conference aims to draw together information systems, practitioners and management experts from all quadrants involved in developing computer technology to improve healthcare quality.

Submissions were accepted under the following 3 main areas in the field of e-health within specific topics:

-Research Issues
-Management Issues
-Applications

The Network period in the evolution of computer technology is very much based on the convergence and integration of three main technologies; computer technology, tele technology and media technology. Telecommunication technology is playing a more and more dominant role in this convergence, especially internet and web technology. Embedded (ubiquitous) computer technology is making the process invisible, and media technologies converge within itself (multimedia and cross media). The convergence process is enforced all the time by smaller, cheaper, and more powerful components.

ICT and its applications are interacting with environments, roles, and processes which can also be modelled by converging circles. The process of social and psychosocial change and ICT from a global perspective is described graphically in the convergence model in figure 1 (Bradley 2006 Routledge) with concepts and their interrelations. Both "convergence" and "interactions" are important features in the model. Read from the left hand side in the model for the titles of some main tracks of the conference:

- Globalisation and ICT:. When technology, economy, norms/values and labour market are converging on a global level, what are the hard questions? When the geographical space in the future will be both global and beyond – including virtual reality (VR) what is the state of art in research? (see the list of key words under ´globalisation´)
- Information and Communication Technology (ICT), next cluster of circles to the left in the figure, what applications contribute to desirable goals in the society?
- When Work Environment, Home Environment, and Public Environment are converging and the work and public issues tend to merge into the private sphere of our homes – what main changes in peoples Life Environment occur?.
- If the Professional Role (Work Life), Private Role (Private Life) and Citizen's Role (Public Life) converge forming a Life Role, what are the main social-psychologial changes?

• Four circles representing Virtual Reality (VR) are marked with dotted lines and are surrounding the set of converging circles. These circles reflect our participation in cyberspace on various levels. To the left part in figure we could talk about Virtual Worlds on the global level. Within the concept of ICT, the steps taken by applied Embedded and ubiquitous technology make technology more hidden to the individual and society as a whole.
Virtual Environments are already a common concept. Finally, we could talk about Virtual Human Roles, which could in more extreme forms be another personality that you play e g avatars. The converging circles are forming a Life Role and new life styles are being shaped.
• Effects on humans become more multi faceted and complex. Research focusing upon the individual is crucial i. e. research on how the use of ICT interacts with and impacts identity, social competence, creativity, integrity, trust, dependency etc. A compass rose (card) for "Effects on Humans" (to the right) is used as a metaphor reminding us of the importance to keep the direction towards desirable human and societal goals and qualities at the development and use of ICT.

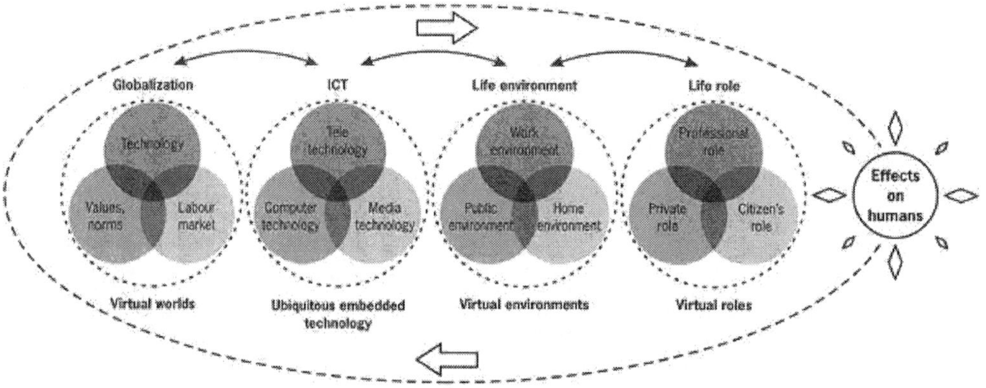

Figure 1. Convergence Model on ICT and Psychosocial Life Environment (Source: Bradley, 2005, 2006)

ICT can provide tools for promoting sustainability (environmental, economic, and social sustainability) but can also be a threat for sustainability. Sustainability as a guiding principle involves system perspective, holism, human aspects, bottom up approach, common good, and equality. A change in focus regarding research and development is taking place. Analysis and design increasingly address both the work process and management connected to the sphere of production life and people's life environment. Analysis and design of ICT and societal systems both at local level and globally become important. What research in the field exists or is needed?

Community research in a broad sense comes to the fore – both physical and virtual communities. There is also a requirement to involve new and additional actors at the deeper and broader integration of ICT in the society (children, elderly, and consumer organisations). Educational programs on Community Informatics and Social Informatics are appearing in many academic institutions. Can a new infrastructure of the society be identified?

The effects of ICT on human beings as well as the interaction between ICT, individuals, and society are all within the focus of this conference. Both analyses of interactions and effects are important. Changes in behaviour, perspectives, values, competencies, human and psychological aspects and feelings are all of interest. Reflections on past, present, and future challenges – especially planning for handling the latter – are encouraged.

Today, computer science and ICT-related disciplines are working more and more together with various behavioural and social sciences including child psychology and developmental psychology. For this reason, the conference pays attention to societal changes, global and more local organisational and institutional changes, changes in values and in lifestyles, as well as individual cognitive effects and changes, motivational and emotional changes. It also appeals to solution-building in terms of desirable goals and actions for reaching a Good Information Society.
In general all types of research strategies are encouraged, and especially cross-disciplinary and multi-disciplinary studies. Case studies, broader empirical field studies, theoretical analyses, cross-cultural studies, scenarios, ethnographic studies, epistemological analyses may all be presented.

The ICT, Society and Human Beings (ICT) 2018 conference addresses in detail seven main aspects:
- Globalization and ICT
- Life environment and ICT
- Life role and ICT
- ICT and effects on humans
- Perspectives on ICT
- Desirable goals and ICT
- Actions for reaching the Good Information Society

The World Wide Web has migrated from information space into opportunities for social communication. Social Media are growing rapidly and play an increasingly important role in the development of Online Communities. They are all about identity, reputation, presence and relationships. Web based communities announce themselves both in your professional and private life through several new media such as LinkedIn, Twitter, Plaxo, etc. In order to keep you up to date with the pace of these new technological developments this Conference offers a dedicated overview and informative discussion on today's most relevant issues in new media for social life on the web.

Social Media are growing rapidly and play an increasingly important role in the development of Online Communities. Social Network Sites and Web-based communities announce themselves both in your professional and private life through new media such as Facebook, LinkedIn, Twitter, Plaxo, etc. Social media allow more dynamic roles in participation, virtual presence and online communities. These new ways to communicate via online social media have great societal effects and are motivating the creation of best practices to help individuals, corporations and authorities to make the best of it. It raises the awareness of the growing impact of social media and the influence of web based communities in today's users / consumers behavior; many organizations spend an increasing share of their budget in online social marketing strategies.

The mission of the Web Based Communities and Social Media (WBC) 2018 conference is to publish and integrate scientific results and act catalytically to the fast developing culture of web communities, while helping to disseminate and understand the latest developments social media and their impact.

Submissions were accepted under the following 6 main topics:
- The History, Architecture and Future of Virtual Communities
- Cyborgs, Teleworking, Telemedicine, Art Games and Learning Communities
- Virtual Communities for People with Special Needs
- Group Processes and Self-Organization
- Expanding Markets through Virtual Communities
- Collaborative Technologies
- Social Media

These conferences received 126 submissions from more than 27 countries. Each submission has been anonymously reviewed by an average of five independent reviewers, to ensure that accepted submissions were of a high standard. Consequently only 20 full papers were approved which means an acceptance rate of 16%. A few more papers were accepted as short papers, reflection papers and poster. An extended version of the best papers may be published in the IADIS International Journal on Computer Science and Information Systems (ISSN: 1646-3692) and/or in the IADIS International Journal on WWW/Internet (ISSN: 1645-7641) and also in other selected journals, including journals from Inderscience.

Besides the presentation of full, short and reflection papers and a poster, the conference also included three keynote presentations from internationally distinguished researchers and one Special Talk by Prof. Eugene Kaluzniacky. We would therefore like to express our gratitude to Associate Professor George (Giorgos) Kontaxakis, Department of Electronic Engineering, Information Processing and Telecommunications Center, Universidad Politécnica de Madrid (UPM), Spain, Prof. Rosa M. Carro, Head of the Department of Computer Engineering, Universidad Autónoma de Madrid, Spain and Ana Fernández-Pampillón Cesteros, Professor, Universidad Complutense de Madrid, Spain for accepting our invitation to be keynote speakers.

This volume has taken shape as a result of the contributions from a number of individuals. We are grateful to all authors who have submitted their papers to enrich the conference proceedings. We wish to thank all members of the organizing committee, delegates, invitees and guests whose contribution and involvement are crucial for the success of the conference.

Last but not the least, we hope that everybody had a good time in Madrid, and we invite all participants for the next edition of these conferences, that will be held in Porto, Portugal.

Mário Macedo, Universidade Atlântica, Portugal
e-Health 2018 Program Chair

Piet Kommers, University of Twente, The Netherlands
ICT, Society and Human Beings 2018 & Web Based Communities and Social Media 2018 Program Chair

Piet Kommers, University of Twente, The Netherlands
Pedro Isaías, The University of Queensland, Australia
MCCSIS 2018 General Conference Co-Chairs

Madrid, Spain
July 2018

PROGRAM COMMITTEE

E-HEALTH PROGRAM CHAIR

Mário Macedo, Universidade Atlântica, Portugal

ICT, SOCIETY AND HUMAN BEINGS
&
WEB BASED COMMUNITIES AND SOCIAL MEDIA PROGRAM CHAIR

Piet Kommers, University of Twente, The Netherlands

MCCSIS GENERAL CONFERENCE CO-CHAIRS

Piet Kommers, University of Twente, The Netherlands
Pedro Isaías, The University of Queensland, Australia

E-HEALTH

COMMITTEE MEMBERS

Abdel-Badeeh M.Salem, Ain Shams University, Egypt
Adel Taweel, Kings College/Birzeit University, United Kingdom/Palestine
Alain Pruski, Lcoms Université de Lorraine, France
Alexander Hoerbst, UMIT - University for Health Sciences, Austria
Andreas Schrader, University of Lübeck, Germany
Anna Bianchi, Politecnico di Milano, Italy
Antonio Lanatà, University of Pisa, Italy
Asa Smedberg, Stockholm University, Sweden
Athina Lazakidou, University of Peloponnese, Greece
Bennoor Kazi, National Institute of Diseases of Chest & Hospital, Bangladesh
Bert-Jan van Beijnum, University of Twente, Netherlands
Bobak Jack Mortazavi, Texas A&M University, USA
Bridget Kane, Karlstad University, Sweden
David Isern, University Rovira I Virgili, Spain
Edward de Quincey, Keele University, United Kingdom
Elizabeth Krupinski, Emory University, USA
Elske Ammenwerth, UMIT - University for Health Sciences, Austria
Eric Campo, LAAS-CNRS, France
Farid Naït-Abdesselam, Paris Descartes University, France
Georg Duftschmid, Medical University of Vienna, Austria
Hermie Hermens, Roessingh Research and Development, Netherlands
Jan Aidemark, Linnéuniversitet, Sweden
Jan Vejvalka, Charles University, Czech Republic
Jaspaljeet Singh, Universiti Tenaga Nasional, Malaysia
Jesuk Ko, Gwangju University, South Korea
José Massa, Universidad Nacional del Centro de la Provincia de Buenos Aires, Argentina
Laurent Billonnet, Ensil - Université de Limoges, France
Lenka Lhotska, Czech Technical University in Prague, Czech Republic
Lina F. Soualmia, LITIS-TIBS EA 4108, France
Lucy Hederman, Trinity College Dublin, Ireland
Malina Jordanova, Bulgarian Academy of Sciences, Bulgaria

Mario Cannataro, University "Magna Græcia" of Catanzaro University, Italy
Maurice Mars, University of KwaZulu-Natal, South Africa
Michael Schumacher, University of Applied Sciences and Arts Western Switzerland, Switzerland
Mohammad Pourhomayoun, California State University Los Angeles, USA
Mohy Uddin, King Abdullah International Medical Research Center, Saudi Arabia
Nadine Vigouroux, IRIT, CNRS 5505, Paul Sabatier Univesrity, Toulous, France
Nathalie Souf, University Paul Sabatier, France
Nikolaos Mastellos, Imperial College London, United Kingdom
Ovidiu Stan, Technical University of Cluj Napoca, Romania
Paivi Hamalainen, The National Institute for Health and Welfare, Finland
Panos Liatsis, The Petroleum Institute, United Arab Emirates
Patricia Sánchez González, Universidad Politécnica de Madrid, Spain
Pedro Pablo Escobar, Intelymec Group, UNCPBA, Argentina
Peter Kokol, University of Maribor, Slovenia
Roberto Hornero, University of Valladolid, Spain
Sandro Scattareggia Marchese, Signo Motus Srl, Italy
Silvia Alayon, Universidad de La Laguna, Spain
Sylvain Giroux, Université de Sherbrooke, Canada
Tacha Serif, Yeditepe University, Turkey
Vaclav Kremen, Czech Institute of Informatics, Robotics and Cyber, Czech Republic
Vicente Traver Salcedo, Universitat Politècnica de València, Spain
Yiqiang Chen, Institute of Computing Technology, China

ICT, SOCIETY AND HUMAN BEINGS

COMMITTEE MEMBERS

Ana Isabel Paraguay, CIDACS/FIOCRUZ-BA, Brazil
Arthur Tatnall, Victoria University, Australia
Cecile K. M. Crutzen, Open University of the Netherlands, Netherlands
Eila Jarvenpaa, Aalto University, Finland
Eleni Berki, University of Tampere, Finland
Elspeth Mckay, Rmit University, Australia
Elvis Mazzoni, University of Bologna, Italy
Fabio Cassano, Università Degli Studi Di Bari, Italy
Farshad Badie, Aalborg University, Denmark
Ilkka Arminen, University of helsinki, Finland
Jan Meyer (Pr.M), Faculty Member, GSB&GL, NWU, South Africa
Jorge Franco, Sao Paulo University, Brazil
Kai Kimppa, University of Turku, Finland
Lech Zacher, Kozminski Business School, Poland
Manuela Repetto, INDIRE, Italy
Monica Banzato, University of Ca' Foscari - Venice, Italy
Nestor Garay-Vitoria, Universidad del País Vasco/Euskal Herriko Uniberts, Spain
Nicola Doering, Ilmenau University of Technology, Germany
Paivi Fadjukoff, University of Jyväskylä, Finland
Pertti Jarvinen, University of Tampere, Finland
Robert Pintér, Corvinus University of Budapest, Hungary
Sarai Lastra, Universidad Del Turabo, Puerto Rico
Stina Giesecke, Aalto University, Finland
Virve Siirak, Tallinn University of Technology, Estonia

WEB BASED COMMUNITIES AND SOCIAL MEDIA

COMMITTEE MEMBERS

KEYNOTE LECTURES

EMPATHY, CREATIVITY, TECHNOLOGICAL INNOVATION: KEYS IN THE PATH OF DESIGNING HEALTHCARE BEYOND 2020

A/ Professor George (Giorgos) Kontaxakis,
Department of Electronic Engineering, Information Processing and Telecommunications Center,
Universidad Politécnica de Madrid (UPM), Spain

ABSTRACT

Barack Obama was here in Madrid on July 6, 2018, invited to the Spain Summit 2018 on Technological Innovation and Circular Economies organized by Advanced Leadership Foundation. The man who was the 44th President of the USA between 2008 and 2016 confessed his surprise by the speed of the digital revolution that global economy is currently witnessing. This global change is comparable in its potential to change the world to the Industrial Revolution. However, while the latter was developed over almost two centuries, this current one occurs in just 10 to 20 years.

Healthcare experiences the same pace of digital revolution and, to paraphrase Obama´s words at the Madrid Summit, it remains uncertain whether most of our current healthcare systems are ready for this. Change through technological innovation is imperative in order to position current healthcare to existing demands for a personalized, preventive and participative care. To achieve this, it is essential that carers better understand who their patients are and then create an entire ecosystem of care based on creative solutions that would drive the changes required.

Empathy to view patients as persons first and then as patients, along with creative mindsets to drive technological innovation, can be powerful tools for looking beyond the status quo and building the healthcare systems beyond 2020. In this talk we will review creative techniques, such as design thinking which, when applied to healthcare, can lead to a human focus and an openness to generate and test lots of ideas to find more innovative, far-reaching solutions to current challenges and help paint clear pictures of the near future in healthcare. We will then outline opportunities for innovation in healthcare, discuss the need of digital transformation in this sector and address the complexities behind delivering high-quality patient treatment outcomes without leaving out economic considerations.

EMOTIONS AND INCLUSION IN E-LEARNING: STUDENT MODELLING AND ADAPTIVE E-TRAINING

Rosa M. Carro,
Head of the Department of Computer Engineering,
Universidad Autónoma de Madrid, Spain

ABSTRACT

In the e-learning context, knowing the students' needs, abilities or preferences is essential to provide personalised learning and training resources and tools. Student modeling and, moreover, the automatic acquisition of information to fill in student models, are key issues in this direction. For example, the students' emotions have an impact on their learning performance. Therefore, acquiring and using this information to adapt some elements to each student (such as the type of activities to be proposed to him at a certain time) may influence on the student's engagement and achievements. Information about the student sentiments and emotions can be extracted from different sources: messages they write, interactions with others in social networks, words they use when writing essays, etc.

The use of adaptation techniques allows the variation of e-learning and e-training resources and strategies automatically to each student to better fit their specific needs during the process, according to the information available in the student model. This is useful for all the students, especially for those with special needs, who can benefit from adaptive e-training specifically designed to favor their social and labour inclusion.

Finally, all the student interactions within e-learning and e-training systems can be stored and analysed with different purposes, such as: tracking the learning processes; predicting failure risk; or getting more information about the student needs or features and, therefore, enriching the student models.

In this conference, Rosa M. Carro will present some advances on automatic student model acquisition, such as the research done in the GHIA group at the Computer Science Department of the Universidad Autonoma de Madrid, in which social networks, written texts or mouse movements, among others, are analysed to get information about the student sentiments, emotions, personality, etc. In addition, she will describe some applications developed to support adaptive e-learning, e-training or e-assistance, either for all or for specific collectives, such as adults with cognitive limitations, children with social adaptation difficulties, people with TDAH or persons with myoelectric prosthesis. Some of these applications have been used by students from Promentor, a program leaded by Prodis Foundation, devoted to helping people with intellectual disabilities in their personal development and labor inclusion. Finally, she will comment on some additional ways to use all the information stored about the user interactions.

CAN THE QUALITY OF MATERIALS REALLY BE MEASURED BEFORE THEY ARE USED?

HOW CAN WE KNOW IF AN EDUCATIONAL MATERIAL IS OF QUALITY BEFORE IT IS USED?

Ana Fernández-Pampillón Cesteros,
Professor, Universidad Complutense de Madrid,
Spain

ABSTRACT

In a digital world that offers an endless number of educational materials, suppose you are a teacher, or a developer of virtual courses, or a person responsible for the acquisition of educational material, or a student. How do you select the best one or the best ones? Or, if you are creating material for a virtual course, how do you ensure that it will have a minimum quality, i.e. that it will be minimally efficient? Basically by measuring its quality.

However, measuring how good a material is as difficult as measuring how good a work of art is: there is a significant subjectivity factor. Then, is it possible to measure objectively how good a digital educational material is for teaching and learning in a digital environment?

This keynote presentation will present an answer, the answer developed, over three years, by Working Group 12 of National Technical Committee 71 of the Spanish Association for Standardization (UNE). This answer has been, since June 2017, a national standard: UNE 71362 "Standard on Quality of Digital Educational Materials". It was created on the basis of consensus among representatives of the academic, business and public administration sectors. It has also been tested for validity, reliability and usability. It is, therefore, a good starting point for further reflection, testing and evolution towards increasingly effective learning materials.

SPECIAL TALK

21ST CENTURY ICT PERSONNEL - HUMANS OR HUMAN BEINGS?

Prof. Eugene Kaluzniacky,
Applied Computer Science, University of Winnipeg, 515 Portage Avenue, Winnipeg, Manitoba, Canada

ABSTRACT

The widespread, originally European personal development program developed by PRH International (Personnalite et Relations Humaines) firstly identifies three human psychological components: the intellect, the feelings and the body (with its sensations). However, it also defines another, a source of deep, pure and only positive energy which PRH has called the *inner being*. The goal of life for a *human being*, according to this view, is for the three *human* components to connect to the *being* component and transmit the being's vivifying energy through the body, feelings and intellect according to one's unique personality.

PRH has further declared that at this point in human evolution, "our humanity is extremely underdeveloped in the area of in-depth riches", instead focusing sometimes nearly exclusively on intellectual functioning. American Yale University surgeon and health teacher, Dr. Bernie Siegel, says "most of us are in touch with this inner self intermittently, if at all. I got to know this perfect core self through meditation..however you get there, you'll know you've arrived". Other respected scientists and psychologists have commented and promoted this deeper reality. Australian management consultant Sebastian Salicru in fact promotes for the current business world the development of a deeper, *spiritual* consciousness that supersedes Emotional Intelligence. He explains Intellectual Intelligence is about thinking, Emotional Intelligence is about feeling, and a secularly defined Spiritual Intelligence is about *being*. More extensive information, even a Web portal by Richard Griffiths, is increasingly available on Spiritual Intelligence.

The world of ICT and more specifically information systems development has also alluded to this perspective. Kent Beck, the founder of Extreme Programming, has dismissed the idea that IT work is "some Vulcanic world of pure rationality" and emphatically declared that, in IT work, "to be successful *we must learn to grow our whole selves"*. In his textbook Introduction to Information Systems —Essentials for the Internetworked Enterprise_(1998), author James O'Brien tells his readers, *"May you love the Light within you, and in everyone you meet, and in everything you experience."*

Information and Communication Technology (ICT) professionals encompass a variety of persons in an ever-expanding variety of positions. There is a variety of work in this area that requires not only linear, sequential thinking. Web site design, tutorial development, animation output may involve a synergy of different thinking styles as well as emotional input. In addition, teamwork efforts are increasingly commonplace, necessitating effective communication. End-user interaction, to be productive may often require a combination of thinking, feeling and intuition. As well, a psychologically well-balanced ICT worker is in a position for ongoing creativity, tolerance of ambiguity and more effective stress management. Thus, it is not hard to see that ICT requires a fully human approach to study and work in the discipline. However, with recent advances, we can now recognize that such an approach can be most effective if the human approach transcends the one-tier awareness of the human person: that of intellect, emotion and body and rests in a much more complete two-tier awareness comprising intellect, emotion and body rooted in the inner being (i.e, O'Brien's 'light within'). Thus, the ICT professional is being called to development as a full *human being* and not just a "human doing".

With this perspective, the speaker will first outline the two-tier vision of a *human being* with quotations from respected scholars and professionals in various fields promoting such a level of psychological awareness. He will then show how such a deeper awareness can be discovered and cultivated by ICT professionals worldwide. He will show notable advantages in various areas of ICT work that will be enjoyed by an ICT professional engaged in his work from his fullest self. Also, he will suggest and encourage areas for academic and action research regarding the impact of two-tier self awareness on various areas of ICT work. Finally, he will encourage ICT professional organizations to initiate a group approach to psychological self-awareness within ICT work, possibly leading to a new area called *Psycho-Informatics*.

KEYWORDS

Personal Growth, Spiritual Intelligence, Whole Self, Inner Being, Two-Tier Awareness, ICT Professionals

Full Papers

EVALUATION OF SATISFACTION AND USABILITY OF A CLINICAL DECISION SUPPORT SYSTEM (CDSS) TARGETED FOR EARLY OBSTETRIC RISK ASSESSMENT AND PATIENT FOLLOW-UP

Ever A. Torres Silva[1], Ivan F. Luna Gomez[1], Jose F. Florez Arango[1,2], Jack W. Smith[1,2],
Sebastian U. Ocampo[1] and Junior E. Hidalgo[1]
[1]Universidad Pontificia Bolivariana, Medellín, Colombia
[2]Texas A&M University, USA

ABSTRACT

Introduction: Usability of computer systems, and satisfaction of their users, has been related to user performance, especially for clinical decision support systems (CDSS). The purpose of this paper is to present an evaluation of satisfaction and usability of a web-based CDSS called HADA. HADA was developed to improve antenatal healthcare by assisting in obstetric risk assessment.
Methods: This research used an observational study of blocks of user interactions with multiple treatments. Twenty physicians of two rural hospitals were HADA was tested performed eleven independent tasks, using it. Subjects answered the NASA TLX self-assessment questionnaire1 after performing each task. Once they finished the set of tasks, they answered the Post-Study System Usability Questionnaire (PSSUQ) 2,3.
Results: The data was not normally distributed; therefore non-parametric approaches were used for analysis. Workload was measured to be low and satisfaction to be high. Main tasks related to CDSS scored in the lowest range of findings. There was a statistically significant difference of Overall NASA TLX cognitive load medians between tested users at both hospitals. Median differences among types of physicians were not statistically significant. PSSUQ results showed a high overall acceptance. System usefulness was highest on the independent subscales, followed by interface quality and then information quality.
Discussion: Experiences of users does not seem to be a factor that affected user evaluations of HADA. The reduction on workload on similar tasks suggested a learning effect. HADA's match to current workflow seems to be a factor in perception of workload and acceptance of the tool. The users at the hospital that had no comparable processes experienced a higher workload and lower acceptability.
Conclusions: HADA proved to be a tool that generally results in low additional workload for clinicians independent of their experience, and is rated high in acceptability. The design and development techniques used for HADA have resulted in a tool with low workload and high acceptability. The evaluation methods used have proven useful in this clinical system context and confirm previous research that usability of computer systems, user satisfaction levels and user performance, are related.

KEYWORDS

Maternal Mortality, Perinatal Mortality, Clinical Decision-Support System, Observational Study, Usability of Computer Systems, User Satisfaction Levels, User Performance

1. INTRODUCTION

Maternal and neonatal mortality in developing countries, are still above United Nations and World Health Organization Goals (OMS 2013; WHO 2017). Main causes of maternal deaths in these countries are hemorrhage, sepsis and hypertension. Most complications and deaths are related to poor quality of care during pregnancy and childbirth (Dalaba et al. 2015). The Pan-American Health Organization's traditional methods for prenatal risk identification are paper-based and depends highly on providers' memory (CLAP/SMR n.d.). Regardless of multiple efforts, early identification of prenatal risks and subsequent management is still challenging (Horton & Levin 2016).

Clinical Decision Support Systems (CDSS) are tools that offer a great opportunity to improve clinical care by providing evidence-based information, recommendations and alerts for patient specific clinical features (Berner 2009). Our research team developed and evaluated a CDSS called HADA that was designed to improve clinical outcomes in antenatal care. HADA was designed for being embedded within the workflow of healthcare providers.

HADA stands for "toolset for antenatal assistance and diagnosis" or "Herramientas para la Asistencia y el Diagnóstico Antenatal" in Spanish. HADA is a web-based application for providers (primary care and specialists), developed following using an evidence-based human-centered design method UFuRT (Zhang & Butler 2007) , UFuRT stands for User, Functions, Representation and Tasks Analysis. Such human-centered methods are intended to improve usability and user acceptance of systems with minimum additional resource investment required before and during system development over development without such methods. HADA development was publically funded and it is not a commercial system. There is the potential for the platform to be used by other entities; interested parties should contact the lead author.

HADA is intended to connect primary care providers and specialists using a knowledge-driven computational engine that aids in obstetric risk identification and patient specific care plans. The system can generate recommendations for lab tests and medications and, requests additional data needed for better decisions. In addition, HADA enables telemedicine asynchronous encounters when required limited only by the capacity of data interchange with participating institution's electronic medical records (EMRs).

The main functions of HADA are: Patient's Admission/Registration, Scheduling, Historical view of clinical encounters, dynamically generated data entry forms, Computerized physician order entry (CPOE) supported by a CDSS, Risk scoring tool and Asynchronous communication with specialized care providers.

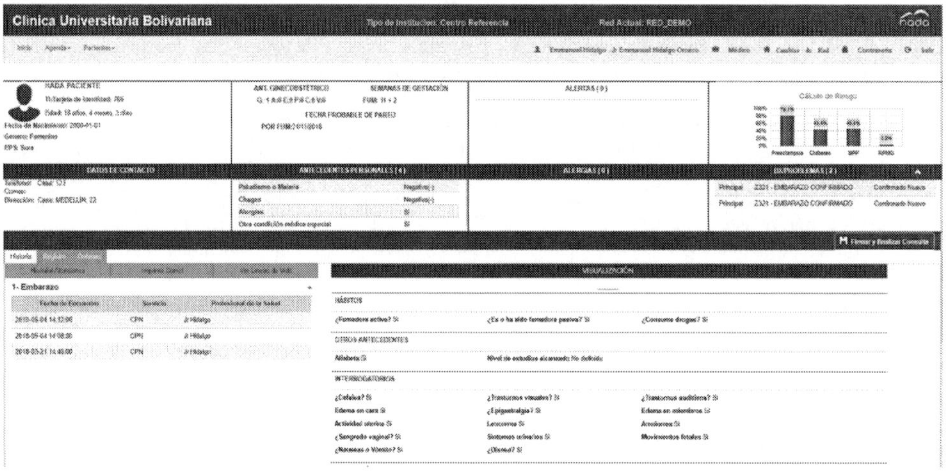

Figure 1. Historical View of Clinical Encounters

Figure 1 shows the layout of the HADA patient specific web-based dashboard. This provides an overview of the essential past and potentially useful future clinical care information for this specific synthetic patient used for testing. This demonstrates some of the functionality described above for the system. Note on the upper right the display of patient specific risk information generated by the DSS component of the system. Some other information on this dashboard is generated specifically for this patient based on the DSS components of the system. For example, the DSS component recommends potential follow-up testing and procedures.

Figure 2 is another patient specific web-based dashboard showing some overlap with the information and functions in Figure 1. However, this dashboard now includes additional interface elements enabling detailed order entry for this specific synthetic patient.

Usability is an attribute of a product and related to user performance. A usable product needs to present a consistent user interface, provide user control, appropriate visual presentation, error-handling and correction, reduce user memory-load, have tasks matching user expectations and provide proper guidance and support (Keinonen n.d.). There are several approaches to measure usability and this paper will use a task centered observational approach based on user perceptions of the system on a structured set of fixed tasks.

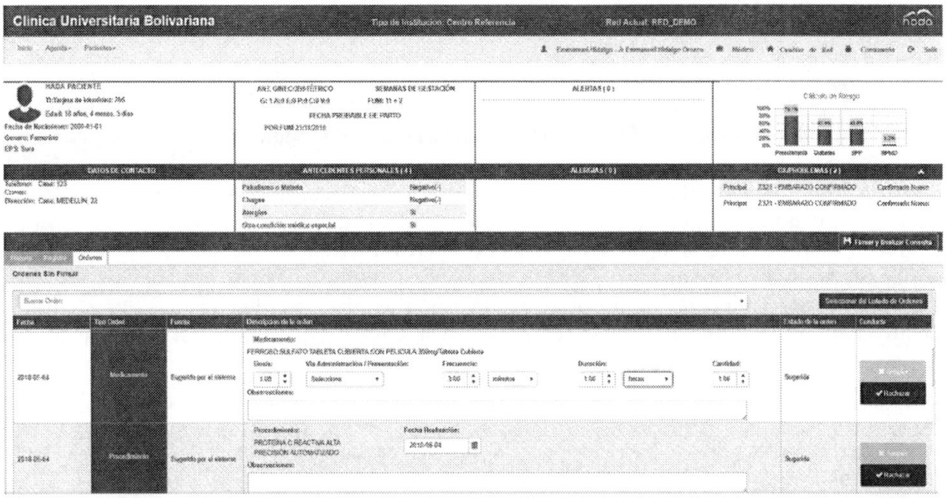

Figure 2. Computerized Provider Order Entry Layout

The purpose of this paper is to present the evaluation results of usability and acceptance observations of our CDSS in pilot tests with a group of intended health care professional users in two rural hospitals in Colombia, S.A. under controlled study conditions. The methods used for the evaluation have proven consistently valuable for these situations and the results presented should generalize when these methods are appropriately applied to other systems. The general relationships between usability, user satisfaction and user performance are also likely to generalize to other systems using the same development and evaluation methods.

2. METHODS

This research was an observational study using blocks of tasks in two settings with multiple treatments (fixed tasks). Sample size was established by convenience. Physicians of two rural hospitals (task blocks) performed eleven fixed independent tasks (treatments), using the information system for obstetric risk assessment and clinical decision support called HADA (intervention). In the evaluation subjects answered the NASA TLX (task load index) self-assessment questionnaire(Hart 1988) after performing each task, and, once they finished performing the set of tasks, answered the Post-Study System Usability Questionnaire (PSSUQ) (Lewis 1995; Lewis n.d.). The ethics committee for health research at the Universidad Pontificia Bolivariana approved the human study protocol.

A standalone version of the HADA system was installed on an Intel® core i5 @ 2,5 GHZ laptop computer with 8GB in RAM, running Windows 10 (Microsoft Corp.), with screen size 13" at 1366 x 768 pixels' resolution. This was the only machine used to run all evaluations.

User inclusion was constrained to be those available physicians at the time of the investigators visit to an institution. Only physicians that were in service at the time of the investigator visit were included. Subjects were relieved of additional clinical activities at the time of the study.

The study takes into considerations that hospitals may have several types of physicians caring for our target population of pregnant women: physicians on social service and fully registered physicians. In Colombia, physicians on social services cannot practice medicine freely, since they have no medical license; they serve between 6 months and a year after graduation of medical school providing limited health services in their social service position before obtaining medical license.

The blocks of tasks used in the study were performed at:

- Hospital 1: is in the southwest part of Antioquia, about 157 km away from Medellin, it is a 4-hour drive. This hospital attends a population over 45.000 inhabitants. The hospital has ten registered physicians and nine doctors in social service. In addition, the hospital serves three small rural health centers. The hospital has an electronic health record (EHR) without a module for antenatal healthcare.

- Hospital 2: is located also in the southwest part of Antioquia, about 97 km away from Medellin it is a 2-hour drive. It serves over 20.653 people. The hospital has one fully licensed physician and seven doctors in social service. The hospital has an EHR with a module for antenatal healthcare.

Evaluations took place in a dedicated room at each hospital. Researchers obtained informed consents from every physician before the initiation of each evaluation. Each subject conducted each task a single time. Subjects had one hour to complete the full set of fixed tasks and the associated evaluations. All the subjects were shadowed while performing the set of tasks by two researchers.

Tasks (treatments) performed using HADA were classified into two domains: administrative and clinical. Administrative tasks are not part of the principal domain of clinical decision, but are a prerequisite to be able to identify and enroll patients. In Table 1 is the task list and the domains of the tasks used in the evaluation.

Table 1. List of tasks and domain of tasks evaluated in HADA

Task	*Domain*
1. **Patient creation**: user fills all the required fields to add a new patient to HADA.	Administrative
2. **Edit the data of a patient:** user edits at least one parameter of a previously created patient.	
3. **Schedule a patient appointment**: user uses agenda component of HADA, then selects a date and time to create an appointment for a previously created patient and assigns himself or herself as a physician of the patient.	
4. **First clinical encounter and clinical data entry**: user goes to the monitoring screen and clicks the "attend" button on the patient, after that, completes the entire questionnaire of the first prenatal encounter.	Clinical
5. **Review of suggested medical orders**: Once user saves the first questionnaire, HADA will recommend some medication and procedure orders, which the user reviews and then decides from the possible actions which to accept or reject.	
6. **Follow-up clinical encounter and data entry**: user attends the same new patient for a second time after finalization of their first encounter and signing of the encounter registry.	
7. **Lab result entry**: user is required to fill lab results accordingly to the generated orders in the first encounter.	
8. **Modify a prescription**: user saves the second encounter, in the screen of orders, makes a new one, and modifies the default prescription fields.	
9. **Request a second opinion/expert consultation**: evaluator instructs user to generate a request for a second opinion by a specialist within HADA's CPOE tool. User is required to generate a clinical question regarding patient status for the specialist.	Clinical
10. **Review answer of the second opinion/expert consultation**: a simulated answer to the user´s question to the specialist is generated and the user requests, searches for and verifies the answer in the encounters list.	
11. **Identify current patient risks**: evaluator requests user to identify the risk bar graphic in the top of the screen, near basic patient data and the user is asked to think aloud regarding the interpretation of the graph.	

Each subject answered the NASA TLX self-assessment questionnaire after each task. NASA TLX identifies the factors associated with subjective workload variation within and between different types of tasks. The instrument evaluates six independent aspects: mental demand (MD), physical demand (PD), temporal demand (TD), performance (OP), effort (EF), and frustration (FR) level. A discrete scale of 20 steps was established for each NASA TLX dimension. In addition, subjects were required to compare a pair of factors, selecting the member of each pair that provided the most significant source of workload variations in the tasks, there are 15 possible pairs comparison. NASA TLX was calculated with the standard equation:

$$NASATLX = \frac{MD * w_{MD} + PD * w_{PD} + TD * w_{TD} + OP * w_{OP} + EF * w_{EF} + FR * w_{FR}}{15}$$

Equation 1. NASA TLX

In Equation 1, W_x represents the number of times that x workload dimension was selected on the pair comparisons. W_x ranges goes from 0 to 5, and the sum of all W_x equals 15. Maximum possible value for NASA TLX is 20. Maximum value for the weighted source of workload (x times $W_{x)}$ is 100. The lower the value of NASA TLX the lower the workload generated by the task, and for the analysis, a lower value means a better result.

At the end of the set of tasks, the subjects answered the PSSUQ. PSSUQ consist of 19 questions about usability, each question use a seven-level Likert scale. In the scale, 1 = "totally agree", and 7 = "totally disagree". PSSUQ consist of questions grouped in three subscales and a global value.

1. System Usefulness (SYSUSE) is calculated as the average of questions 1 -8
2. Information Quality (INFOQUAL) is calculated as the average of questions 9-15
3. Interface Quality (INTERQUAL) is calculated as the average of questions 16-18
4. Global value (Overall) is calculated as the average of all questions 1-19

Lower values on PSSUQ means better usability. To present values in direct relation as percentage of usability the following formula was applied:

$$PSSUQ_{subscale\%} = -(100 * \text{subscale value} - 700)/6$$

Equation 2. PSSUQ Subscale Calculation on Percentage

Both NASA TLX and PSSUQ were answered by the participants using paper forms and then transcribed by the research team using double data-entry into an electronic file.

Researchers used software R 3.2.3 GUI 1.66 Mavericks build 7060 for data analysis. Statistical significance was established as $p<0.05$. Given the sample size of 20 subjects, researchers used non-parametric methods for these calculations

3. RESULTS

All participants (n=20) enrolled in the study that provided informed consent completed the evaluation. Physicians on social service correspond to 80 % of the sample (n=15). Males represented 55 % of the sample, while females represented 45 %. All subjects completed the task set and evaluations within the 1-hour time limit. See Table 2 for details of the general characterization of the health professionals in both hospitals that were sampled.

Table 2. Characterization of Subjects (n=20)

	Sex		*Age range*			*Type of Physician*	
	Male	**Female**	**20-25**	**26-30**	**>30**	**Social Service**	**Fully Licensed**
Hospital 1	7	6	5	5	3	8	5
Hospital 2	4	3	4	3	0	7	0
Total	11	9	9	8	3	15	5

3.1 NASA TLX

Observed values of NASA TLX were not normally distributed. Overall NASA TLX values had skewness of 2.44 and kurtosis 9.70. Medians were used as a central tendency statistics. Kruskal-Wallis and Dunn's test were used to identify median differences and source of difference.

See Figure 3 NASA TLX score was 3.67. There was a statistically significant difference of Overall NASA TLX medians between both hospitals of 3.71 (chi-squared = 13.781, df = 1, p-value = 0.0002054). Median difference among type of physician was not statistically, significant; it was 0.84 (chi-squared = 0.6984, df = 1, p-value = 0.4).

Workload analysis by task suggested statistical independence of task results. Accordingly with the Kruskal-Wallis, test results were calculated (chi-squared = 24.901, df = 10, p-value = 0.005536). Task analysis is summarized in Figure 4 and Figure 5. The Dunn Test was the method used for comparisons of paired medians. Applying the Dunn test finds that all tasks are statistically different from other tasks, but there is not a single task different to all others.

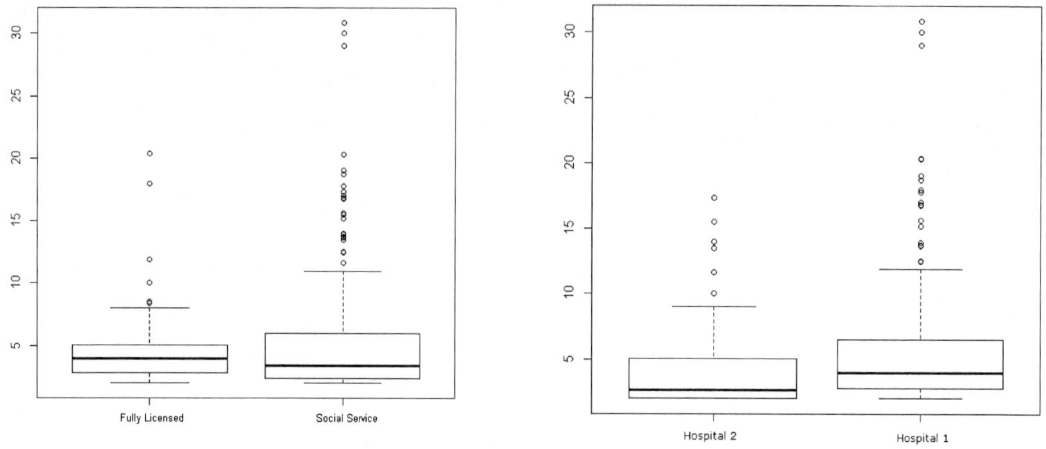

Figure 3. Boxplot Overall NASA TLX by Type of Physician (a) and Hospital (b)

The most demanding tasks were task 4 "**First clinical encounter and clinical data entry" (5.07)** and task 5 **"Suggested medical orders review" (5.03).** It is important to notice that task 6 "**Follow-up clinical encounter and clinical data entry**" uses the same interface as task 4, and there is a significant reduction on workload by 1.57 (z= 1.7; p =0.045) compared with task 4. Task 4 is also the most distinct task of all tasks; the tasks that did not show statistical difference between task 4 are task 5 (z= 0,75; p =0.47) and task 7 (z= 1,6; p= 0.056). See Figure 4.

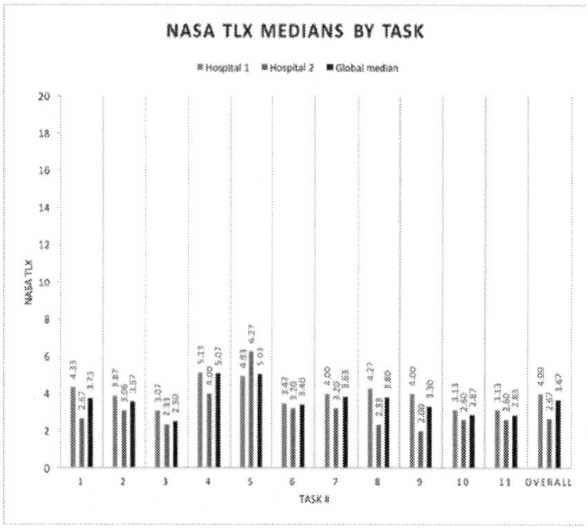

Figure 4. NASA TLX Medians by Task

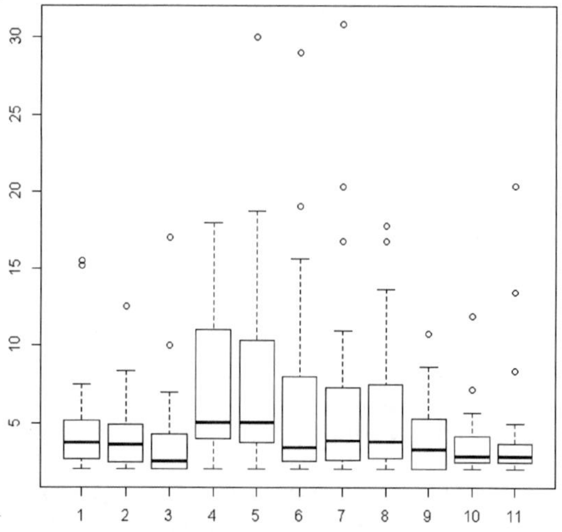

Figure 5. Boxplot of Overall NASA TLX Medians by Task

The task with lower workload was task 3 "**Schedule a patient appointment**" (2.5). The clinical task with lower workload was task 11 "**Identify current patient risks**" (3.67). There is no statistical significant difference (z= 0.51; p=0.31).

The only task that generated more workload at Hospital 2 compared with Hospital 1 was task 5 "**Suggested medical orders review**" but with no statistical difference (chi-squared = 0.1571, df=1, p = 0.6918).

Analyzing individual sources of workload by weighted values showed that the main workload source is temporal demand (TD) 6. Additional dimensions were scored lower: mental demand (MD) 5, effort (EF) was 4, observed performance (OP) 3, Physical demand (PD) 2 and frustration (FR) 0.5. TD is most noticeable at task 4 with value 20.

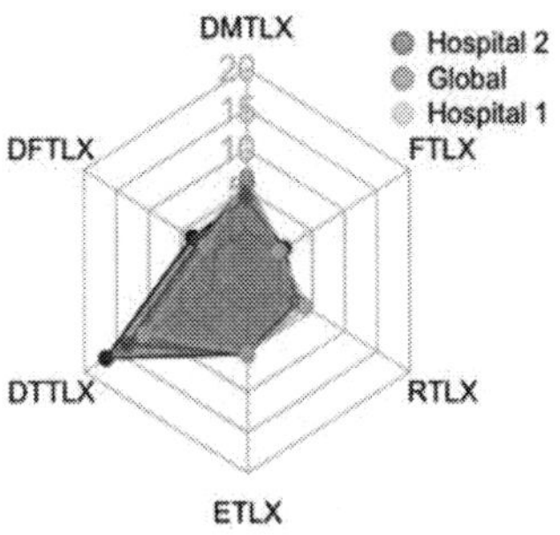

Figure 6. NASA TLX Weighted Workload Sources for Task 4 by Hospitals

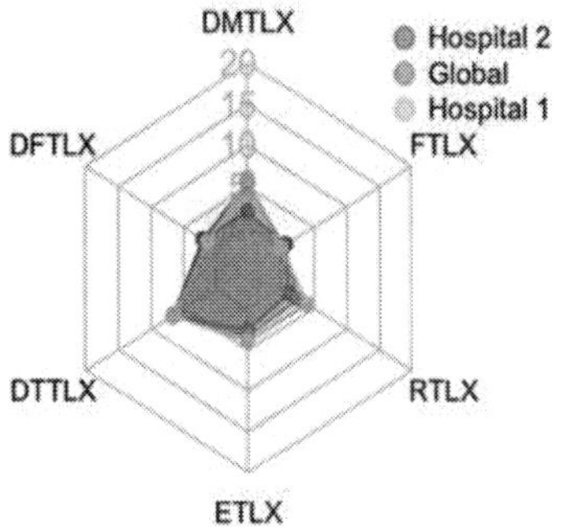

Figure 7. NASA TLX Weighted Workload Sources for Task 6 by Hospitals

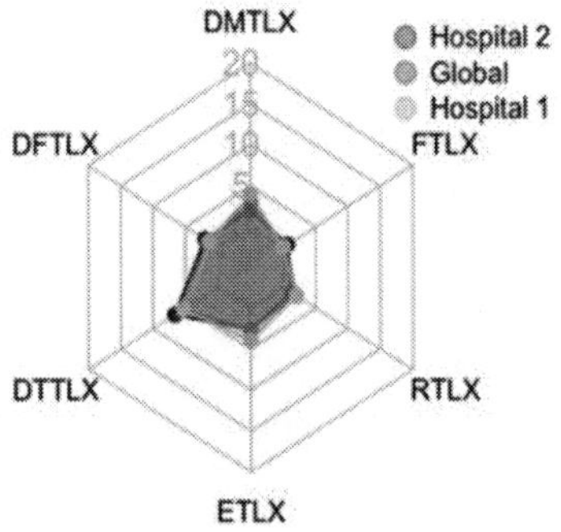

Figure 8. NASA TLX Weighted Workload Sources for Task 7 by Hospitals

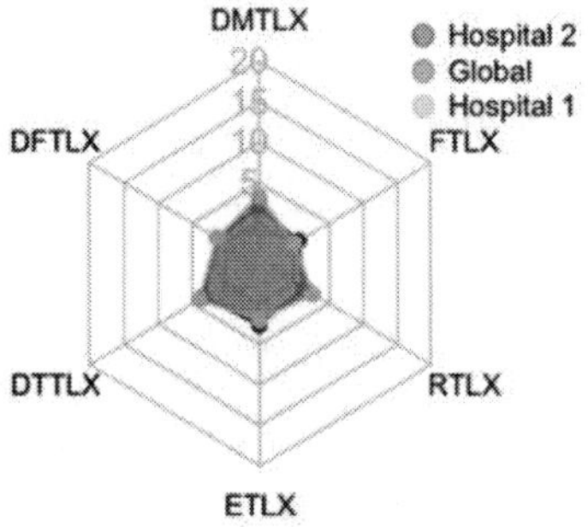

Figure 9. NASA TLX Weighted Workload Sources for Task 11 by Hospitals

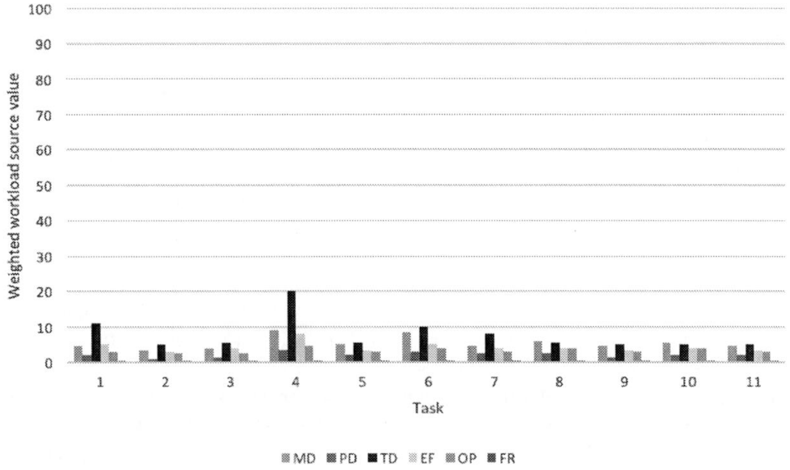

Figure 10. Medians of Weighted Individual Workload Sources by Task

3.2 PSSUQ

Values of PSSQUQ are not normally distributed; overall the PSSQU results had a skewness of 1.75 and kurtosis of 5.87. Medians and non-parametric analysis where used. PSSUQ values showed a high overall acceptance of the tool by scoring 1.52 (91.28%). System usefulness is the higher independent subscale with a score of 1.25 (95.83%), followed by interface quality with a score 1.67 (88.89%), finally information quality had a score 1.86 (85.71%). There are no differences between physicians' types on scores. There is a statistical significant difference between hospitals in the OVERALL scale of 0.952 (chi-sqr=4.79, df=1, p=0.028), but not in any of the subscales. See Table 3.

Table 3. PSSQU results global and subscales, by physicians' type and hospital

		OVERALL (%)	*SYSUSE (%)*	*INFOQUAL(%)*	*INTERQUAL(%)*
	Median	1.52 (91.28)	1.25 (95.83)	1.86 (85.71)	1.67 (88.89)
	Skewness	1.75	2.21	1.56	1.09
	kurtosis	5.87	7.73	5.38	2.99
Type of Physician	K-W chi-sqr,	0.325	0.241	0.049	0.908
	df	1	1	1	1
	p	0.57	0.6235	0.82	0.34
Hospital	Hospital 1 median	2.052	1.625	2.42	2
	Hospital 2 Median	1.1	1.125	1.14	1
	K-W chi-sqr,	4.79	2.62	3.73	3.59
	df	1	1	1	1
	p	0.028	0.105	0.053	0.058

4. DISCUSSION

The present study was a pilot study exploring acceptability and usability of HADA for prenatal care embedded in care workflow. The size of the user sample was limited by availability of subjects. There were no dropouts among participants. A single subject behaved as an outlier, a contextual analysis allowed us to identify external factors from those in the experiment affecting her performance.

Human centered design and its impact on human performance when using systems following these design and development methods is of rising interest (Harte et al. 2017). Our use of such an approach to HADA design and development resulted in positive evaluations and confirmation of the relation between performance, usability and satisfaction with the system. UFuRT as a human center design methodology for improving usability, was useful in the system development based on the statistical analysis, since the rate of acceptability and usability of HADA on evaluated tasks is above 90 %. A more detailed analysis will be present in a separated paper on predictive factors of success using human centered design reflected in human performance for the application HADA.

Source of variation on overall NASA TLX results could be explained by different modules installed on the information systems from both hospitals under the evaluation. Also, each hospital has a unique care workflow. A larger study with different hospitals and different settings is required to identify clearly the source of the differences as suggested by Brunner (Brunner et al. 2017).

The reduction on workload on similar tasks (they share the same basic interface and most of the same content) suggests a learning factor. This is evident in task 4 (**First clinical encounter and clinical data entry**) and task 6 (**Follow-up clinical encounter and data entry**). In future studies an experimental design with repeated measures will be needed to isolate these effects.

A higher workload on task 5 (**Review of suggested medical orders**), in hospital 2, could be related to the lower clinical experience of subjects, and a more conservative approach, since all in hospital 2 were nonregistered physicians on a rural service, while in hospital 1 there were both registered and nonregistered types of physicians. A larger study will allow the isolation of this potential effect. However, there were no differences in global usability scores comparing both physician groups. User experience does not seem to be a factor in the reported evaluations of use of HADA.

Overall, all workload sources scored low, reflecting the effects of the human center design method used for system development. Principal workload source was time demand. Future studies will benefit from timing measures for each task and a contrast analysis with these findings.

According to the results, both workload and usability are in desirable ranges.

5. CONCLUSION

HADA proved to be a tool that generates low additional workload for clinicians independent of their experience, with high acceptability. The participants in the research evaluated interfaces and information quality positively. This is promising for future research, since findings suggest a lower risk of system adoption failure, and increased success rate on system adoption for systems with better usability, satisfaction and user performance. A larger study is needed to isolate sources of observed differences and identify sources of variation in the measured parameters of the system.

ACKNOWLEDGEMENTS

Universidad Pontificia Bolivariana, Ruta N, Sistema general de regalías, Proyecto "Desarrollo de Soluciones en CTi para Telesalud en el Departamento de Antioquia".

REFERENCES

Berner, E., 2009. Clinical Decision Support Systems: State of the Art, Eta S. Berner, Ed.D. ed, No. 09-0069-EF. Alabama at Birmingham, Maryland: Agency for Healthcare Research and Quality.

Brunner, J., Chuang, E., Goldzweig, C., Cain, C.L., Sugar, C., Yano, E.M., 2017. User-centered design to improve clinical decision support in primary care. Int. J. Med. Inf. 104, 56–64. https://doi.org/10.1016/j.ijmedinf.2017.05.004

Dalaba, M.A., Akweongo, P., Aborigo, R.A., Saronga, H.P., Williams, J., Blank, A., Kaltschmidt, J., Sauerborn, R., Loukanova, S., 2015. Cost-Effectiveness of Clinical Decision Support System in Improving Maternal Health Care in Ghana. PLOS ONE 10, e0125920. https://doi.org/10.1371/journal.pone.0125920

Gong, Y., Zhang, J., 2005. A human-centered design and evaluation framework for information search. AMIA Annu. Symp. Proc. AMIA Symp. AMIA Symp. 281–285.

Harte, R., Quinlan, L.R., Glynn, L., Rodríguez-Molinero, A., Baker, P.M., Scharf, T., ÓLaighin, G., 2017.

Human-Centered Design Study: Enhancing the Usability of a Mobile Phone App in an Integrated Falls Risk Detection System for Use by Older Adult Users. JMIR MHealth UHealth 5, e71. https://doi.org/10.2196/mhealth.7046

Hart, S.G., 1988. Development of NASA-TLX (Task Load Index): Results of Empirical and Theoretical Research. North-Holland.

Horton, S., Levin, C., 2016. Cost-Effectiveness of Interventions for Reproductive, Maternal, Neonatal, and Child Health, in: Black, R.E., Laxminarayan, R., Temmerman, M., Walker, N. (Eds.), Reproductive, Maternal, Newborn, and Child Health: Disease Control Priorities, Third Edition (Volume 2). The International Bank for Reconstruction and Development / The World Bank, Washington (DC).

Keinonen, T., n.d. Arteology: Usability Of Artefacts [WWW Document]. URL http://158.132.155.107/posh97/private/ResearchMethods/158.htm (accessed 6.13.17).

Lewis, J., n.d. Computer System Usability Questionnaire: Psychometric Evaluation and Instructions for Use [WWW Document]. URL http://garyperlman.com/quest/quest.cgi (accessed 5.18.17).

Lewis, J.R., 1995. IBM computer usability satisfaction questionnaires: Psychometric evaluation and instructions for use. Int. J. Hum.-Comput. Interact. 7, 57–78. https://doi.org/10.1080/10447319509526110

OMS, 2015. Objetivo de Desarrollo del Milenio 5: Mejorar la salud materna [WWW Document]. WHO. URL http://www.who.int/maternal_child_adolescent/topics/maternal/mdg/es/ (accessed 12.3.15).

OPS/OMS CLAP - Tecnologías Perinatales [WWW Document], n.d. URL http://www.paho.org/clap/index.php?option=com_content&view=article&id=80:tecnologias-perinatales&Itemid=240&limitstart=1&lang=es (accessed 5.18.17).

WHO | Global Strategy for Women's, Children's and Adolescents' Health, 2016-2030 [WWW Document], n.d. . WHO. URL http://www.who.int/life-course/partners/global-strategy/global-strategy-2016-2030/en/ (accessed 5.18.17).

Yuan, M.J., Finley, G.M., Long, J., Mills, C., Johnson, R.K., 2013. Evaluation of User Interface and Workflow Design of a Bedside Nursing Clinical Decision Support System. Interact. J. Med. Res. 2. https://doi.org/10.2196/ijmr.2402

Zhang, J., Butler, K.A., n.d. UFuRT: A Work-Centered Framework and Process for Design and Evaluation of Information Systems.

Zhang, J., Walji, M., n.d. For EHR Usability Research Goals for TURF.

Zhang, J., Walji, M.F., 2011. TURF: toward a unified framework of EHR usability. J. Biomed. Inform. 44, 1056–1067. https://doi.org/10.1016/j.jbi.2011.08.005

ELDERLY PATIENTS WITH COGNITIVE IMPAIRMENT: APPLICATION OF A NEW SERVICE AND ORGANIZATIONAL MODEL SUPPORTED BY DIGITAL SOLUTIONS

Pietro Davide Trimarchi[1], Paolo Locatelli[2], Federica Cirilli[2], Emanuele Tomasini[1], Pedro A. Moreno[3] and Rodrigo Pérez-Rodríguez[3]
[1]*Fondazione IRCCS Don Carlo Gnocchi, Italy*
[2]*Fondazione Politecnico di Milano, Italy*
[3]*Fundación para la Investigación Biomédica - Hospital Universitario de Getafe, Spain*

ABSTRACT

Ageing has significant impacts on the organization of healthcare services, especially for elderly people affected by cognitive impairment. Recent findings suggest that physical exercise, a controlled diet, and intellectual and social stimulation are modifiable factors to prevent and slow down cognitive decline. Moreover, having information on the disease, creating strategies to minimize cognitive and functional disabilities, managing behavioral disturbances and reducing psychological distress can reduce the patients' and caregiver burden. This paper defines a new service and organizational model enabled by a digital solution that meets these needs, and describes the related application within the DECI initiative, a Horizon 2020 project focused on elderly patients affected by Mild Cognitive Impairment and Mild Dementia. The paper describes the application of the model in two countries (Italy and Spain), which have similar characteristics, and reports the main lessons learnt, giving some advices on how successfully implement the model in other similar healthcare systems.

KEYWORDS

Service Model, Organizational Model, Cognitive Impairment, Digital Solution, Application

1. INTRODUCTION

Cognitive impairment and chronic disease are the main causes for disability among the elderly. Recent findings suggest that some cognitive functions remain stable across the life-span while others gradually decrease. Working memory, processing information speed and the capacity for encoding end retrieval seem to suffer from the passing of time (Borson, 2010). On the other hand, cognitive functions like short-term memory, autobiographical memory, semantic knowledge and emotional processing remain unchanged during late life.

The term dementia describes a chronic and progressive dysfunction of the cognitive and functional abilities due to a pathology on the cortical and/or sub-cortical brain structures. Alzheimer disease is the most common type of dementia (Alzheimer's Association, 2018). In 2011, people with dementia accounted for 35.6 million; the World Health Organization (WHO) estimates that in 2030 will be about the double. In 2018, the estimated cost to manage dementia is to be $277 billion. No treatments are currently available to cure or even alter the progressive course of dementia. Moreover, Mild Cognitive Impairment (MCI) (Alzheimer's Association, 2018) has a minimal impact on autonomy in daily life (Albert et al., 2011). Recent findings suggest that physical exercise, a controlled diet, and intellectual and social stimulation are modifiable factors to prevent and slow down cognitive decline (Livingston et al. 2017).

Cognitive impairment is associated with a lower quality of life for both patients and their caregivers, who may suffer from an increase care burden. To reduce this burden, it has been identified that having information on the disease (Miranda et al., 2013), creating strategies to minimize cognitive and functional disabilities, managing behavioral disturbances and reducing psychological distress are of major relevance. It seems that the key needs in the management of people with cognitive problems are more linked to the ability to cope with the disease rather than handling instrumental problems (Prince et al., 2016).

These needs can be met through a new service and organizational model based on shared information management, coordination of different care stakeholders in planning and executing services, empowerment of citizens in their homes while keeping them monitored. Moreover, digital solutions are key to enable new service models (Christensen et al., 2009), indeed an emerging approach perceives the relevance of digital environments and their capability to surround the patient at home, directly providing features or acting as a virtual "extension" of the care providers. However, the success of the application of digital solutions is determined by the adherence to the specific and real needs of the patients and their caregivers (Killin LOJ, Russ TC, Surdhar SK, et al., 2018).

This paper describes an innovative service and organizational model supported by a digital solution for elderly people affected by cognitive impairment and its application in two pilot sites, underlining the challenges that have arisen with the implementation of this innovative model. Obtained findings have allowed eliciting a set keys to successfully implement the DECI approach in other healthcare systems.

2. RESEARCH OF A NEW MODEL

The research of this new service and organizational model supported by a digital solution has been conducted in the scope of the DECI (Digital Environments for Cognitive Inclusion) project, an Innovation Action funded by the European Commission (H2020-PHC-2014-2015/H2020-PHC-2014-single-stage). The core focus of the project revolves around the definition of an innovative model for elderly people affected by MCI and MD, granting higher levels of quality of life. MCI and dementia among elderly population has been recognized as a huge clinical, social and financial burden (Wimo et al. 2017), and new models of intervention that could assist patients and caregivers in their path as well as reducing assistance load, mainly in charge of family member caregivers (Alzheimer's Association 2018), are crucial today, also due to the lack of effective therapeutic solutions.

The research process started with an analysis of the state-of-the art related to the project horizon: clinical literature on MCI and MD, organizational models for the care and assistance to these patients. Moreover, the researchers analyzed the state-of-art of digital solutions for improving MCI and MD care and assistance, case studies on innovative organizational models and advanced implementation of digital technologies for improving inclusion and for cognitive stimulation. Then, they conducted a deep field analysis of the current leading practices in care and assistance processes, organizational models and use of digital technologies in the project-specific settings.

Starting from the results of these research activities, all partners involved in the DECI project (herein after, the Consortium) organized some co-creation workshops to map the stakeholder's current roles and responsibilities regarding the care of elderly patients affected by MCI and MD and to create an inventory of identified gaps and needs of the target population in the current care models. The main gap in the care services that emerged during these workshops was the lack of a central figure to follow and support the patients throughout the overall care process, especially when they are at home.

For this reason, the Consortium focused the attention in designing a new service and organizational model to meet this need. This model includes the role of the Case Manager (Reilly et al. 2015), a figure to lead the care process and collaborate with the rest of the clinical team in following and monitoring the patients during the whole care process, and providing support in different daily life activities, such as the physical and cognitive training that is part of the therapy. Specifically, the Case Manager: communicates continuously with the patients and the caregivers; shares information about the patient with all the clinical staff that is involved in the care process; organizes and carries out the implementation of the activity protocol (physical and cognitive exercises) defined by the clinical staff; monitors the adherence to the treatment periodically; monitor periodically the level of physical activity through analyzing the data provided by an actigraphy smartwatch.

To translate this model into real scenarios, the Consortium discussed how this new service and organizational model could be supported by digital technologies. The result is the DECI digital solution, composed of different integrated modules that allow remote activity monitoring, physical coaching and cognitive training as well as provide the needed means for the communication between patients, caregivers and clinicians. This solution supports the Case Manager, the clinical staff, the patients and the caregivers through:

1. A Web-based interface for clinical professionals to share information about the patients (e.g. clinical data, assistance project, relevant clinical events). This interface also allows following-up the evolution of the patents in terms of physical activity, cognitive performance and adherence to the treatment;

2. A mobile application for the patients to: interact with the Case Manager and the care staff or access to up-to-date educational material about cognitive decline and other relevant topics; follow a computer-based cognitive stimulation program with several exercises covering Memory, Attention, Executive functions, and other mental abilities. The system records the patients' activity in terms of duration of the sessions and performance; follow a computer-based physical activity program (e.g. stretching, balance, and strength improvement) delivered by a video tutorial. The system records the patients' activity in terms exercises performed and duration;

3. An activity monitoring system worn by the patients, based on an actigraphy smartwatch which allows registering the number of steps, intensity of the physical activity and where it took place (in house or out of the house).

The integrated digital solution has been implemented in four pilot countries involved in the DECI project (Italy, Spain, Israel and Sweden). The following section describes the results emerged from the implementation of DECI in real environments.

3. APPLICATION IN REAL CONTEXTS

The DECI Consortium implemented a comparative, multi-centre, randomized and prospective study to investigate the effect of the application of the DECI model in the assistance of those older adults suffering from MCI and MD and their caregivers. According to the inclusion criteria, patients older than 60, with a diagnosis of MCI (Petersen 2004 criteria) or MD (DSM 5 criteria) and living at home have been recruited. Exclusion criteria were: previous or current major psychiatric illness (e.g. schizophrenia or recurrent major depression episodes), presence of previous or current major neurological illness other than MCI or MD (e.g. stroke or multiple sclerosis), other serious comorbidity (e.g. severe chronic obstructive pulmonary disease or severe heart disease), drug history or alcohol abuse, severe sensory impairments (mainly visual and auditory), history of intellectual disability and/or other developmental diseases, and a life expectancy lower than one year as judged by the clinician. The study was approved by the local Ethical Committee at each participating centre. All participants have given oral and written informed consent to participate. The recruitment phase lasted from June 2017 to January 2018. At the end of the recruitment, a total number of 511 patients were enrolled. The intervention time has been defined to 6 months, so accordingly to the recruitment dates, the last patient to leave the study will do it in August 2018.

Stratification by diagnosis (MCI, MD), age group (60-75, 76-58, +85), gender (male, female) and education level (primary school, secondary school, high school and university degree) was applied for the randomization of the participants in 3 different groups: Control Group (CG), and two intervention groups, namely Intervention Group 1(IG1) and Intervention Group 2 (IG2). Figure 1 shows the general structure of the DECI study.

Patients in the CG are followed according to the usual clinical practice delivered to elderly patients suffering from MCI or MD in the clinical services of the four pilot sites. For the two intervention groups, patients are involved in several activities delivered through the ICT tools available at the patient home.

In IG1, the following activities are implemented:

- An organizational model including the Case Manager and clinical staff specialized in treating elderly patients suffering from cognitive impairment;
- The care staff uses the Web-based interface of the platform to share information about the patient;
- The patient and her/his caregiver have the possibility to use the mobile application to interact with the Case Manager and the care staff and, if needed, to access to up-to-date educational material about cognitive decline.

In IG2, apart from the services offered to IG1, the following are also accessible:

- The patient has the possibility to use the mobile application to follow a program of computer-based cognitive stimulation. The activity protocol is defined by a clinical professional and and adherence monitored by the Case Manager;
- The patient has the possibility to use the mobile application to follow computer-based physical activity program. The activity protocol is defined by a clinical professional and adherence monitored by the Case Manager;
- The patient has the possibility to use an activity monitor based on an actigraphy smartwatch. The recorded data are monitored by the Case Manager through the Web-based access to the platform.

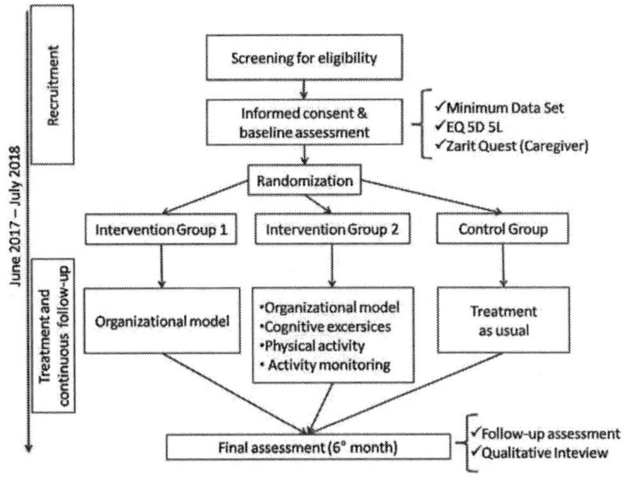

Figure 1. General Structure of DECI Study

The primary endpoints defined for the evaluation of the DECI model:

- Improve the quality of life of the patients, measured using EuroQL-5D-5L (Herdman et al. 2011).
- Improve the quality of life of the informal caregivers, measured using the Zarit questionnaire (Zarit et al. 1986)
- Improve the patients' autonomy in daily living activities, measured using the B-ADL (Katz et al. 1970) and I-ADL (Lawton & Brody 1969).
- Improve the satisfaction of the care staff about information sharing regarding patient management, measured by a specifically designed qualitative interview.
- Improve the patients' cognitive functions or prevent cognitive deterioration, measured with MMSE (Folstein et al. 1975) and Clock Drawing Test (Shulman 2000).

In the following sub-sections, the detailed application of the DECI model in Spain and Italy is described. The choice to report these two currently ongoing clinical pilot implementations, instead of all the four, is related to the following reasons: (1) Italy and Spain have a quite similar Health Services, so that the frame where the pilots are being implemented can be thought as uniform; and (2) Italy and Spain are both Mediterranean countries with comparable elderly populations in terms of age, educational level, lifestyle and digital literacy.

3.1 Spanish Application

One-hundred and fifteen patients have been recruited in Spain to participate in the validation of the DECI solution: 38 were assigned to the CG, 38 to the IG1 and 39 to the IG2. Up today, 13 participants voluntarily left the study. Technology delivered to both IG1 and IG2 consisted in a Vodafone Smart Tab N8 with a 4G SIM card to guarantee connectivity regardless of the location (no Internet connection required in the homes of the participants). Participants assigned to IG2 were also provided with the smartwatch to monitor physical activity. Five clinical professionals, apart from an engineer, are involved in the application of the DECI solution in Spain: 1 geriatrician, 1 psychiatrist, 1 neuropsychologist, 1 psychologist and 1 nurse. These professionals are not full-time dedicated to the application of the solution but integrate it in their daily clinical routine.

In Spain, the application of DECI is currently being developed in Madrid Region, under the collaboration between the application leader, Getafe University Hospital (GUH, in Getafe, Madrid Region), and the Centre for the Prevention of Cognitive Impairment (CPCI, in Madrid city). From an organizational perspective, the Case Manager is based at GUH; a geriatrician is in charge of the general management of all the patients that use the solution. The Case Manager is in direct contact with all the clinical professionals involved in the care process regardless the clinical team (GUH or CPCI). From the CPCI, a multidisciplinary team composed of a psychiatrist, a neuropsychologist and a psychologist oversee the prescription of tailored cognitive stimulation treatments; from GUH, a geriatrician designs physical activity programs depending on the specific needs of the patients. The Case Manager checks the incoming messages in a daily basis and reply them; in case it is required, she contacts the specific clinical professional to perform certain actions (update treatments, etc.). Technology issues are managed by the GUH engineering team: service deployment, supervision of the correct performance of all the modules that compose DECI, monitoring of the battery status of the smartwatches and replacement whenever necessary. The engineering team also coordinates the training provided to patients and caregivers. There are specific sessions for patients participating in IG1 and IG2, with an attendance limit per session of 5 and 3 patients (with caregivers) respectively; IG1 training sessions last 1 hour, while those devoted for patients in IG2 last 1 hour and a half. The training sessions are given by trained a nurse from GUH who is experienced in treating older patients to facilitate learning; patients and caregivers can attend sessions whenever if they feel they need a refreshment. The scheduling of these educational activities is fixed to take place every Wednesday at the CPCI; for refreshment sessions patients must call in advance to book a place. Furthermore, patients and caregivers have a helpdesk available (Monday to Friday from 8:00h to 15:00h), also managed by the engineering team, that they can use (via phone call) in case they have a technical query.

The overall workflow to include a new patient in the DECI care loop is the following: (1) pre-screening visit for the application of the inclusion/exclusion criteria and signature of the informed consent; (2) inclusion visit (clinical anamnesis, neuropsychological and functional examination); (3) randomization; (4) training and delivery of the material; (5) follow-up; and (6) exit visit (clinical and satisfaction assessment). The workflow to apply after the piloting phase, when DECI is integrated within the general healthcare delivery would be very similar, taking out the randomization and substituting the exit clinical visit for periodic face-to-face follow up visits. Figure 2 depicts the Spanish DECI workflow (dashed lines indicate that those actions and transitions only take place during the piloting phase).

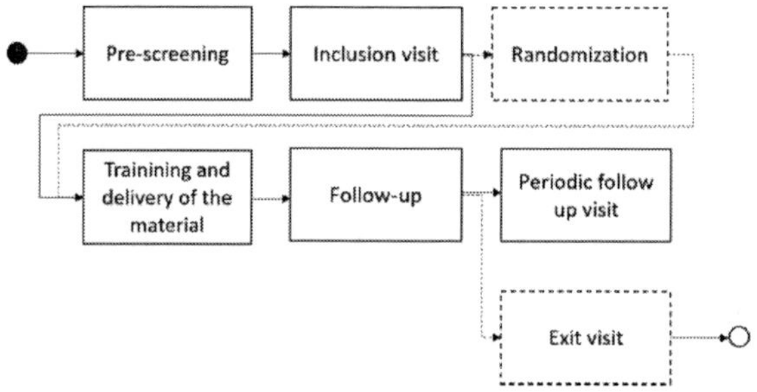

Figure 2. DECI Workflow in Spain

3.2 Italian Application

The Italian pilot site is running in the Istituto Palazzolo–Fondazione Don Gnocchi (IP–FDG), based in Milan. One-hundred and sixty-one older adults have been enrolled to participate in the validation of the DECI model: 54 were assigned to the CG, 54 to the IG1, and 53 to the IG2. Stratification of participants has been applied as previously described. Up today, 18 participants voluntarily left the study, but13 of them remained available for the follow-up assessment at the end of the study. Technology delivered to both IG1 and IG2 consisted in a 10.1 inches screen tablet with a 4G SIM card to guarantee connectivity regardless of the location (no Internet connection required in the homes of the participants). If participants already had and used a personal technological device (both tablet or PC) with Internet connection, they were allowed to use them in order to avoid the need to learn the use of a new device. Participants assigned to IG2 were also provided with a smartwatch to monitor physical activity. The Italian DECI team is composed by 5 professionals: 1 geriatrician, 2 neuropsychologists, 1 social worker and 1 engineer. The social worker, who has the role of the Case Manager, is full-time dedicated. The other professionals are not full-time dedicated to the application of the solution but integrate it in their daily clinical routine.

Participants have been mainly enrolled from the memory clinic of the IP–FDG (application leader). Further enrolment has been carried out in collaboration with the memory clinics of two other hospitals in the Milan area (Ospedale Policlinico and Istituto Redaelli). From an organizational perspective, the Case Manager is based at IP–FDG: a social worker in charge of the general management of all the patients that use the solution. The Case Manager is in direct contact with all the clinical professionals involved in the care process. The geriatrician designs physical activity programs depending on the specific needs of the patients. The neuropsychologist designs the tailored cognitive stimulation programs. The Case Manager checks the incoming messages in a daily basis and reply them and coordinates the specific clinical professional to perform certain actions (update treatments, modify exercises, etc.) in case it is required. Moreover, the Case Manager checks the adherence to the prescribed therapies on a weekly basis and, if long lasting inactivity is detected (more than two weeks), a kind reminder is sent (reasons for this lack of adherence are currently being investigated). The Case Manager also monitors the battery status of the smartwatches and organizes a replacement whenever necessary. Technology issues are managed by the DECI engineer: tablet configuration, service deployment, supervision of the correct performance of all the modules, solution of technical problems related to the Internet connection and updates of the tablets operating system and apps. The Case Manager also coordinates the training provided to the patients and the caregivers belonging to IG1 and IG2. There are specific sessions, on a single patient basis (patient and her/his caregiver if present), for patients participating in each intervention group. IG1 training sessions last 45 minutes, while those devoted for patients in IG2 last 75-90 minutes. IG1 training sessions are given by the Case Manager while those for the IG2 are given by the neuropsychologist. In order to facilitate learning, the methodology followed during the training have been previously defined with a neuropsychologist expert in cognitive decline and dementia. The DECI team remains available if patients and caregivers need further training sessions to refresh learned information. Furthermore, patients and caregivers have a helpdesk available (Monday to Friday from 10:00h to 17:00h), managed by the Case Manager, that they can use (via phone call) in case they have a technical or procedural query. Assistance is provided in case of technical problems using the delivered devices. If a specific technical problem cannot be solved by the Case Manager, the engineer is involved. The assistance can be required sending a message through the DECI integrated platform or by a dedicated phone number.

In Italy, the overall workflow to include a new patient in the DECI care loop is the following: (1) screening of the IP–FDG memory clinical database to find patients meeting the inclusion criteria; (2) contact of the pre-selectedpatients by phone or during a clinical follow-up visit, to show them the study and propose participation; (3) inclusion visit (signature of the informed consent, clinical anamnesis, neuropsychological and functional examination) to those patients that accept to participate (inclusion and exclusion criteria arefurther checked); (4) randomization; (5) training and delivery of the material; (6) follow-up (if needed); and (7) exit visit (clinical and satisfaction assessment). The workflow to apply after the piloting phase, when DECI will be integrated within the general healthcare delivery, would be very similar, taking out the randomization and substituting the exit clinical visit for periodic face-to-face follow up visits. Figure 3 depicts the Italian DECI workflow (dashed lines indicate that those actions and transitions only take place during the piloting phase).

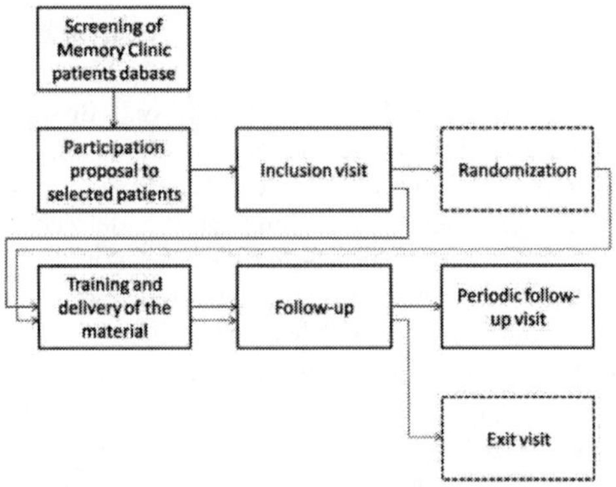

Figure 3. DECI Workflow in Italy

3.3 Results and Discussion

Preliminary feedback in both the Spanish and the Italian experiences is promising for the IG2 scenario, where patients are full-equipped. Indeed, patients assigned to IG2 have good adherence, regularly using the system and much appreciating the possibility of having a tailored cognitive and physical therapy designed and supervised by a clinical professional. Also, the clinicians are expressing satisfaction in the application of this innovative model, highlighting the feature that allows them remotely to analyze how patients are performing and update the prescription accordingly; clinical professionals also appreciate having fresh information from the Case Manager about the evolution of the patients.

Preliminary feedback from participants assigned to IG1 is not that satisfactory: in general, these patients started using the system, checking the available online information and using the messaging service to communicate with the case manager, but use decreases with time.

The experimental validation of the DECI solution has allowed the research and innovation team learn what the most important points are to bring this new organizational model to other healthcare systems.

The first lesson learnt comes from the low satisfaction expressed by participants assigned to IG1, where they were only benefiting from the organizational model without having access to the cognitive and physical stimulation activities. This low satisfaction is discouraging the use of the DECI system, which implies that the new ways offered to communicate with the clinical professionals and having up-to-date information about the own disease is not enough for the patients to feel they are getting a substantial benefit from using the DECI solution (see Killin et al. 2018 for similar results).

The second lesson learnt is related to the crucial role of the Case Manager. In a complex and differentiated system with several actors that deliver services to patients suffering from cognitive impairment, the Case Manager can support both patient/caregiver dyads and clinical staff. For patients and caregivers, the Case Manager organizes, supports and follows the care journey, helping to avoid the drop-out from the clinical path, and working as a pivot between patient/caregiver dyads and the stakeholders that deliver the clinical and social services. For the clinical staff, the Case Manager is a reference figure that plans, coordinates and harmonizes the different interventions, assures continuity of care and promptly detects eventual deteriorations that need intervention.

Finally, the third and last of the main lessons learnt relates to the importance of the training sessions and the helpdesk. Nowadays, technology literacy among the elderly is usually enough to self-learn how to use new mobile applications. For that reason, it is of major importance to structure the training sessions addressing the key aspects that present the highest challenges for the target population in using the provided services; MCI and MD may affect learning abilities, reducing the acquisition of new competencies and discouraging the use of the tools needed for the intervention. For this reason, the training sessions need to be delivered to the caregivers too, who play a very important role in supporting the use of the technological tools. Training can be delivered either by a professional with a technical or clinical profile but understanding gained during the piloting of DECI indicates that previous experience in working with the target population allows increasing the learning efficiency. Furthermore, it is also a priority to have available a helpdesk to solve the most common problems of the older adults when using the material.

4. CONCLUSION

In conclusion, the application of the DECI model in Italy and Spain has allowed defining a series of suggestions for other sites with similar healthcare systems, to successfully implement this innovative model.

The first point to consider is the need of using the full digital solution (organizational platform and cognitive and physical stimulation programs) to support the delivery of the service, since experience indicates that not providing access to therapeutic contents severely impacts adherence. The second suggestion is the enablement of a Case Manager to coordinate the delivered interventions, acting as a reference for both patients and clinical personnel. And the third and last recommendation is the establishment of a competent training system and a helpdesk to support patients and clinicians in using the new digital solution. Only when all these three aspects act together, the potential of this innovative service and model supported by digital technologies for elderly people affected by MCI and MD will be a reality.

DECI is an ongoing initiative that has not yet come to the end. Quantitative data demonstrating the impact of the proposed innovative model in terms of clinical and economic outcomes will be released at the end of the undergoing experimentations in the 4 clinical sites. These results will be used to define next evolutions of the DECI digital solution in order to launch it on the market. This quantitative evaluation will also analyze the implementation of DECI in other healthcare models, different to the Spanish and Italian ones, based on the DECI experience in Sweden and Israel. Moreover, an economic sustainability of the implementations will be analyzed in the four countries. Dedicated papers on the final results will be published by the partners in the next months.

ACKNOWLEDGEMENT

The authors would like to thank the whole Consortium of the DECI project: Fondazione Politecnico di Milano, Consoft Sistemi SpA, Fondazione Don Carlo Gnocchi Onlus (Italy), Centre for Healthcare Improvement–Chalmers University of Technology, Västra Götalandsregionen (Sweden); Hospital Universitario de Getafe–Servicio de Geriatría (Spain); Maccabi Healthcare Services (Israel); Roessingh Research and Development (The Netherlands).

REFERENCES

Albert, M. S., DeKosky, S. T., Dickson, D., Dubois, B., Feldman, H. H., Fox, N. C., ...& Snyder, P. J. (2011). The diagnosis of mild cognitive impairment due to Alzheimer's disease: Recommendations from the National Institute on Aging-Alzheimer's Association workgroups on diagnostic guidelines for Alzheimer's disease. *Alzheimer's & dementia: the journal of the Alzheimer's Association*, *7*(3), 270-279.

Alzheimer's Association. (2018). 2018 Alzheimer's disease facts and figures. *Alzheimer's & Dementia*, *14*(3), 367-429.

Borson, S., 2010. Cognition, aging, and disabilities: conceptual issues. *Physical Medicine and Rehabilitation Clinics*, *21*(2), pp.375-382.

Cesari, M., Demougeot, L., Boccalon, H., Guyonnet, S., Abellan Van Kan, G., et al. 2014. A Self-Reported Screening Tool for Detecting Community-Dwelling Older Persons with Frailty Syndrome in the Absence of Mobility Disability: The FiND Questionnaire. *PLoS ONE,* 9(7), e101745. doi: 10.1371/journal.pone.0101745.

Christensen, C.M., Grossman, J.H. and Hwang, J., 2009. *The innovator's prescription*. McGraw Hill, New York, USA

Folstein, M., Folstein, S.E., McHugh, P.R. 1975. "Mini-Mental State" a Practical Method for Grading the Cognitive State of Patients for the Clinician. *Journal of Psychiatric Research*, 12(3); 189-198.

Hendrich, A.L. Bender, P.S. &Nyhuis, A. 2003. Validation of the Hendrich II Fall Risk Model: A large concurrent case/control study of hospitalized patients. *Applied Nursing Research*, 16(1), 9-21.

Herdman, M., Gudex, C., Lloyd, A., Janssen, M., Kind, P., Parkin, D., Bonsel, G., Badia, X. 2011. Development and preliminary testing of the new five-level version of EQ-5D (EQ-5D-5L). *Quality of Life Research,* 20: 1727–1736.

Katz, S., Down, T.D., Cash, H.R., &Grotz, R.C. 1970. Progress in the development of the index of ADL. *The Gerontologist*, 10(1), 20-30.

Killin LOJ, Russ TC, Surdhar SK, et al. Digital Support Platform: a qualitative research study investigating the feasibility of an internetbased, postdiagnostic support platform for families living with dementia. BMJ Open 2018; 8:e020281. doi:10.1136/ bmjopen-2017-020281.

Lawton, M.P., & Brody, E.M. 1969. Assessment of older people: Self-maintaining and instrumental activities of daily living. *The Gerontologist*, 9(3), 179-186.

Livingston, G., Sommerlad, A., Orgeta, V., Costafreda, S. G., Huntley, J., Ames, D., & Cooper, C. (2017). Dementia prevention, intervention, and care. *The Lancet*, 390(10113), 2673-2734.

Miranda-Castillo, C., Woods, B., &Orrell, M. (2013). The needs of people with dementia living at home from user, caregiver and professional perspectives: a cross-sectional survey. *BMC health services research*, *13*(1), 43.

Pistorio, A., Locatelli, P., Cirilli, F., Gastaldi, L., Solvi, S., 2017. A Business Model for Digital Healthcare Environments: An Organic Approach and a Use Case for Handling Cognitive Impairment. Proceedings of the 10th International Joint Conference on Biomedical Engineering Systems and Technologies *HEALTHINF, BIOSTEC*, pp. 340–347.

Prince, M., Comas-Herrera, A., Knapp, M., Guerchet, M., &Karagiannidou, M. (2016). *World Alzheimer report 2016: improving healthcare for people living with dementia: coverage, quality and costs now and in the future.*

Reilly, S., Miranda- Castillo, C., Malouf, R., Hoe, J., Toot, S., Challis, D., & Orrell, M. (2015). *Case management approaches to home support for people with dementia*. The Cochrane Library.

Reynolds, T., Thornicroft, G., Abas, M., Woods, B., Hoe, J., Leese, M. and Orrell, M., 2000. Camberwell Assessment of Need for the Elderly (CANE): Development, validity and reliability. *The British Journal of Psychiatry*, *176*(5), pp.444-452.

Shulman, K. I. 2000. Clock-drawing: is it the ideal cognitive screening test? *International Journal of Geriatric Psychiatry*, 15, 548–561.

Zarit, S.H., Todd, P.A. and Zarit, J.M., 1986. Subjective burden of husbands and wives as caregivers: a longitudinal study. *The Gerontologist*, *26*(3), pp.260-266.

Wimo, A., Guerchet, M., Ali, G. C., Wu, Y. T., Prina, A. M., Winblad, B., ... & Prince, M. (2017). The worldwide costs of dementia 2015 and comparisons with 2010. *Alzheimer's & dementia: the journal of the Alzheimer's Association*, 13(1), 1-7.

EXTRACTING FRAILTY STATUS FOR POST SURGICAL MORTALITY PREDICTION

Qing Zeng-Treitler[1,2], Yijun Shao[1], Yan Cheng[1], Kristina Doing-Harris[3], Rashmee U. Shah[4], Charlene R. Weir[5,6] and Bruce E. Bray[5,4]
[1]*Biomedical Informatics Center, George Washington University, Washington, DC, USA*
[2]*Washington DC VA Medical Center, Washington, DC, USA*
[3]*KDH-NLP Consulting, kdh-nlp.com, Salt Lake City, UT, USA*
[4]*Division of Cardiovascular Medicine, University of Utah, Salt Lake City, UT, USA*
[5]*Department of Biomedical Informatics, University of Utah, Salt Lake City, UT, USA*
[6]*Veterans Affairs Medical Center, Salt Lake City, UT, USA*

ABSTRACT

Frailty is increasingly recognized as a leading indicator of worsening health outcomes, even death. Frailty is also used implicitly by many clinicians for decision-making, but rarely in a systematic way. We developed a novel ontology-based natural language processing approach to extract frailty status from clinical notes and support retrospective clinical studies and prospective clinical decision making. The natural language processing classification of frailty description achieved an accuracy of 80.3%. We used the extracted frailty information along with other clinical variables to predict mortality after major cardiovascular procedures. The area under the curve for mortality prediction was 78.3% on the test data for a deep neural network model, compared to 74.9% using a support vector machine.

KEYWORDS

Frailty, Natural Language Processing, Predictive Modeling

1. INTRODUCTION

Physicians can now offer invasive and aggressive treatments including implantable cardioverter defibrillators, mechanical support devices, valve replacements and other surgical procedures.(American Heart Association, 2015, Kadish and Mehra, 2005, Slaughter et al., 2009) These interventions, are invasive, involve risk, and can be painful; careful patient selection is imperative.(Morken et al., 2014, Ladwig et al., 2008, Magid et al., 2015, Rowe et al., 2014) Ideally, patients who receive these treatments should have a reasonable life expectancy in order to benefit from aggressive treatment and should have sufficient reserves to mitigate risk. The treatment intensity should match the expected patient outcome, yet providers do not have a reliable method to estimate prognosis as most clinical trials do not include all types of patients.

Frailty assessments are particularly applicable to patients referred for cardiac interventions. Older patients represent a growing segment of the population referred for invasive procedures, especially as availability of transcatheter technology grows, and providers and patients need better ways to decide which patients will benefit from (or be harmed by) interventions12. Current risk prediction models are known to underestimate risk in frail older patients, yet these patients have up to five times the risk of complications plus increased mortality risk.(Dasgupta et al., 2009) Frail patients have 20 times higher risk to be discharged to institutional care compared to non-frail patients.(Makary et al., 2010) In 2015, the United States National Institute on Aging cited frailty assessment as a key priority in the perioperative approach to cardiac surgery.(Robinson et al., 2015)

Frailty is a key factor in cardiac surgical decision making, but current methods are limited by: 1) inconsistent frailty definitions 2) poor specificity in predicting outcomes 3) inability to scale measurement for large populations and 4) difficulty in obtaining repeated measures over time. A consensus conference defined frailty as " ... a state of increased vulnerability to stressors due to age-related decline in physiologic reserve across neuromuscular, metabolic and immune systems."(Rodriguez-Manas et al., 2013) Despite its importance in clinical decision making, frailty assessment in its current form poses several limitations to widespread use. Although there is consensus regarding the underlying concept of frailty – a series of accumulated deficits – different tools vary in terms of dimensions measured. Some measures incorporate weakness, mobility problems, cognitive limitations, impaired hearing or vision and poor nutritional status as components, whereas other measures include impaired cognition, poly-pharmacy, social connectedness, or the inclusion of activities of daily living. In addition to different dimensions, some instruments are based on self-report (e.g. Groningen Frailty Indicator), while others are based on clinician judgment (e.g. Canadian Study of Health and Aging (CSHA) or the Fried frailty phenotype)(Fried et al., 2001, Jones et al., 2005) or structured data from the EHR. For instance, the Frailty Index, based on medical history, is unique in that it is simply a sum of chronic conditions reflecting overall burden plus age.(Mitnitski and Rockwood, 2015)

In addition to variation in the components used to measure frailty, current instruments are inadequate to inform clinical decision making. Current instruments have relatively low specificity and may result in many false positives, making the cost-effectiveness of mass screening unappealing.(Pijpers et al., 2012) Another important limitation of frailty assessment in its current state is the inability to assess change over time. Pijpers, et al and others have argued extensively for the necessity of assessing all individuals over time in order to understand the trajectory and implications of frailty.(Pijpers et al., 2012, Pijpers et al., 2009) Frailty assessments would therefore need to be systematically conducted at all encounters, a requirement that would add substantial burden to the clinician and cost to the healthcare system.

Finally, a population-based assessment is essential in order to fully support routine clinical care and pragmatic trials, in which measurement and outcome assessments are embedded in routine clinical care.(Tunis et al., 2003, Ware and Hamel, 2011) In summary, current frailty assessment in clinical practice and clinical research is variable and incomplete. The study we described here is intended to address current limitations, allow a broader use of frailty assessment in clinical practice, and ultimately reduce patient morbidity and mortality by allowing smarter selection of patients for invasive procedures.

2. BACKGROUND

We developed a frailty ontology with a focus on physical deficits (Figure 1). An ontology consists of a group of concepts (ideas) and the relationships between them. Each concept is associated with a set of terms. These term groups are often referred to as synonyms for the concept, because the concept name is taken from the term list. For example, the concept urinary incontinence has a term list including "urinary incontinence" and "incontinence of urine." A detailed description of the ontology has been submitted and is currently under review. We briefly describe the ontology development process in Figure 1.

Terms from 10 different validated functional status assessments and 3 frailty assessment instruments were collated. There are a number of official measures of functional status used by clinicians in both real clinical care settings as well as in research and they overlap significantly. Many of these tools have been validated through extensive psychometric testing: SF-36(McHorney et al., 1994, McHorney et al., 1993, Ware and Sherbourne, 1992), Barthel(Mahoney and Barthel, 1965), Katz ADL(Katz, 1983), FIM(Tinetti, 1986), HAQ(Fries et al., 1980), PSMS(Lawton and Brody, 1969), RDRS(Linn and Linn, 1982), NYHA(Raphael et al., 2007), DASI(Hlatky et al., 1989) and FACIT(Tamburini, 2001). We added to these terms items commonly used in the GFI, the CSHA, and the Fried phenotype for men and women.(Fried et al., 2001) The items in the instruments vary in their level of abstraction, scope and uniqueness. Terms from the World Health Organization's International Classification of Functioning, Disability and Health (ICF) were also included in the analysis.

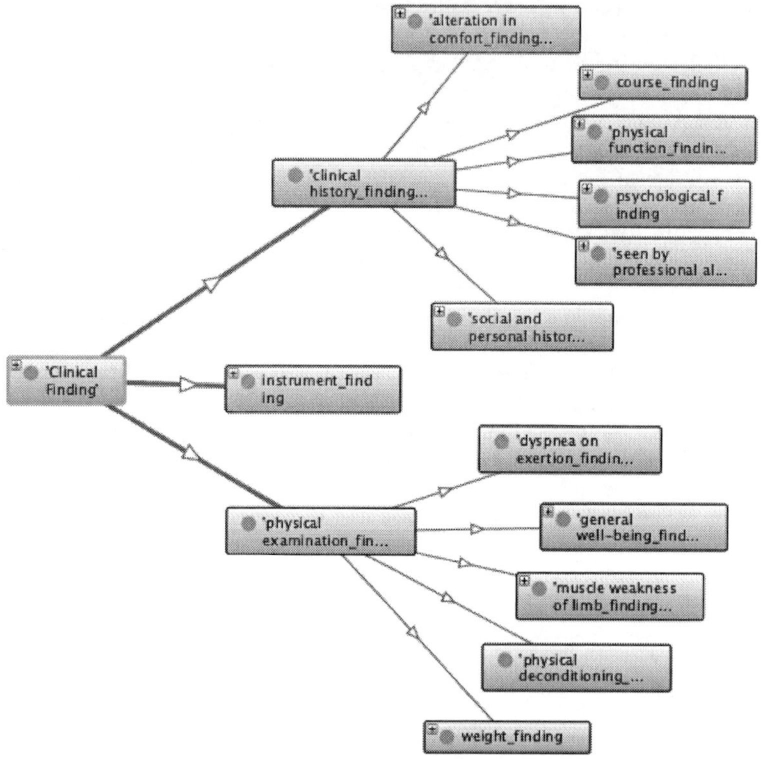

Figure 1. Snapshot of Part of the Frailty Ontology

These terms underwent an initial sorting by the research team in order to identify explicit synonyms. Terms were found to correspond to the general deficit categories of toileting, mental health, social functioning, working, exercise, walking, eating, general health, bathing, dressing & grooming, transfers, and modifiers (i.e., body locations and qualifiers). We had an initial set of about 110 unique terms.

Our research team met regularly to iteratively find terms and map them onto the UMLS meta-thesaurus(Aronson, 2001). When collecting terms into their associated concepts, their potential relations to surgical decision making were taken into consideration. For example, terms relating to housework
(e.g., "dusting," "washing dishes," and "vacuuming") were grouped into a single concept ability to perform domestic activities. In addition, we included terms extracted from note review from a prior studies we conducted(Kuang et al., 2015). Terms identified in this way included "stagger" as indicative of the concept impairment of balance.

To establish the relationship of the concepts to frailty, three members of the team and 3 participant cardiologists mapped the concepts to frailty scores in Rockwood's Clinical Frailty Scale(Rockwood et al., 2007, Rockwood et al., 2005, Howlett and Rockwood, 2013). This mapping made clear that some concepts were associated with three levels of severity of impairment (able, assisted, and unable). The concept ability to walk, for example, has these three levels and each level is associated with a frailty score. The clinicians in the team also characterized the constructs related to their likely cardiac origin or not and their relevance in cardiac decision making.

The identified frailty concepts were arranged in hierarchical relationship by mapping them to SNOMED CT equivalents using SNOMED CT concepts within the category finding, where possible. The goal for this process was to restrict the SNOMED CT mappings to as small a selection as possible, while maintaining correspondence, which means our ontology is compatible with SNOMED CT.

3. METHODS

In this study, we developed an ontology-based natural language processing (NLP) method to extract and aggregate frailty findings from EHR. We then created models using frailty status and other risk factor to predict all-cause mortality following cardiac interventions.

3.1 Data Sources

Data was collected from the US national veteran health administration (VHA) electronic health record (EHR) repository called VINCI with 17 million unique patients.

3.2 Inclusion Criteria

Patients were considered eligible for inclusion if they have a diagnosis of cardiac disease as identified by ICD codes (inpatient or outpatient) and have >= 1 cardiology visit or cardiac surgical consult. We identified ~1.6 million Veterans (including deceased) who meet this criteria. The predictive modeling experiments used a randomly selected set of 20,000 patients who received a major cardiovascular procedure (MCVP) between 2010 and 2015.

3.3 Frailty Extraction

The frailty status extraction method was developed through 4 steps: 1) Identify frailty indicator terms to retrieve potential descriptions of frailty; 2) Manually annotate potential descriptions of frailty; 3) train a machine learning model to classify potential descriptions as frailty deficits; and 4) categorize positive deficit descriptions into major frailty groups and assign severity score.

3.3.1 Identify Frailty Terms

Clinicians typically describe different aspects of frailty such as fatigue, mobility, and daily activity rather than explicitly saying "frail". The specific terms and descriptors they document are largely related to their decision-making processes, their belief in how well the patient will recover, and how patient preferences might be significant for different types of interventions. We first compiled a list of frailty terms from the frailty ontology described in the Background and from a prior study (Shao et al., 2016). The list was jointly reviewed and revised by the research team. Finally, 1000 text snippets were extracted and a manual review of the term list added new terms (e.g. 'WC' as an acronym of "wheel chair") and removed terms that occurred relatively frequently but never used to indicate frailty (e.g., "dressing"). This process resulted in 146 keywords.

3.3.2 Annotation

We extracted 2460 snippets centered at the frailty keyword terms. We define a text snippet as a piece of text containing n words or symbols, in this case, 20 words before and 20 words after frailty keyword. The rationale for using text snippets instead of entire notes for annotation is that human reviewers can usually classify frailty based on a small number of words surrounding a frailty-related term (e.g. "memory loss"). Narrowing the review down from entire notes to these short snippets can have a dramatic impact on annotation speed and make it feasible to obtain much larger annotated datasets: an experienced annotator is able to review 50 to 100 snippets per hour.

A group of clinicians (including 2 cardiologist, 1 registered nurse, and 1 physical therapist) and two informatics researchers reviewed the snippets. They categorized each into one of three classes: a) Yes Deficit; b) No Deficit; and c) Irrelevant. Frailty deficit classification is not simply a negation detection process. "Unable to walk" and "does not need wheelchair" are both negated but the former describes a positive deficit and the latter does not. Disagreements in the team was resolved by consensus. To facilitate the annotation, we used a user-friendly and open source tool called the Visual Tagging Tool (VTT)(Lu et al., 2010). VTT was developed by the National Library of Medicine and optimized by us to accommodate larger data sets assess inter-rater agreement.

3.3.3 Text Classification

We trained a classifier on the snippets using support vector machine (SVM). A linear kernel was used for the SVM. The 3 classes were merged into two categories: "Yes Deficit" and "Other", so that the classifier would classify a snippet into one of the two classes. The features we extracted for learning included 1 and 2 grams (words) before and after the keyword as well as the keyword itself. The weights for the words before and after the keyword were assigned a value between 0 to 1, with lower values for words more distant from the keyword.

3.3.4 Ontology-Based Categorization

The frailty terms that were classified as "Yes Deficit" were mapped to concepts in the frailty ontology. Since the frailty terms were first identified based on the ontology, this is a straight forward step. In the ontology (Figure 1), concepts are categorized into 7 main deficit groups: therapy, medical findings, exercise, mobility, living activity, self care and social function. In the ontology, each concept is associated with a severity level (1-9), with the higher level indicating more severe frailty status. We aggregated the frailty concepts by group and selected the maximum severity level in each group.

3.4 Mortality Risk Assessment

On a cohort of 20,000 heart failure patients who underwent a MCVP, we collected 2-year clinical history data before the first MCVP for each patient. The data features included hospitalizations, ICD9 codes for diagnosis, medications, and frailty deficits from the clinical notes.

We represented the temporal data as a rectangular image, with x-axis representing time (measured in weeks) and y-axis representing the features. Positive findings are represented as a grey scale pixel. Except frailty, all other findings are binary (i.e. either black or white). The color of frailty variables represent the severity. A darker color is assigned to more severe frailty status. An example of the representation is shown in Figure 2.

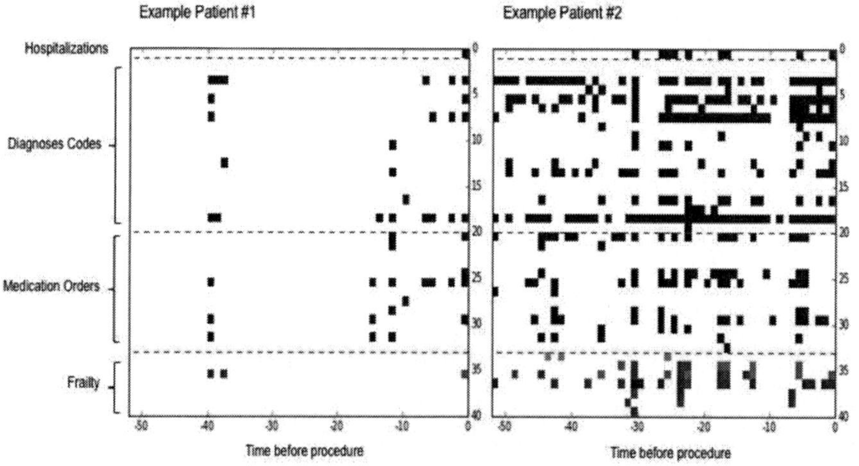

Figure 2. Temporal Graphic Representation of Patient Medical History Including Frailty Status

We built a predictive model using a deep neural network (DNN) for prediction of mortality within the first year after MCVP. The DNN had two branches: one branch took the temporal images as inputs, followed by two convolutional layers and two max pooling layers; the other branch took non-temporal data (e.g., age, gender, race, specific MCVP) as inputs (Figure 3). The two branches were joined to combine the data for final outcome prediction. The cohort was partitioned into 3 sets: train (70%), validation (10%) and test (20%). Parameters in DNN were fined tuned to maximize performance on the validation dataset. We then calculated area under the curve on the testing data set. For comparison, a SVM with linear kernel was trained, validated, and tested on the same datasets.

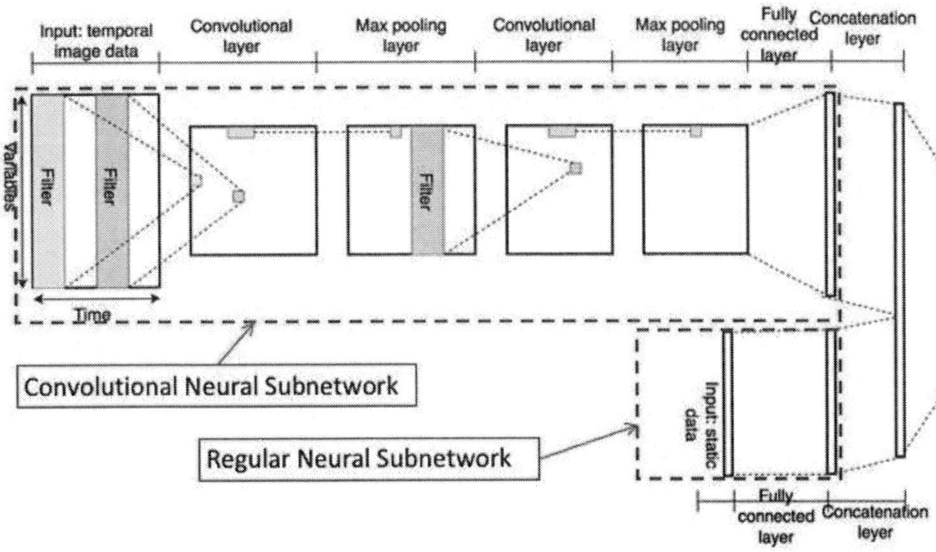

Figure 3. Deep Neural Network System Architecture

4. RESULTS

NLP training and testing was performed using the 2460 annotated snippets. The distribution of the categories included 1080 "yes deficit", 493 "no deficit", and 887 "irrelevant" instances. "No deficit" and "irrelevant" categories were grouped into one class because we were mainly interested in the positive (yes) deficits.

In ten-fold cross validation, the average SVM performance was: accuracy = 80.5%, positive predictive value (PPV) = 77.9%, negative predictive value (NPV) = 82.5%, sensitivity = 77.5%, and specificity = 82.8%. Since frail patients typically have multiple frailty descriptions, the accuracy was deemed to adequate.

The characteristics of the MCVP patients are shown Table 1. We identified 2.5 million frailty text snippets in this patient population. Amongst these, 855,000 snippets were classified as positive by the SVM classifier.

Table 1. Characteristics of MCVP Patients, Grouped by Mortality Status

Characteristic	Death	Non-death
Age, mean years (SD)	71.5 (10.7)	66.3 (9.9)
Female sex, n (%)	27 (1.9%)	540 (2.7%)
White race	1,061 (73.5%)	15,474 (77.7%)
# Hospitalizations in prior year, mean (SD)	1.8 (2.4)	0.8 (1.6)

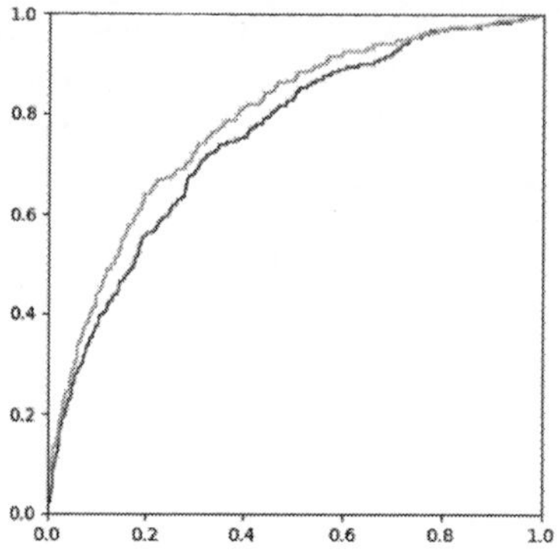

Figure 4. Receiver operator curves of the DNN and SVM models (DNN – light grey, SVM – dark grey) for post MCVP mortality risk

The frailty descriptions and other risk factors were utilized by the DNN and SVM risk prediction models. The area under the curve (AUC) for mortality prediction was 78.3% (95% CI 77.1% to 79.5%) on the test data for the DNN model, compared to 74.9% (95% CI 73.6% to 76.2%) for the SVM model. (Figure 4)

5. DISCUSSION

We have developed a novel ontology-based NLP approach to extract frailty information from EHR. This approach involved several steps: first we identified frailty-related keywords based on a frailty ontology, then we classified the frailty snippets to identify descriptions of positive deficits, and finally we grouped the frailty descriptions into deficit groups and assigned severity scores based on the ontology. In prior work75, we used a topic modeling approach to extract frailty, which does not differentiate positive from negative findings. In comparison, the ontology-based approach offers more precise identification of frailty status.

This approach was validated in two ways. The text classification method was evaluated using annotated data and achieved 80% accuracy. The frailty deficits were then used along with other risk factors to predict mortality and achieved 78.3% AUC with DNN and 74.9% with SVM. Since the prediction of mortality post MCVP is a challenge, these results are promising.

In utilizing the DNN, we created temporal graphics to represent frailty and other medical history data. This provided a more granular data representation and we believe is part of the reason that DNN outperformed SVM. DNN can recognize non-linear patterns while SVM with linear kernel can recognize linear patterns. This is another part of the reason for DNN's superior performance. Since DNN typically performs better on big data, the gap between DNN and SVM performance will likely widen if we utilize more training data.

The frailty extraction reported in this study focused on physical deficits. Cognitive deficits and general wellbeing are two areas that we hope to add in the future. Signs of cognitive deficit and general welling are documented in clinical notes and in structured data. The integration of NLP results and structured data relating to frailty is another task we plan to undertake.

The ultimate goal of our research is support more in-depth research on frail patients and to support physician decision making. We have demonstrated that it is feasible to extract detailed frailty information from clinical notes and incorporate the extracted information in predictive models. For clinical adoption, we need to further improve the prediction accuracy and provide explanations of the models.

6. CONCLUSION

Ontology-based NLP approach is very helpful to extract frailty status from clinical notes and support retrospective clinical studies. Both of the DNN and SVM algorithms are promising method to predict mortality, with the DNN a little bit superior to the SVM.

ACKNOWLEDGEMENTS

This work is funded by the NIH grant R56 AG052536-01A1 and grants from the US Department of Veterans Affairs, Office of Research and Development, Health Services Research and Development including CHIR HIR 08-374, HIR 08-204, CRE 12-315 and the CREATE: A VHA NLP Software Ecosystem for Collaborative Development and Integration.

REFERENCES

American Heart Association. 2015. Prevention & Treatment of Heart Failure [Online]. Available: http://www.heart.org/HEARTORG/Conditions/HeartFailure/PreventionTreatmentofHeartFailure/Prevention-Treatment-of-Heart-Failure_UCM_002048_Article.jsp [Accessed June 3 2015].

Aronson, A. R. 2001. Effective mapping of biomedical text to the UMLS Metathesaurus: the MetaMap program. Proc AMIA Symp, 17-21.

Dasgupta, M., Rolfson, D. B., Stolee, P., Borrie, M. J. & Speechley, M. 2009. Frailty is associated with postoperative complications in older adults with medical problems. Arch Gerontol Geriatr, 48, 78-83.

Fried, L. P., Tangen, C. M., Walston, J., Newman, A. B., Hirsch, C., Gottdiener, J., Seeman, T., Tracy, R., Kop, W. J., Burke, G., Mcburnie, M. A. & Cardiovascular Health Study Collaborative Research, G. 2001. Frailty in older adults: evidence for a phenotype. J Gerontol A Biol Sci Med Sci, 56, M146-56.

Fries, J. F., Spitz, P., Kraines, R. G. & Holman, H. R. 1980. Measurement of patient outcome in arthritis. Arthritis Rheum, 23, 137-45.

Hlatky, M. A., Boineau, R. E., Higginbotham, M. B., Lee, K. L., Mark, D. B., Califf, R. M., Cobb, F. R. & Pryor, D. B. 1989. A brief self-administered questionnaire to determine functional capacity (the Duke Activity Status Index). Am J Cardiol, 64, 651-4.

Howlett, S. E. & Rockwood, K. 2013. New horizons in frailty: ageing and the deficit-scaling problem. Age Ageing, 42, 416-23.

Jones, D., Song, X., Mitnitski, A. & Rockwood, K. 2005. Evaluation of a frailty index based on a comprehensive geriatric assessment in a population based study of elderly Canadians. Aging Clin Exp Res, 17, 465-71.

Kadish, A. & Mehra, M. 2005. Heart failure devices: implantable cardioverter-defibrillators and biventricular pacing therapy. Circulation, 111, 3327-3335.

Katz, S. 1983. Assessing self-maintenance: activities of daily living, mobility, and instrumental activities of daily living. J Am Geriatr Soc, 31, 721-7.

Kuang, J., Mohanty, A. F., H., R. V., Weir, C. R., Bray, B. E. & Zeng-Treitler, Q. 2015. Representation of Functional Status Concepts from Clinical Documents and Social Media Sources by Standard Terminologies. AMIA Fall Symp Proc, In press.

Ladwig, K.-H., Baumert, J., Marten-Mittag, B., Kolb, C., Zrenner, B. & Schmitt, C. 2008. Posttraumatic stress symptoms and predicted mortality in patients with implantable cardioverter-defibrillators: results from the prospective living with an implanted cardioverter-defibrillator study. Archives Of General Psychiatry, 65, 1324-1330.

Lawton, M. P. & Brody, E. M. 1969. Assessment of older people: self-maintaining and instrumental activities of daily living. Gerontologist, 9, 179-86.

Linn, M. W. & Linn, B. S. 1982. The rapid disability rating scale-2. J Am Geriatr Soc, 30, 378-82.

Lu, C. J., Divita, G. & Browne, A. C. 2010. Development of Visual Tagging Tool. AMIA 2010 Symposium Proceedings

Magid, K. H., Matlock, D. D., Thompson, J. S., Mcilvennan, C. K. & Allen, L. A. 2015. The influence of expected risks on decision making for destination therapy left ventricular assist device: An MTurk survey. The Journal Of Heart And Lung Transplantation: The Official Publication Of The International Society For Heart Transplantation.

Mahoney, F. I. & Barthel, D. W. 1965. Functional Evaluation: The Barthel Index. Md State Med J, 14, 61-5.

Makary, M. A., Segev, D. L., Pronovost, P. J., Syin, D., Bandeen-Roche, K., Patel, P., Takenaga, R., Devgan, L., Holzmueller, C. G., Tian, J. & Fried, L. P. 2010. Frailty as a predictor of surgical outcomes in older patients. J Am Coll Surg, 210, 901-8.

Mchorney, C. A., Ware, J. E., Jr., Lu, J. F. & Sherbourne, C. D. 1994. The MOS 36-item Short-Form Health Survey (SF-36): III. Tests of data quality, scaling assumptions, and reliability across diverse patient groups. Med Care, 32, 40-66.

Mchorney, C. A., Ware, J. E., Jr. & Raczek, A. E. 1993. The MOS 36-Item Short-Form Health Survey (SF-36): II. Psychometric and clinical tests of validity in measuring physical and mental health constructs. Med Care, 31, 247-63.

Mitnitski, A. & Rockwood, K. 2015. The rate of aging: the rate of deficit accumulation does not change over the adult life span. Biogerontology.

Morken, I. M., Bru, E., Norekvål, T. M., Larsen, A. I., Idsoe, T. & Karlsen, B. 2014. Perceived support from healthcare professionals, shock anxiety and post-traumatic stress in implantable cardioverter defibrillator recipients. Journal Of Clinical Nursing, 23, 450-460.

Pijpers, E., Ferreira, I., Stehouwer, C. D. & Nieuwenhuijzen Kruseman, A. C. 2012. The frailty dilemma. Review of the predictive accuracy of major frailty scores. Eur J Intern Med, 23, 118-23.

Pijpers, E., Ferreira, I., Van De Laar, R. J., Stehouwer, C. D. & Nieuwenhuijzen Kruseman, A. C. 2009. Predicting mortality of psychogeriatric patients: a simple prognostic frailty risk score. Postgrad Med J, 85, 464-9.

Raphael, C., Briscoe, C., Davies, J., Ian Whinnett, Z., Manisty, C., Sutton, R., Mayet, J. & Francis, D. P. 2007. Limitations of the New York Heart Association functional classification system and self-reported walking distances in chronic heart failure. Heart, 93, 476-82.

Robinson, T. N., Walston, J. D., Brummel, N. E., Deiner, S., Brown, C. H. T., Kennedy, M. & Hurria, A. 2015. Frailty for Surgeons: Review of a National Institute on Aging Conference on Frailty for Specialists. J Am Coll Surg, 221, 1083-92.

Rockwood, K., Abeysundera, M. J. & Mitnitski, A. 2007. How should we grade frailty in nursing home patients? J Am Med Dir Assoc, 8, 595-603.

Rockwood, K., Song, X., Macknight, C., Bergman, H., Hogan, D. B., Mcdowell, I. & Mitnitski, A. 2005. A global clinical measure of fitness and frailty in elderly people. CMAJ, 173, 489-95.

Rodriguez-Manas, L., Feart, C., Mann, G., Vina, J., Chatterji, S., Chodzko-Zajko, W., Gonzalez-Colaco Harmand, M., Bergman, H., Carcaillon, L., Nicholson, C., Scuteri, A., Sinclair, A., Pelaez, M., Van Der Cammen, T., Beland, F., Bickenbach, J., Delamarche, P., Ferrucci, L., Fried, L. P., Gutierrez-Robledo, L. M., Rockwood, K., Rodriguez Artalejo, F., Serviddio, G., Vega, E. & Group, F.-C. 2013. Searching for an operational definition of frailty: a Delphi method based consensus statement: the frailty operative definition-consensus conference project. J Gerontol A Biol Sci Med Sci, 68, 62-7.

Rowe, R., Iqbal, J., Murali-Krishnan, R., Sultan, A., Orme, R., Briffa, N., Denvir, M. & Gunn, J. 2014. Role of frailty assessment in patients undergoing cardiac interventions. Open Heart, 1, e000033.

Shao, Y., Mohanty, A. F., Ahmed, A., Weir, C. R., Bray, B. E., Shah, R. U., Redd, D. & Zeng-Treitler, Q. 2016. Identification and Use of Frailty Indicators from Text to Examine Associations with Clinical Outcomes Among Patients with Heart Failure. AMIA Annu Symp Proc, 2016, 1110-1118.

Slaughter, M. S., Rogers, J. G., Milano, C. A., Russell, S. D., Conte, J. V., Feldman, D., Sun, B., Tatooles, A. J., Delgado, R. M., 3rd, Long, J. W., Wozniak, T. C., Ghumman, W., Farrar, D. J. & Frazier, O. H. 2009. Advanced heart failure treated with continuous-flow left ventricular assist device. The New England Journal Of Medicine, 361, 2241-2251.

Tamburini, M. 2001. Health-related quality of life measures in cancer. Ann Oncol, 12 Suppl 3, S7-10.

Tinetti, M. E. 1986. Performance-oriented assessment of mobility problems in elderly patients. J Am Geriatr Soc, 34, 119-26.

Tunis, S. R., Stryer, D. B. & Clancy, C. M. 2003. Practical clinical trials: increasing the value of clinical research for decision making in clinical and health policy. JAMA, 290, 1624-1632.

VA Informatics and Computing Infrastructure [Online]. Available: http://www.hsrd.research.va.gov/for_researchers/vinci/#.USPMOqURCnY [Accessed April 7 2017].

Ware, J. E., Jr. & Sherbourne, C. D. 1992. The Mos 36-item short-form health survey (SF-36). I. Conceptual framework and item selection. Med Care, 30, 473-83.

Ware, J. H. & Hamel, M. B. 2011. Pragmatic trials--guides to better patient care? The New England Journal Of Medicine, 364, 1685-1687.

EXTRACT CLINICAL MEASUREMENT VALUES USING A REGULAR EXPRESSION PATTERN DISCOVERY ALGORITHM VS SUPPORT VECTOR MACHINE

Douglas Redd[1], Bryan Gibson[2], Maureen A. Murtaugh[3], Joseph Goulet[4] and Qing Zeng-Treitler[1]
[1]Department of Clinical Research and Leadership, George Washington University, Washington, DC, USA, and Washington DC VA Medical Center, Washington, DC, USA
2Department of Biomedical Informatics, University of Utah, Salt Lake City, UT, USA
[3]Division of Epidemiology, University of Utah, Salt Lake City, UT, USA
[4]VA Connecticut Healthcare System, West Haven, CT, USA, and Yale University School of Medicine, New Haven, CT, USA

ABSTRACT

Background: Clinical measurements are commonly embedded in free-text clinical notes. These can be extracted using natural language processing, but this can be resource intensive with limited generalizability. We demonstrate a new approach using regular expression discovery for extraction (REDEx), a supervised machine learning algorithm that we have developed that automatically generates regular expressions to extract measurements with reduced effort.
Results: We compare this approach to that of a support vector machine (SVM) in the task of body weight extraction. 968 weight values were annotated in 300 clinical notes and used for training of the REDEx and SVM models. 98 regular expressions were automatically generated by REDEx. In 10-fold cross validation the REDEx model consistently outperformed the SVM model, with precision .99 vs .85, recall .98 vs. .87, f1-score .99 vs .86, and accuracy .98 vs. .82.

KEYWORDS

Medical Informatics, Clinical Informatics, Natural Language Processing, Machine Learning, Regular Expressions

1. BACKGROUND

Many numeric and categorical measurements are embedded in clinical notes within Electronic Health Records (EHR). Since clinical notes are free text, it is not possible to retrieve these measurement values for use in either individual or population level analyses without performing natural language processing (NLP). For example, estimated oxygen consumption (VO2, estimated during a cardiac stress test), and aortic aneurysm diameter (measured by ultrasound or CT scan) are routinely collected in individuals with known or suspected cardiovascular disease. These data are important diagnostic and prognostic variables that could be used by decision support systems, in quality improvement projects, or comparative-effectiveness re- search. The extraction of measurement values is also important to research such as pragmatic randomized controlled trials (pRCTs) (Raymond, 2014; Patsopoulos, 2011). pRCTs use clinical populations that are more like the population who might receive the intervention (Chalkidou, 2012; Thorpe, 2009; Zwarenstein, 2008). As such, pRCTs depend upon the ability to extract the clinical variables of interest from EHRs. However, since many numeric and categorical measurements are currently stored as text they cannot be readily accessed for these purposes.

Prior natural language processing (NLP) studies have extracted important clinical variables from medical records with excellent accuracy (often reaching 90% or higher) but with limited capacity to generalize to other variables. Investigators have extracted categorical and continuous measures such as tumor staging (TNM) (Liu, 2005), ejection fraction (EF) (Garvin, 2012), blood pressure (BP) (Turchin, 2006), Gleason Score (GS) (D'Avolio, 2008), and medication dosage, frequency, or duration (Xu, 2010).

However, developing these NLP tools can take months, or even years. For each new numerical/categorical variable of interest, clinicians and informaticians need to manually develop case-specific rules and regular expressions and annotate a reference standard for training and testing. The development of NLP tools for numerical/categorical variables is further complicated by the presence of templates (e.g. slot-value pairs). In large healthcare systems the variety of different templates in use limits the scalability of NLP tools. Regular expressions and rules developed for one template often do not work for a different template. This makes the manual development of NLP methods for the extraction of numerical and categorical values inefficient.

We have developed a new approach of machine-learning-based regular expression discovery for extraction (REDEx). Machine learning is a modern approach to NLP that has become increasingly popular in and outside of the biomedical domain as it reduces most of the human effort required for understanding and representing patterns in clinical text. Several of our early experiments have demonstrated the feasibility of this approach by generating regular expressions from annotated text and achieving >90% precision, recall, and F-measure.

One question regarding the development of REDEx and other regular expression-based pattern discovery is how they compare with traditional machine learning methods. In this paper we describe an experiment that applied REDEx to the task of weight extraction and compare REDEx with SVM.

1.1 Weight Value

It is somewhat surprising that weight and height measurements are not always available as structured data (Das, 2005; Littman, 2012). Researchers at the Group Health Cooperative, testing the ability to use EHR data to calculate cardiac risk, found that amongst the records of 122,270 individuals, 11.5% were missing data for either height, weight, or both (Green, 2012). Similarly, Das et al. reported that among 1.8 M Veterans who received outpatient care at Veterans Affairs (VA) facilities in the year 2000, 50.4% had no height or weight recorded as structured data (Das, 2005). More recently, Littman et al. reported that 32.8% of records of 173,127 veterans in the northwestern US were missing structured data for weight or height (Littman, 2012). In VA system, patient weight should be found exclusively in the vital signs table. However, it is often recorded in free text instead of the vitals table.

1.2 Challenge in Extracting Numerical and Categorical Values from Text

Using current methods to extract the wide range of variables recorded in clinical text would be prohibitively time consuming and expensive. This is because most existing dictionary/knowledge-based NLP pipelines or systems like METAMAP (Aronson 2001), CTAKES (Savova, 2010), or v3NLP (Divita, 2016) focus on concepts/findings rather than numerical and categorical values. While a single general concept extraction pipeline may be used to identify many concepts (e.g. symptoms), each numerical and categorical variable often requires their own unique NLP module. For each unique NLP module, clinicians and informaticians need to manually develop specific regular expressions and rules, as well as annotate the training and testing data. Machine learning is increasingly used in NLP development because models created through machine learning can be scaled up or adapted using additional or alternative training samples and reduce the time and energy requirement of intensive human knowledge engineering. The proposed automation of regular expression development is consistent with the modern NLP trend and is a promising approach to improve development efficiency.

An additional challenge involving numerical and categorical values is that they are just as likely to be documented in templates (e.g. tables and questionnaires) as in narratives. Take pain score for example: we have seen a large number of templates used in clinical notes including "Pain - 7 (0-10)", "Pain Scale: [ ] 0 [ ] 1 [ ] 2 [] 3 [ ] 4 [x] 5 [ ] 6 [ ] 7 [ ] 8 [ ] 9 [ ] 10)" and "Pain: 3/10". While it is easy to craft parsing rules for a specific template, in large healthcare systems such as the VA there are numerous templates to deal with. Regular expressions and rules developed for one template do not tend to be generalizable to other templates. Automated regular expression generation provides a better approach for handling templates than manual development because machine-generated expressions can be updated by augmenting the annotated learning sample while manually developed expressions must be revised by human experts.

1.3 Regular Expressions in NLP

NLP extraction of numerical and categorical values usually involves regular expressions (McCart, 2012; Frenz, 2007; Stenner, 2012; Denny, 2009; Piwowar, 2008; Xu, 2011; Matheny, 2009; De Bruijn, 2008; Dorr, 2006; Zeng, 2006; Chapman, 2001; Huang, 2007). A regular expression is a sequence of characters describing a pattern of text (e.g. \d+ represents an integer and 0[1- 9]|1[012] represents a month number). Software developers typically work with domain experts to create regular expressions for a specific NLP use case. Since there is currently no standardized method to generate or test regular expressions, the maintenance and extension of regular expression libraries is also a challenge.

To overcome these challenges, algorithms have been developed to automatically formulate regular expressions from training text samples. A key challenge in algorithmic learning of regular expressions is the huge search space of candidates. Two main learning approaches have been employed to reduce the search space. The top-down approach starts with domain experts supplying seed patterns; they are then transformed or modified until they satisfy predefined evaluation metrics. Such algorithms have been developed for email addresses (Li, 2008; Xie, 2008), URLs (Li, 2008; Xie, 2008), software names, notebook names (Brauer, 2011), course numbers (Li, 2008), phone numbers (Li, 2008; Brauer, 2011; Babbar, 2010), invoice numbers (Brauer, 2011), and date-time (Babbar, 2010) patterns. The bottom-up approach starts by finding similarities or patterns among textual data that are then generalized to build regular expressions. For example, Prasse et al (Prasse, 2012) used this approach to discover campaign message templates from predefined batches of emails.

A major issue with both bottom-up and top down algorithms is their generalizability. The performance of the top-down approach is dependent on seed patterns, the creation of which requires significant human effort. The bottom-up approach is totally automated, however the algorithms described in the literature have targeted entities like phones numbers, URLs, HTML tags, and swift codes which have relatively fixed patterns. Thus, the context surrounding the entities is typically ignored by the algorithms (Brauer, 2011). Clinical use cases, in contrast, require a context sensitive approach; a value of "5-6" cannot be interpreted on its own, the related unit of measure must also be extracted (e.g. 5-6 METs). In addition, clinical texts often refer to multiple measures in the same sentence (e.g. "hr 96, bp 136/78, rr 18"). These texts therefore require co-reference resolution so that individual measures and values are appropriately parsed. The REDEx algorithm that we have developed addresses these issues by learning from contextual clues and will be generalizable to the extraction of any numeric or categorical measures in clinical text. We have applied preliminary versions of REDEx to measurement extraction previously (Murtaugh, 2015; Redd, 2015; Redd, 2016) with good results, but have not compared directly to traditional machine learning methods, such as SVM, as we do in this study.

2. METHODS

2.1 Text Sample

We collected a random sample 300 clinical notes from the VA EHR clinical data repository using "wt", "weight", "wgt", "#", "lb", and "kg" as keywords. Two human reviewers manually reviewed the 300 text samples and annotated body weight values. A third reviewer evaluated the annotations, which had a very good agreement with a kappa score of 99.43%. The few disagreements were adjudicated by the third reviewer. The text was split into snippets of text containing the annotated weight values and the textual context surrounding it, resulting in 968 text snippets. Each snippet contained 20 tokens before and after each weight value. Examples of text snippets are shown in table 1.

2.2 REDEx Extraction

We used the REDEx algorithm to automatically create a set of regular expressions that extract body weight values. REDEx is an iterative process for discovering regular expressions for extraction of values from annotated text snippets and is summarized by the pseudo-code in table 2.

Table 1. Examples of snippets retrieved using the keywords and the regular expressions resulting from REDEx training. Annotated weight values are shown boxed

Sample Snippets	Sample Regular Expressions
No need for pain med. He has gained about 35 lbs. since discharged from active duty and has lost weight already as he is starting to exercise more and count calories	N/A
4 times a week, including swimming and cardio machine (walking hurts his back) 3) goal of at least 13 pound wt. loss in 4 months 4) continue Ketaprophen with food 5) heat to hips at night 15 minutes 6) stretch hips	N/A
terazosin 5 mg one capsule q.h.s., Depends briefs. PHYSICAL EXAM: Vital signs: 97.6, 81, and 20. Blood pressure 27/80, and weight 93.1. Fundi: No AV nicking. Eyes: PERLA. EOMI. Neck: Carotids equal bilaterally without bruits. No thyromegaly. Lungs: Clear to P&A	(?i)weight\s{1,2}(\d+\p{Punct}\d+)\p{Punct}
BP: 121/71 (03/27/99 11:25) T: 98 (03/27/99 11:25) P: 7O (03/27/99 11:25) R: 21 (03/27/99 11:25) HT: 75 (03/27/99 11:25) WT: 335 (03/27/99 11:25) PHYSICAL EXAM FINDINGS: Pt alert and oriented x3. Psoriatic lesions around hairline. Pleasant and cooperative during apt.	(?i)wt\p{Punct}\s{1,2}(\d+)\s{1,2}\p{Punct}
Most recent visit: oct 7, 2005, actinic changed, face, cyro ear/forearm, plan: surg tx in about one mo, b/p 155/75 wt 187 2005 visits: multiple for lesions, 7/2005 b/p 145/75, 5/2005: surg txs: 'suspect, squam cell CA' b/p 130/55	(?i)wt\s{1,2}(\d+)\s{1,2}
polyp was excised and found to be benign. He should have follow-up sigmoid after 5 years. BP = 133/88 P = 68/min Wt = 154 lbs Good carotid upstroke without bruits, no oral lesions, ear canals have some wax but TM's are clear.	(?i)wt\s{1,2}\pPunct\s{1,2}(\d+)\s{1,2}
The patient has no other specific complaints. He would like to stop smoking. PHYSICAL EXAMINATION: VITAL SIGNS: Weight 142 , temperature 97.7, pulse 72, blood pressure 126/89, height 73 inches.	(?i)weight\s{1,2}(\d+)\p{Punct}\s{1,2}
Had colon cancer at 58 S doing well gen : no c/o GI : no BRBPR no constipation BP 152/86 P 84 Wt 166.9 gained wt heart s1s2 ext no edema neck no ln palpable, lungs clear.	(?i)\d+\s{1,2}wt\s{1,8}(\d+\p{Punct}\d+)\s{1,2}[a-z]{1,3}
PAST MEDICAL HISTORY: Please see prior note from April of 2006 for full details. PHYSICAL EXAMINATION: Weight is 182 pounds which is down 4 pounds. Blood pressure is 132/80, pulse of 69 and regular.	(?i)is\s{1,2}(\d+)\s{1,2}pounds

Table 2. Pseudocode describing the REDEx Algorithm

```
let S = the set of annotated text
snippets. let w = the labeled weight
value.
let b = the text of the snippet before
w. let a = the text of the snippet
after w.
let t = a test function that tests a regular expression against all annotated snippets
   and fails if the regular expression produces any matches that are not actual
for each i in [w, b, a] do
loop
   for s_i in S do
   loop
      replace all whitespace in s_i with regular expression \s{1,50}
      replace all punctuation in s_i with regular expression \p{Punct}
      replace all digits in s_i with regular expression \d+
      while t(s_i) does not fail
      loop
         remove a term from the front of the snippet, or the end of the snippet if a term
            was removed from the front the previous iteration.
      end loop
      restore the previous version of s_i that did not cause t(s_i) to fail.
      for each x in X do
      loop
         replace all occurrences of x in s_i with regular expression \S+ if it does not cause
          t(s_i) to fail
      end loop
   end loop
end loop
```

2.3 SVM Extraction

To provide a control method for REDEx, we trained SVM models for weight extraction using the same annotated samples. The SVM models were used to classify the context surrounding numerical values in the text sample, with the weight value serving as the label. After experimenting with several different windows of context (n tokens before and after), the best performance we obtained from SVM was with bag of words features and a 5 token before and after window. The Weka SMO package was used as the SVM implementation (Bartoli, 2014).

2.4 Comparison

10-fold cross validation was used in the case of REDEx and also SVM evaluation. In the case of REDEx we determined a weight prediction to be a true positive (TP) if a weight was extracted and it matched the actual weight annotated in the snippet. We determined false positives (FP) to be cases where a weight was extracted, but it did not match the actual weight annotated in the snippet, or there was no weight annotated in the snippet. False Negatives (FN) were considered to be cases where no weight was extracted, but there was a weight annotated in the snippet. And true negatives (TN) were cases where no weight was extracted, and there was no weight annotated in the snippet.

3. RESULTS

REDEx produced 98 distinct regular expressions that provided a precision of 0.99, recall of 0.98, F1-score of 0.99, and accuracy of 0.98. SVM evaluation on the 938 examples resulted in a precision of 0.85, recall of 0.87, F1-score of 0.86, and accuracy of 0.82. These metrics are presented in table 3. Even though REDEx had to both identify the span and determine context, it achieved higher precision, recall, F1-score and accuracy.

Table 3. REDEx and SVM Accuracy Comparison

	REDEx	SVM
Precision	**0.99**	0.85
Recall	**0.98**	0.87
F1-score	**0.99**	0.86
Accuracy	**0.98**	0.82
# Reg Ex	98	N/A

4. DISCUSSION

In our prior studies, we demonstrated that REDEx can extract values with 90% accuracies in most cases. In this study, we compared REDEx against SVM for weight extraction. The weight value itself is relatively straightforward, but the context is more complex, requiring us to distinguish past weight from current weight and patient goals from patient status. When unit (lb or kg) is not present, we also need to use context to determine if a value is weight or another variable like heart rate or blood pressure. Since SVM is fundamentally a classification algorithm, we trained an SVM model focusing on the context classification. The comparison of SVM and REDEx result is actually in favor of SVM because SVM is only used to classify context, while REDEx is used to identify both the span and classify context. The results show that REDEx clearly outperformed SVM.

Software developers are fully capable of creating regular expressions and they often do when performing NLP tasks. Developer's time, however, can be expensive. In addition, maintaining and/or extending regular expressions can be quite difficult. This is why the development of algorithms to learn regular expressions from text is an active area of research (Brauer, 2011; Bartoli, 2012; Moreo, 2013). Our unique contribution lays in the generation of patterns not only for variables of interest but also for the surrounding context.

One of the limitations of REDEx is that it is designed to extract local patterns (i.e. patterns in small snippets of text e.g. <50 tokens) rather than global patterns in an entire document. Sometimes contextual cues may be located far from the target variable. In such cases, REDEx is not an efficient solution as it optimizes the search strategy by assuming tokens close to a variable are more likely to be relevant. We have experimented with other optimization strategies, e.g. assuming tokens with lower frequency and/or less distinguishing power are the less relevant context. Each strategy has some trade-offs. For example, in a relatively small dataset, some low frequency words are actually very good contextual features.

For future work, we are considering wrapping REDEx up as an open source software module. We also would like to integrate it with a clinical chart annotation tool for end users who have no NLP expertise, simplifying the extraction of categorical or numerical variables. Since there is an increasing demand for NLP by biomedical researchers, there is a need for simple tools that allow biomedical researchers (including research assistants) to annotate text and develop NLP functions without relying on NLP experts.

5. CONCLUSIONS

REDEx is a promising new approach for the extraction of categorical and numerical values from clinical notes. It can outperform SVM and has the potential for efficient maintenance and adaption to new or expanding corpora.

REFERENCES

Aronson, A.R., 2001. Effective mapping of biomedical text to the UMLS Metathesaurus: the MetaMap program. In *Proceedings of the AMIA Symposium* (p. 17). American Medical Informatics Association.

Babbar, R. and Singh, N., 2010, October. Clustering based approach to learning regular expressions over large alphabet for noisy unstructured text. In *Proceedings of the fourth workshop on Analytics for noisy unstructured text data* (pp. 43-50). ACM.

Bartoli, A., Davanzo, G., De Lorenzo, A., Mauri, M., Medvet, E. and Sorio, E., 2012, July. Automatic generation of regular expressions from examples with genetic programming. In *Proceedings of the 14th annual conference companion on Genetic and evolutionary computation* (pp. 1477-1478). ACM.

Bartoli, A., De Lorenzo, A., Medvet, E. and Tarlao, F., 2014, July. Playing regex golf with genetic programming. In *Proceedings of the 2014 Annual Conference on Genetic and Evolutionary Computation* (pp. 1063-1070). ACM.

Brauer, F., Rieger, R., Mocan, A. and Barczynski, W.M., 2011, October. Enabling information extraction by inference of regular expressions from sample entities. In *Proceedings of the 20th ACM international conference on Information and knowledge management* (pp. 1285-1294). ACM.

Chalkidou, K., Tunis, S., Whicher, D., Fowler, R. and Zwarenstein, M., 2012. The role for pragmatic randomized controlled trials (pRCTs) in comparative effectiveness research. *Clinical trials*, *9*(4), pp.436-446.

Chapman, W.W., Bridewell, W., Hanbury, P., Cooper, G.F. and Buchanan, B.G., 2001. A simple algorithm for identifying negated findings and diseases in discharge summaries. *Journal of biomedical informatics*, *34*(5), pp.301-310.

D'Avolio, L.W., Litwin, M.S., Rogers Jr, S.O. and Bui, A.A., 2008. Facilitating clinical outcomes assessment through the automated identification of quality measures for prostate cancer surgery. *Journal of the American Medical Informatics Association*, *15*(3), pp.341-348.

Das, S.R., Kinsinger, L.S., Yancy, W.S., Wang, A., Ciesco, E., Burdick, M. and Yevich, S.J., 2005. Obesity prevalence among veterans at Veterans Affairs medical facilities. *American journal of preventive medicine*, *28*(3), pp.291-294.

De Bruijn, B., Carini, S., Kiritchenko, S., Martin, J. and Sim, I., 2008. Automated information extraction of key trial design elements from clinical trial publications. In *AMIA Annual Symposium Proceedings* (Vol. 2008, p. 141). American Medical Informatics Association.

Denny, J.C., Miller, R.A., Waitman, L.R., Arrieta, M.A. and Peterson, J.F., 2009. Identifying QT prolongation from ECG impressions using a general-purpose Natural Language Processor. *International journal of medical informatics*, *78*, pp.S34-S42.

Divita, G., Zeng, Q.T., Gundlapalli, A.V., Duvall, S., Nebeker, J. and Samore, M.H., 2014. Sophia: a expedient UMLS concept extraction annotator. In *AMIA Annual Symposium Proceedings* (Vol. 2014, p. 467). American Medical Informatics Association.

Dorr, D.A., Phillips, W.F., Phansalkar, S., Sims, S.A. and Hurdle, J.F., 2006. Assessing the difficulty and time cost of de-identification in clinical narratives. *Methods of information in medicine*, *45*(03), pp.246-252.

Frenz, C.M., 2007. Deafness mutation mining using regular expression based pattern matching. *BMC medical informatics and decision making*, *7*(1), p.32.

Garvin, J.H., DuVall, S.L., South, B.R., Bray, B.E., Bolton, D., Heavirland, J., Pickard, S., Heidenreich, P., Shen, S., Weir, C. and Samore, M., 2012. Automated extraction of ejection fraction for quality measurement using regular expressions in Unstructured Information Management Architecture (UIMA) for heart failure. *Journal of the American Medical Informatics Association*, *19*(5), pp.859-866.

Green, B.B., Anderson, M.L., Cook, A.J., Catz, S., Fishman, P.A., McClure, J.B. and Reid, R., 2012. Using body mass index data in the electronic health record to calculate cardiovascular risk. *American journal of preventive medicine*, *42*(4), pp.342-347.

Huang, Y. and Lowe, H.J., 2007. A novel hybrid approach to automated negation detection in clinical radiology reports. *Journal of the American Medical Informatics Association*, *14*(3), pp.304-311.

Li, Y., Krishnamurthy, R., Raghavan, S., Vaithyanathan, S. and Jagadish, H.V., 2008, October. Regular expression learning for information extraction. In *Proceedings of the Conference on Empirical Methods in Natural Language Processing* (pp. 21-30). Association for Computational Linguistics.

Littman, A.J., Boyko, E.J., McDonell, M.B. and Fihn, S.D., 2012. Evaluation of a weight management program for veterans. *Preventing chronic disease*, *9*.

Liu, K., Mitchell, K.J., Chapman, W.W. and Crowley, R.S., 2005. Automating tissue bank annotation from pathology reports–comparison to a gold standard expert annotation set. In *AMIA Annual Symposium Proceedings* (Vol. 2005, p. 460). American Medical Informatics Association.

Matheny, M.E., FitzHenry, F., Speroff, T., Hathaway, J., Murff, H.J., Brown, S.H., Fielstein, E.M., Dittus, R.S. and Elkin, P.L., 2009. Detection of blood culture bacterial contamination using natural language processing. In *AMIA Annual Symposium Proceedings* (Vol. 2009, p. 411). American Medical Informatics Association.

McCart, J.A., Finch, D.K., Jarman, J., Hickling, E., Lind, J.D., Richardson, M.R., Berndt, D.J. and Luther, S.L., 2012. Using ensemble models to classify the sentiment expressed in suicide notes. *Biomedical informatics insights*, *5*, pp.BII-S8931.

Moreo, A., Eisman, E.M., Castro, J.L. and Zurita, J.M., 2013. Learning regular expressions to template-based FAQ retrieval systems. *Knowledge-Based Systems*, *53*, pp.108-128.

Murtaugh, M.A., Gibson, B.S., Redd, D. and Zeng-Treitler, Q., 2015. Regular expression-based learning to extract bodyweight values from clinical notes. *Journal of biomedical informatics*, *54*, pp.186-190.

Patsopoulos, N.A., 2011. A pragmatic view on pragmatic trials. *Dialogues in clinical neuroscience*, *13*(2), p.217.

Piwowar, H.A. and Chapman, W.W., 2008. Identifying data sharing in biomedical literature. In *AMIA Annual Symposium Proceedings* (Vol. 2008, p. 596). American Medical Informatics Association.

Prasse, P., Sawade, C., Landwehr, N. and Scheffer, T., 2012. Learning to identify regular expressions that describe email campaigns. *arXiv preprint arXiv:1206.4637*.

Raymond, J., Darsaut, T.E. and Altman, D.G., 2014. Pragmatic trials can be designed as optimal medical care: principles and methods of care trials. *Journal of clinical epidemiology*, *67*(10), pp.1150-1156.

Redd, D., Shao, Y., Yang, J., Divita, G. and Zeng-Treitler, Q., 2015. Automated Learning of Temporal Expressions. *Studies in health technology and informatics*, *216*, pp.639-642.

Redd, D., Kuang, J., Mohanty, A., Bray, B.E. and Zeng-Treitler, Q., 2016. Regular Expression-Based Learning for METs Value Extraction. *AMIA Summits on Translational Science Proceedings*, *2016*, p.213.

Savova, G.K., Masanz, J.J., Ogren, P.V., Zheng, J., Sohn, S., Kipper-Schuler, K.C. and Chute, C.G., 2010. Mayo clinical Text Analysis and Knowledge Extraction System (cTAKES): architecture, component evaluation and applications. *Journal of the American Medical Informatics Association*, *17*(5), pp.507-513.

Stenner, S.P., Johnson, K.B. and Denny, J.C., 2011. PASTE: patient-centered SMS text tagging in a medication management system. *Journal of the American Medical Informatics Association*, *19*(3), pp.368-374.

Thorpe, K.E., Zwarenstein, M., Oxman, A.D., Treweek, S., Furberg, C.D., Altman, D.G., Tunis, S., Bergel, E., Harvey, I., Magid, D.J. and Chalkidou, K., 2009. A pragmatic–explanatory continuum indicator summary (PRECIS): a tool to help trial designers. *Journal of clinical epidemiology*, *62*(5), pp.464-475.

Turchin, A., Kolatkar, N.S., Grant, R.W., Makhni, E.C., Pendergrass, M.L. and Einbinder, J.S., 2006. Using regular expressions to abstract blood pressure and treatment intensification information from the text of physician notes. *Journal of the American Medical Informatics Association*, *13*(6), pp.691-695.

Xie, Y., Yu, F., Achan, K., Panigrahy, R., Hulten, G. and Osipkov, I., 2008. Spamming botnets: signatures and characteristics. *ACM SIGCOMM Computer Communication Review*, *38*(4), pp.171-182.

Xu, H., Stenner, S.P., Doan, S., Johnson, K.B., Waitman, L.R. and Denny, J.C., 2010. MedEx: a medication information extraction system for clinical narratives. *Journal of the American Medical Informatics Association*, *17*(1), pp.19-24.

Xu, H., Jiang, M., Oetjens, M., Bowton, E.A., Ramirez, A.H., Jeff, J.M., Basford, M.A., Pulley, J.M., Cowan, J.D., Wang, X. and Ritchie, M.D., 2011. Facilitating pharmacogenetic studies using electronic health records and natural-language processing: a case study of warfarin. *Journal of the American Medical Informatics Association*, *18*(4), pp.387-391.

Zeng, Q.T., Goryachev, S., Weiss, S., Sordo, M., Murphy, S.N. and Lazarus, R., 2006. Extracting principal diagnosis, co-morbidity and smoking status for asthma research: evaluation of a natural language processing system. *BMC medical informatics and decision making*, *6*(1), p.30.

Zwarenstein, M., Treweek, S., Gagnier, J.J., Altman, D.G., Tunis, S., Haynes, B., Oxman, A.D. and Moher, D., 2008. Improving the reporting of pragmatic trials: an extension of the CONSORT statement. *Bmj*, *337*, p.a2390.

E-HEALTH DEPLOYMENT BUSINESS MODEL TELE-EXPERTISE CASE STUDY

Angela Martin[1], Corentin Canu[2], Gabriele Breda[1] and Camille Jean[2]
[1]*ALTRAN Research Department, 2 Rue Paul Dautier, Vélizy, France*
[2]*ENSAM, 151, bd de l'Hôpital, Paris, France*

ABSTRACT

The main focus of this study is to provide a new framework for business model design by using complex system modelling. The tool offers both entrepreneurs and investors a comprehensive view of value streams by the comparison of different business models based on the economic shared value creation during the go to market process.

KEYWORDS

ICT, Tele-Expertise, Business Model, Deployment, Innovation, Financial Flow Modelling

1. INTRODUCTION

To survive in today's competitive economy entrepreneurs need to find new markets to explore and new customer demand to capture. The mix between innovation and a good business model has proved to be effective. Yet, on the complex eHealth market, start-ups are facing major difficulties. Although many innovative projects are successfully tested, only a few lead to an economic return. Therefore, designing robust business models is an important challenge for entrepreneurs.

The goal of this paper is to provide a framework for designing multiple business models and to select the best ones based on economic shared value creation. Our tool tackles the complexity of business models by using complex system modelling. The paper is organized as follows: in the next section, a literature review examines business model design and challenges. Then, the methodology is presented. The case study is described in the fourth section. Then, the deployment model of the case study is explained and the main results are discussed. Finally, a conclusion ends this paper.

2. THEORETICAL BACKGROUND

Business models and innovation are inter-linked. Innovation is the tool that creates value while business model provides the explanation of how a company creates, delivers and captures value (Osterwalder and Pigneur, 2010). As a consequence, innovation triggers changes in all company activities, thus changing the business model itself. Just like the business model, it is essential to implement innovation (Chesbrough, 2002). Business models are vital for any start-up that intends to raise funds and are widely used especially in go to market strategy of innovations (Chanal, 2011).

A review of recent studies reveals a wide variety of business models. First, economists listed and described business models in use, like the franchise model or low cost model (Zott *et al.*, 2011). Second, they defined the components (Krumeich *et al.*, 2012). According to the literature review, business models are classified into four categories: value proposition, value network, value finance and value architecture (Al-Debei *et al.*, 2008). However, few business model design methods exist. Among the frameworks provided, the business model canvas introduced by Osterwalder and Pigneur (2010) is one of the first attempts to create a tool that specifically helps entrepreneurs to design business models. Based on creativity methods such as design thinking, business model canvas are effective especially in the early stage of projects or in start-up environment.

Created by Riley (2013), the modelH is an adaptation of the business model canvas. It adds on Porter's theory of creating shared value (Porter and Kramer, 2011) and Christensen's jobs to be done (JTBD) (Christensen, 2013). Furthermore, it is specifically designed for the health sector. Indeed, health care business models must deal with the multiplicity of stakeholders and their strong interdependency.

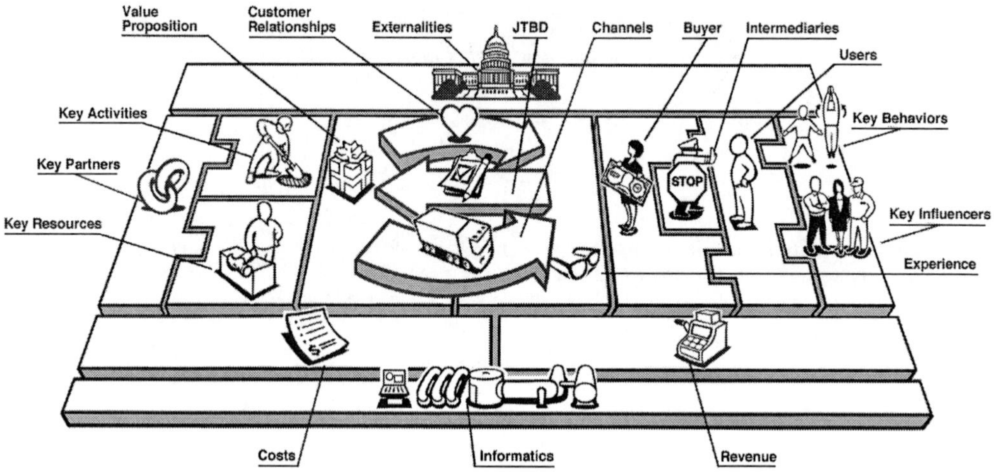

Source: modelH Business Model Canvas for Healthcare by Kevin Riley & Associates. Drawings by Mike Werner.

Figure 1. ModelH: Business Model Canvas For Healthcare (Riley at al., 2013)

In addition to business model canvas, Osterwalder and Pigneur (2010) proposed a business model design that offers entrepreneurs many tools to help design their business models and improve the innovation processes. Visual thinking, storytelling, and prototyping are some of them. Most of these methods are useful during early stages but there is still a lack of tools to test and select business models at the end of the design stage. Furthermore, the organization of the health industry is unique. Stakeholders are particularly dependent on each other and their multiplicity enhances the complexity of business models. Thus, the ecosystem of an innovative project / start-up can be seen as a complex system. Therefore, we look for complex system modelling tools useful for the implementation of an e-health business model.

3. METHODOLOGY

Jean's studies provide an example of the use of a complex system modelling method known as the financial flow modelling method (FFM) to model the implementation of an innovative telehealth system (Jean et al., 2016). FFM method identifies the added value of an innovative product/system. The main idea is to compare the existing scenario to the new one taking into account the new product/system. The key indicator of the FFM method is the annual margin balance i.e. the balance of profits and losses of each stakeholder when the new business system goes to the market, as follows:

$$Margin\ M_{before} = Revenues_{before} - Costs_{before}$$
$$Margin\ M_{after} = Revenues_{after} - Costs_{after}$$
$$Margin\ balance = M_{after} - M_{before}$$

The FFM method consists of seven stages. The first stage is to identify all the stakeholders involved. The second is to identify the existing financial flows. The third is to make hypothesis about new financial flows. The fourth is to collect data to quantify them. The fifth is to make hypothesis on product volume. The sixth is to estimate the workload. The last stage is to analyse the margin balance of each stakeholder. At the end, the modelling implementation scenario is either valid or not. If it is not valid, new hypotheses about financials flows are set, back to stage 3 until a valid/sustainable business model is created.

FFM method is proven to be effective when quantifying the added value of an innovative system and the shared value among stakeholders. Up to date, no direct link with any business model framework has been established. Thus, the research goal of this paper is to use, during the design process, Financial Flow Modelling (FFM) alongside business model canvas (BMC) in order to build up the design of business models and offer more inputs to decision makers. Hence, the use of FFM can help entrepreneurs/ project leaders overcome three major challenges of the business model design process: finding the right model, testing the model before a full-scale launch, and continuously adapting the model in response to market feedback.

4. CASE STUDY

4.1 Project EclairAge

The methodology presented above was tested on EclairAge, an innovative project conducted by the Research Department of Altran Group. EclairAge was selected and financed by the French Government as an eHealth demonstrator on the French territory. The project aimed to create and test innovative services that intend to improve elderly permanence of care in long-term dependency care centres (called EHPAD).

At the present time, the continuity of care is provided by the French Emergency Services System (SAMU) that faces two major challenges. Firstly, SAMU affiliated physicians are not specialized in elderly care. Secondly, SAMU can quickly become congested. To face these challenges, the tele-expertise service EclairAge is a geriatric hot line for EHPAD's residents during weekdays. In case of an alarming situation, the nurse contacts EHPAD's coordinator physician or the resident's attending physician. If available, the case is treated. If not, the nurse calls the emergency service or EclairAge.

Two main benefits arise from the utilization of EclairAge geriatric hotline. First, based on the immediate transmission of EclairAge Emergency Files (containing the complete medical record of all EHPAD' residents), the on-call geriatrician is able to make personalized medical decisions to treat the case. As a consequence, transfers to the hospital or SAMU are avoided. In the case of a confirmed emergency, EclairAge can initiate the intervention of SOS Doctor (private emergency service). As a result, a physician is sent to the caller EHPAD. So, the case is treated without transferring the resident to the hospital. Second, when transfers are unavoidable, the pre-diagnosis performed by EclairAge's geriatrician and the transmission of resident's emergency file may shorten the waiting time at the hospital reception desk. In the case of a differed emergency, the on-call geriatrician can schedule a hospitalization and help find a hospital bed in geriatrics saving time and improving residents' reassurance.

4.2 Local Experimentation

To assess the feasibility and viability of EclairAge, we conducted an *in situ* and *in vivo* non randomized trial in 5 French EHPADs situated in the South of the Essonne department (Ile de France region) for 18 months (August 2014- February 2016). EclairAge geriatrician provided remote diagnosis during the experimental period. Emergency first aid and geriatric training were offered to EHPAD medical staffs for the use of EclairAge platform. The experiment data useful for EclairAge deployment is summed up in the following table.

Table 1. EclairAge Experimentation Data

Data	Value
Number of EHPADs	5
Number of EclairAge calls	51
Number of emergency transports	11
Number of on-site interventions	17

5. DEPLOYMENT MODELLING

The aim here is to provide a go-to-market strategy in order to implement EclairAge at the regional level. Thus, the methodology presented in the third section is tested on this study case. Using FFM, we model the financials streams of each business model canvas and compare them to the financial streams of the current economic system. The aim is to choose the best business model based on the added value created by innovation before the go to market stage. FFM is adapted to fit in the business model design process so that project leaders can adapt the business model according to the market response.

The first step is to model the existing environment where the innovation is to be implemented. The second step is to model the business model canvas designed. The third step is to collect data to quantify the financial flows invested by the business models. Then, hypotheses on the volume of products and the deployment scenarios are done. Next, the model is run and the margin balance is computed. Finally, different business models are compared based on the shared value created by the innovation.

5.1 Modelling the Existing Economic Environment

We start by modelling the current business model (Model 0). Before the implementation of EclairAge, no other system in place avoided the transferring of patients. In case of an alarming situation, if the EHPAD's coordinating physician and the patient's attending physician are not available, two options are presented. Either the patient is transferred to the emergency service or a physician comes to the EHPAD (SOS Doctors). When patients are transferred, they usually go to the emergency department of the nearest hospital (GHT). EHPAD's residents are reimbursed by the National Health Insurance Fund (CNAM) and their private health insurance. We mapped the financial flows underlying the current permanence of elderly care in figure 2.

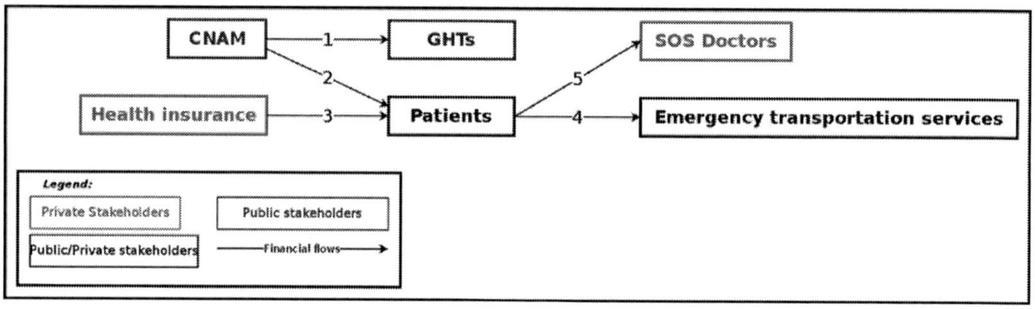

Figure 2. Financial Flows of the Current Business Model

Every arrow represents a parametrized flow. For example, flow 1 = Average Number of SAMU Calls per EHPAD * Hospitalizations Rate * Average Hospitalization Costs.

5.2 Modelling the New Business Model Canvas

In order to build different scenarios for EclairAge business models, we conducted two workshops with different stakeholders of the local experimentation: EclairAge Project Manager, the CEO of the software company that developed EclairAge solution, a geriatrician, a physician from SOS Doctors, a SAMU representative, an economic expert and an eHealth specialist. The ModelH (Riley, 2013) canvas was used to help synthetize the new organization. The first workshops result is the new economic environment model with the new financial flows (figure 3).

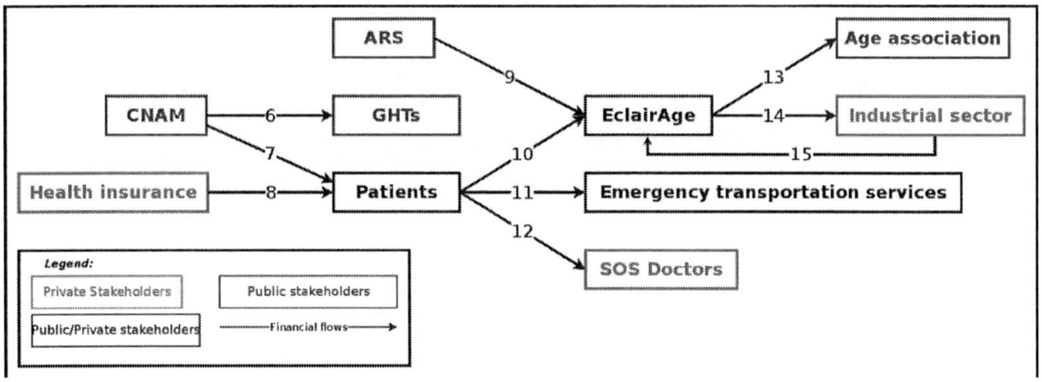

Figure 3. Financial Flows of the New Business Models

Four new stakeholders appear: ARS (Regional Health Agency), EclairAge, Age (geriatric association) and Industrial sector. As the previous graph, every arrow represents a parametrized flow associating equations with the 35 quantitative parameters such as the average number of patients per EHPAD or the Public Insurance Reimbursement Rate for Transportation.

The second result of the two workshops is 3 business models scenarios (BM 1, BM 2, and BM 3). In the first two ones, we assume that ARS invests in EclairAge in order to cover the implementation costs while the industrial sector covers them in the third business model. In the first and the third business model, we assume that EclairAge pays Age for every training session while in the second business model; a monthly salary is payed to Age. The following table sums up the differences between the three business models.

Table 2. New Business Models

	Implementation Stage Stakeholder Strategy	Training Management
Model 1	ARS	Pay per training session
Model 2	ARS	Monthly payment
Model 3	Industrial sector	Pay per training session

5.3 Collect Data to Quantify Financial Flows

The third stage of FFM is to collect data to quantify the parameters. Anonymized data was collected from: EHPAD's annual reports, SAMU of Essonne data base, SOS Doctors data and excel sheets and EclairAge dashboard of geriatric calls. We also conducted semi-structured interviews with relevant stakeholders to get an estimation of the parameter (integrating confidence intervals) when quantitative data was not available. To better understand the key variables and to improve our model, we conducted sensitivity analysis. Due to the confidentiality of some data, we cannot publish the parameters in this paper (but they were included in simulations).

5.4 Compute the Model and Calculate the Margin Balance

All modelled scenarios are now simulated so we can compare the margin balance of each stakeholder present in the financial flows. The following graphs (4, 5, and 6) show the sum of margin balances over time.

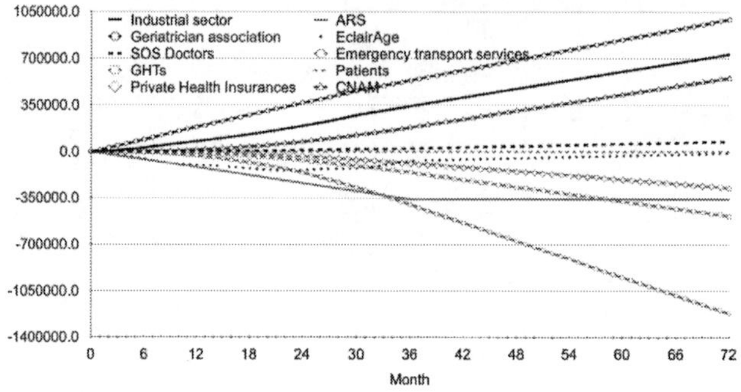

Figure 4. Total Margin Balances for BM1

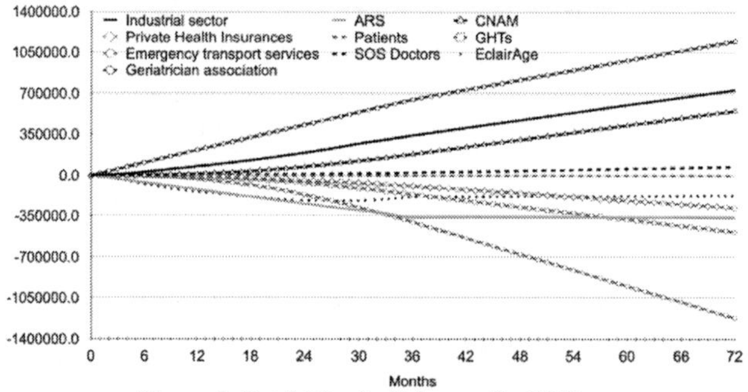

Figure 5. Total Margin Balances for BM2

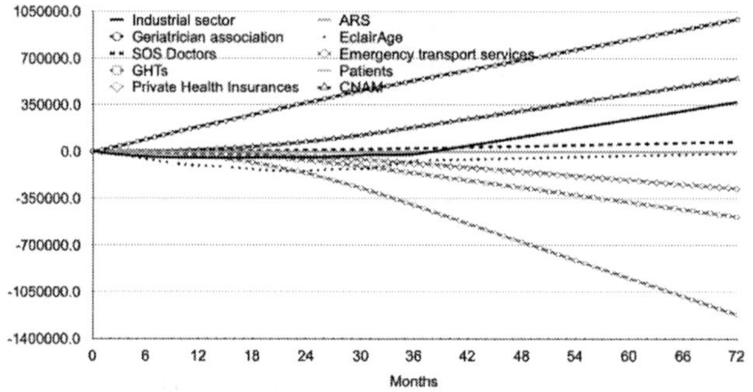

Figure 6. Total Margin Balances for BM 3

The figures show that the shared value is the same for at least the following stakeholders: SOS Doctors, GHTs, Patients, CNAM, Emergency Transport Services, and Private Health Insurance. In all cases, CNAM, SOS Doctors and the geriatric association Age have a positive margin balance if EclairAge is implemented. However, emergency transport services, private health insurances, GHTs are losing money in the process. The business models are designed so that patients don't spend any money, so their margin balances are null over time. Some differences are to be noted for the following stakeholders: Industrial sector, ARS, Age association, and EclairAge.

5.5 Scenario Comparison

The comparison of BM scenarios can provide some inputs for decision makers. Here, we focus on attracting new investors and managing key resources and activities.

Concerning the first scenario, modelling shared value created by innovation is a good way to attract investors. In fact, figure 7 shows that the industrial sector – especially software companies – has a certain interest in the deployment of EclairAge. In fact, in the current business model of care permanence, they are not at all implicated. Whatever new business model is chosen to be implemented, they will make profit. In the first two business models, they earn money as soon as the service is implemented in an EHPAD while in the last scenario they must wait about 3 years to have a return on their investment.

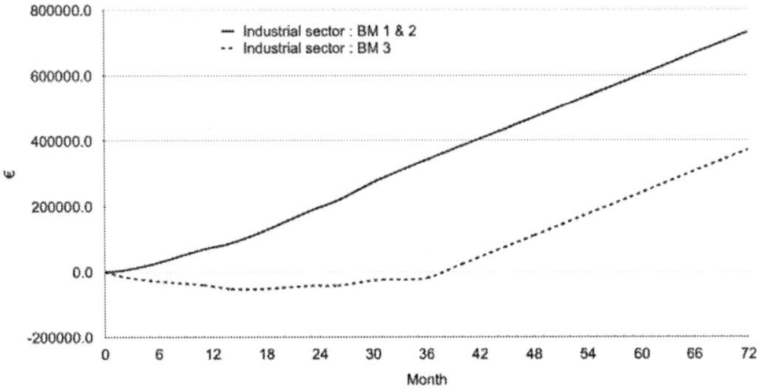

Figure 7. Total Margin Balances of the Industrial Sector

The French health system is not used to private investments. However, they can be extremely useful in the deployment of innovative services or products on the E-health market at a lower cost for the government and in the same time improving the quality of care services.

When managing key resources and activities, our tool can also be useful to compare training costs (see figure 8). As a result, on the one hand, it is more relevant to pay a monthly salary for EHPADs' training sessions during the first 3 years of EclairAge implementation. On the other hand, once EclairAge is entirely implemented, it is more pertinent to pay per training session. As a consequence, our tool provides inputs for cost forecasts thus adjusting the key resources and activities.

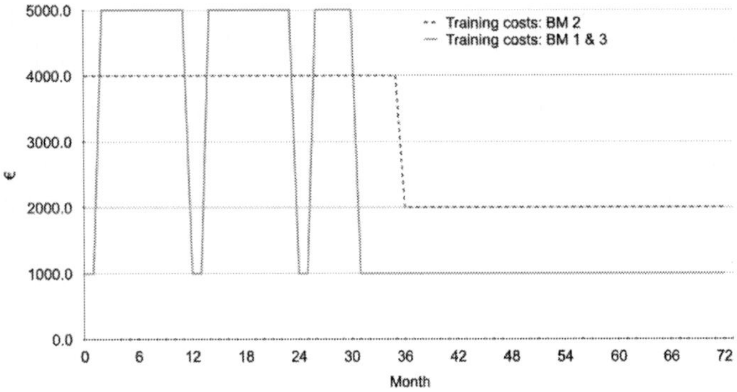

Figure 8. Training Costs

6. CONCLUSION

Our framework combining Financial Flow Modelling (FFM) and Business Model Canvas (BMC) provides 5 main benefits.

1) The framework reinforces the consistency of business models. Through scenarios simulation, the cost and revenue structure are in line with the resources and activities.

2) Systemic approach is useful to design business models for early stage innovative projects when entrepreneurs must react quickly. In fact, the FFM method allows entrepreneurs to test a wide variety of

business models and to choose the best one in accordance with the implementation scenario and market response.

3) The business model developed for the case study is both flexible and complete. Thus, we are able to model 3 coherent business models and select the most appropriate for the implementation of EclairAge tele-expertise service.

4) The results of empirical studies show that detailed business models are more pertinent for investors (Chanal, 2011). The detailed financial structure quantifying long and short term benefits can help convince some stakeholders to invest in innovative projects.

5) BMC together with FFM can be widely used for key business activities including: resource allocation, short and long-term planning, decision-making support, business strategy, and profit optimization.

This tool helps entrepreneurs compare and select the best business model canvas based on financial shared value. Further work is needed to implement a selection tool based on social value created. In fact, according to Porter (2011), generating social and medical value is an advantage for companies. Furthermore, the development and use of a second tool provides a multi-perspective approach that enhances both the quality and the validity period of the business model (Fendt, 2002).

REFERENCES

Al-debei, M., El-Haddadeh, R. and Avison, D. 2008. Defining the Business Model in the New World of Digital Business, *In Proceedings of the Americas Conference on Information Systems (AMCIS)*, pp. 1–11.

Chanal V., 2011. *Business Models dans l'innovation : pratiques et méthodes*. Presses Universitaires de Grenoble.

Chesbrough, H., and. Rosenbloom, R. 2002. The role of the business model in capturing value from innovation : evidence from Xerox Corporation' s technology spin-off companies, *In Ind. Corp. Chang.*, vol. 11, no. 3, pp. 529–555.

Christensen, C. and Raynor. M., 2013. *The innovator's solution: Creating and sustaining successful growth*. Harvard Business Review Press.

Fendt C., 2002. *Meaning, legitimacy and impact of business models in fast moving environments*, University of East London.

Jean, C., Duong, T., Cardinal, J., Jankovic, M., Bocquet J., and Espinoza P., 2016. Sharing economic value between the stakeholders of a telehealth project: methodological issues?, *In Eur. Res. Telemed.*, vol. 5, pp. 37–44, 2016.

Krumeich, J., Burkhart T., Werth D., and Loos, P., 2012. Towards a Component-based Description of Business Models- A State-of-the-Art Analysis, *In AMCIS 2012 Proc.*, p. 19, 2012.

Porter, M., and Kramer, M., 2011. Creating shared value - How to reinvent capitalism and unleash a wave of innovation and growth, *In Harvard Business Revue.*, vol. 39, pp. 62–77.

Osterwalder, A. and Pigneur, Y., 2010. *Business Model Generation*. John Wiley & Sons.

Zott, C., Amit, R. and Massa, L., 2011. The Business Model: Recent Developments and Future Research, *In J. Manage.*, vol. 37, no. 4, pp. 1019–1042.

"GDPR" IMPACT ON HEALTH DATA EXCHANGE IN EUROPEAN DIGITAL ENVIRONMENT

Paola Aurucci, Eleonora Ciceri, Mariet Nouri Janian, Andrea Micheletti and Alberto Sanna
Center for Advanced Technology in Health and Wellbeing, IRCCS Hospital San Raffaele, Italy

ABSTRACT

The aim of this paper is to provide practical guidance to ensure compliance of the SHiELD project with the General Data Protection Regulation (EU), 2016/679 also known as "GDPR". The SHiELD project is a H2020 project whose objective is to support the lawful exchange of clinical information across Europe and is built upon the epSOS project. GDPR remains in the path defined in 1995 and does not change the main elements of the model provided by the Directive. Notice and consent remain an important legal ground for the processing of sensitive data, this is why a two-steps-consent by the data subject to the transfer and processing of health data should be envisaged by the project. However, compared to the Directive 95/46/EC, the GDPR imposes stricter security obligations on data processors and controllers. This new Regulation simultaneously broadens the relevance of the risk - as it is explicitly based on the notion of risk/based approach – and a detailed Data Protection Impact Assessment must be undertaken and documented prior the project is implemented. In line with the new accountability principle, processors and controllers should also be able to demonstrate compliance with the GDPR. This paper, in the conclusions, also underlines that SHiELD project full compliance to the Regulation could be assessed only when all Member States have revised or adapted their legislation in order to comply with the GDPR end to exploit their discretional power, delegated to them by the EU legislator on specific subjects, like processing of genetic, biometric or health data.

KEYWORDS

Health Data, Security, GDPR, Interoperability, Cloud, Exchange

1. INTRODUCTION

Nowadays, European citizens are enabled to move freely across Member States. Freedom of movement helps people in many situations, as they are allowed to move to places where jobs are less scarce, or simply to travel across several countries without having to show passports at internal borders.

However, freedom of movements brings its own challenges, as one would expect personal data to travel freely with people, while this cannot be often guaranteed due to the sensitivity of data and the data protection rules that ensure the safety of those data is not corrupted. This is especially true with health data, that are considered highly sensitive (and thus in need of a high protection degree)[1], as they are not perishable and of great value. In this context, data often risk to stay locked in the country where they are produced, and there is a huge friction in making them travel across borders. This produces two specific issues. On the one hand, as people may be cured in a foreign country, clinical data risk to be scattered across several countries, dispersing the information one has on a person's health in separate sources. On the other hand, when a person travels abroad and incurs in an emergency situation which requires appropriate treatment, doctors find it difficult to access his information instantly and with high recall, so as to have all the needed information at hand and formulate a proper diagnosis. The SHiELD project[2] is a H2020 project whose objective is to support the lawful exchange of information across Europe. SHiELD builds upon the epSOS project[3], adapting its concepts to the new standards and data protection laws, and focuses on all those case studies in which a European citizen needs medical assistance when in another Member State, and physicians need to access data from different Member States.

---

[1]Article 29 Data Protection Working Party, 2007. Working Document on the processing of personal data relating to health in electronic health records (EHR) (WP 131). Available at: http://www.dataprotection.ro/servlet/ViewDocument?id=228 (Accessed: 28 May 2018).
[2] http://project-shield.eu/
[3] www.epsos.eu/

The architecture of SHiELD extends the one of epSOS[4], in which each Member State participates in the exchange of clinical information (in the form of, e.g., Patient Summary) via the usage of National Contact Points (NCPs). SHiELD adds up to this architecture by providing a set of cost-effective components that verify and monitor over time the compliance with General Data Protection Regulation (EU) 2016/679[5] (from now on "GDPR") – which has repealed Directive 95/46/EC (from now on "Directive") with effect from 25 May 2018[6] (Piciocchi, C. et al., 2017). Moreover, it will govern the processing of data contained in medical documentation, in electronic health records and in EHR systems, and will identify automatically potential security risks that would expose sensitive data, with the help of a system modeller exploiting a knowledge base of known threats. This is of vital importance in all those cases in which insecure devices (even mobile ones) are used to exchange or store clinical data.

With this in mind, the paper aims to provide practical guidance to ensure compliance of SHiELD project with the GDPR. After providing a short description of one of the SHiELD use cases (Chapter 2), this paper will focus on the new accountability principle, provided by Article 5 of the GDPR, which implies that this new law explicitly embraces risk/based approach, and in specific circumstances requires the controller to carry out a Data Protection Impact Assessment (from now on "DPIA")[7] (Chapter 3). The last part of the paper shall deal on the new discipline for processing special categories of personal data ("sensitive data") under the GDPR, with the aim to identify the legal basis for the processing of personal data in the framework of the analyzed SHiELD's use case. Finally, we discuss those aspects that will be fully analyzed to guarantee the SHiELD full compliance to the regulation only when all Member States have revised or adapted their legislation in order to comply with the GDPR end to exploit their discretional power, delegated to them by the EU legislator on specific subjects, like processing of genetic, biometric or health data (Pagallo U., 2017).

2. SHIELD USE CASE DESCRIPTION

One of the use cases addressed by SHiELD, stemming from the one proposed during the epSOS project, is a cross-border scenario in which a patient finds himself in an emergency situation while he is abroad, and clinicians need to access her clinical history to treat him properly.

The use case is built around two Member States: i) Country A, where the clinical data are stored; and ii) Country B, where the clinical data need to be retrieved to support the healthcare delivery. Both countries elect a National Contact Point (henceforth, NCP), that is a legal entity assuming nationally all legal duties. Cross-border data exchange will take place between the two NCPs when there is the need from some healthcare facility in Country B of treating a patient from Country A. In this context, a natural person living in Country A may be interested in sharing cross-border her clinical information, to best respond to all those situations in which clinical assistance is needed while abroad. She thus enrols in the SHiELD project and stipulates via the offered tools her first consent to transfer information to some other Member States. In this consent she specifies which information she would like to share with other Member States, and chooses where to share it (i.e., which Member States will be able to access it) and when to share it (i.e., which will be the period of time during which the information will be accessible from other States). In the unlikely event of a clinical emergency raising while the person is in Country B, which may see her unconscious, she is brought to a healthcare centre seeking for assistance. The healthcare staff, after confirming the patient's identity and her second consent to process data for the purpose of this very treatment, issues a data retrieval request to the NCP of Country B, which will operate the exchange of information with the NCP of Country A if it is compliant with the first expressed consent.

[4]Article 29 Data Protection Working Party, 2012. Working Document 01/2012 on epSOS (WP 189). Available at: http://ec.europa.eu/justice/article-29/documentation/opinion-recommendation/files/2012/wp189_en.pdf

[5]General Data Protection Regulation (EU) 2016/679. Available in English: http://ec.europa.eu/justice/data-protection/reform/files/regulation_oj_en.pdf

[6]This act, approved in May 2016, has repealed Directive 95/46/EC with effect from 25 May 2018. This two-year period allowed Member States to revise or adapt their legislation in order to comply with the GDPR.

[7]Article 29 Data Protection Working Party (2007) Guidelines on Data Protection Impact Assessment (DPIA) and determining whether processing is "likely to result in a high risk" for the purposes of Regulation 2016/679. Available at: http://ec.europa.eu/newsroom/article29/item-detail.cfm?item_id=611236 (Accessed: 28 May 2018).

3. ACCOUNTABILITY PRINCIPLE AND DATA PROTECTION IMPACT ASSESSMENT

General principles of data processing lay down in Article 5 of the GDPR and broadly replicate those in the Directive. However, GDPR, beside developing more fully certain concepts, adds a new principle (accountability principle)[8], which provides that "*controller shall be responsible for, and be able to demonstrate compliance with the GDPR*". Furthermore, Article 24 (1) specifies that controllers are required to adopt appropriate and effective protective "*technical and organisational measures*" to ensure and demonstrate their GDPR compliance. These measures have to be chosen by controller "*taking into account the nature, scope, context and purposes of processing as well as the risks of varying likelihood and severity for the rights and freedoms*" of individuals[9].

These accountability provision implies that GDPR broadens the relevance of the risk as it is explicitly based on the notion of risk/based approach, that it is first and foremost an effective tool for ensuring a high level of protection of the rights and freedoms of individuals. It enables all stakeholders to dedicate their resources to the areas where the risks and potential harms for individuals are most significant, to mitigate these risks. This in turn creates better outcomes and effective protection for individuals. This does not mean that the protection of the rights of individual depends on the level of risk. Individual's rights apply full irrespective of the level of risk. However, controllers must build, implement and able to demonstrate a comprehensive data privacy compliance program, that is modulated on several factors including specifically the risk level to the fundamental rights and freedom of individuals (risk-based approach) taking into account the nature, scope, context and purposes of the processing (scalability principle).

This does not absolve the organization from the overall obligation to comply with the GDPR in respect of processing. It only means that the actual controls, specific compliance obligations, verifications and possible future penalties will be more intensive in respect of processing that creates risks and freedoms of individuals. In concrete, referring to the SHiELD project, the incorporation of a risk-based approach to data protection require that prior the processing, the controllers don't have the obligation to notify their national supervisory authority, but must assess the "*likelihood and severity of the risk*" of their personal data processing operation to the fundamental right and freedoms of individuals. This is an operation to do *ex ante* on a case-by-case basis and will vary depending on the complexity of the processing. Furthermore, the accountability principle require that processors must indeed also verify the effectiveness of these measures *ex post*, as demonstrable proof of risk minimization. When the processing is "*likely to result in a high risk to the rights and freedoms of natural persons*" [Article 35 (1) illustrated by Article 35 (3) and complemented by Article 35 (4)] a detailed DPIA[10] must be undertaken and documented. However, Article 35(3) provides a non-exhaustive list examples when a processing is "*likely to result in high risks*" that include the "*processing on a large scale of special categories of data referred to in Article 9 (1)*" [Article 35(3(b)]. In addition, a data controller subject to the obligation to carry out the DPIA "*shall maintain a record of processing activities under its responsibility*" including inter alia the purposes of processing, a description of the categories of data and recipients of the data and "*where possible, a general description of the technical and organizational security measures referred to in Article 32 (1)*"[11] and must assess whether a high risk is likely, even if they ultimately decide not to carry out a DPIA. Keeping in line with the risk-based approached embodied by the GDPR, according to which organizations that control the processing of personal data are encouraged to implement protective measures corresponding to the level of risk of their data processing activities for ensuring compliance with the GDPR's general obligations, where is "*likely to result in a high risk to the rights and freedoms of natural persons*"[12], the SHiELD's use case indeed leads the systematic evaluations or processing special categories of data. For this reason, a detailed DPIA must be undertaken and documented. Article 35 (1) states that "a *single assessment may address a set of similar processing operations that present similar high risks*". Recital 92 adds that "*there are circumstances under which it may be reasonable and economical for the subject of a data protection impact assessment to be broader than a single project, for example where public authorities or bodies intend to establish a common application or processing platform or where several controllers plan to introduce a common application or processing environment across an industry sector or segment or for a widely used horizontal activity*".

[8] Article 5 (2) of the GDPR.
[9] Article 24 (1) of the GDPR.
[10] Ibidem, supra n. 7, p. 18.
[11] Article 30 (1) of the GDPR
[12] Article 35 (1) of the GDPR.

Considering SHiELD project, this means that controllers may use a single DPIA to assess multiple processing operations that are similar in terms of the risks presented, provided adequate consideration is given to the specific nature, scope, context and purposes of the processing. On the other end joint controller in SHiELD (NCPs) need to define their respective obligations precisely. Their DPIA should set out which party is responsible for the various measures designed to treat risks and to protect the rights of the data subjects. A DPIA can also be useful for assessing the data protection impact of a technology product, for example a piece of hardware or software, where this is likely to be used by different data controllers to carry out different processing operations. Of course, the data controller deploying the product remains obliged to carry out its own DPIA with regard to the specific implementation, but this can be informed by a DPIA prepared by the product provider, if appropriate.

4. LEGAL BASES FOR PROCESSING

As set out in the Opinion on the processing of personal data relating to health in electronic health records (EHR)[13] of the Article 29 Working Party (WP29), the independent European Union Advisory Body on Data Protection and Privacy, composed of representatives from each of the EU Member States, the European Data Protection Supervisor, and the representative of the European Commission[14], all data contained in medical documentation, in electronic health records and in EHR systems should be considered to be "sensitive personal data" under the Directive. These data are broadly the same as "special categories of data" in the GDPR[15], that now specifically include - other than data concerning health – *"genetic data and biometric data"* where processed *"to uniquely identify a person"*[16].

Protecting these data is basic requirement to guarantee confidentiality of diagnostic and therapeutic information. Although, since they could be used in a discriminatory way, and is likely to be of a private nature, processing of such data in GDPR is subject - as with the Directive - to more stringent conditions than other forms of personal data. This is why, these data are not only subject to all the general rules on the protection of personal data, but also subject to the special data protection rules on the processing of sensitive information contained in Article 9 of the GDPR. The nature of the data is also a factor to consider in deciding what security is appropriate (See Conclusion). Article 9 (1) expressly prohibit - as a general rule - that processing of "special categories of data". However, considering the importance of using information about a patient in order to provide appropriate medical treatment, there are exemptions to the general prohibition of processing these data. Article 9 (2) sets out the circumstances in which the processing of special categories of data, which is otherwise prohibited, may take place.

Exceptions under Article 9 broadly replicate those under the Directive and considering SHiELD use case framework described in Section 2, either explicit consent [(Article 9 (2) (a)] and vital interest of the data subject [Article 9 (2) (c)] (if the patient is not conscious) shall be considered as appropriate legal basis for the processing of personal data. However, it is important to mention that Article 9 (4) broads ability for Member States to adduce new conditions (including limitations) regarding the processing of genetic, biometric or health data. As such, existing differences in approach on these topics, among Member States, will likely be maintained, and further divergence will be permitted. Since SHiELD platform processes "genetic, biometric or health data of data" controllers should continue to keep the development of relevant national law under review (Pagallo U., 2017).

4.1 Article 9 (2)(a): Explicit Consent and Information

Article 9 of the GDPR provides a list of possible exemptions to the ban on processing special categories of data (Granieri M., 2017). The first exemption listed is the situation where the data subject provides consent to the use of this data.

---

[13] Ibidem, supra n. 2, p. 7.

[14] Article 29 Data Protection Working Party: http://ec.europa.eu/newsroom/just/item-detail.cfm?item_id=50083 (Accessed at: 28 May 2017).

[15] Article 9 of the GDPR.

[16] Article 9 (1) of the GDPR.

The necessary elements composing "regular" consent are listed in Article (4) (11) and elaborated in the brand new WP29 Guidelines on consent[17]. Article 4 (11) of the GDPR stipulates that valid consent of the data subject means any: freely given[18], specific, informed and unambiguous indication of the data subject's wishes by which he or she, by a statement or by a clear affirmative action, signifies agreement to the processing of personal data relating to him or her. "Unambiguous" consent is a new requirement but does not represent a change for practical purposes. Recital 32 suggests that this may be signified that "silence, pre-ticked boxes or inactivity should therefore not constitute consent." GDPR indeed prescribes that "clear affirmative act" is a prerequisite for valid consent. The new Regulation aims to raise consent requirements to a higher standard compared to the Directive, thus providing guidance in Article 7 (2), (3), (4) and in recitals 32, 33, 42, and 43 to clarify which additional arrangements controllers should make to reach these standard and to ensure they obtained, maintained and are able to demonstrate, consent of a data subject in line with the GDPR.

Since serious data protection risk may emerge from the processing of sensitive data and, therefore, of EHR, the GDPR, as the Directive, prescribe that consent under Article 9 (2) (a) should also be explicit. This term refers to the way consent is expressed by the data subject. WP29 Guidelines on Consent[19] specify that opt-out solutions will not meet this requirement, while confirming consent in a written statemen make this consent explicit. However, according to the text of the GDPR, written and signed statements is not requested in all circumstances that require valid explicit consent. In the digital or online environment, for instance, like that of SHiELD use case, a data subject may be able to issue the required statement by filling in an electronic form, by sending an email, by uploading a scanned document carrying the signature of the data subject, or by using an electronic signature.

In line with the general principle of accountability which runs throughout the GDPR, Article 7(1) requires that where consent is relied on a ground for lawful processing, controllers have the explicit obligation to demonstrate a data subject's consent. The rest of Article 7 is dedicated to setting out the conditions for a valid consent.

In the end, Article 9 (2) (a) acknowledges that there may be cases of processing of sensitive data in which not even explicit consent of the data subject should lift the prohibition of processing: Member States are free if, and how to regulate such cases in detail. Furthermore, as seen in Section 4, Member States are entitled, under Article 9 (4) GDPR, to maintain or impose further conditions (including limitations) in respect of genetic, biometric or health data.

4.1.1 SHiELD Use Case: Two-steps-Consent

According to this analysis, as for epSOS project a two-steps-consent by the data subject to the transfer and processing of health data should be envisaged. Firstly, an explicit consent should be given to the participation of the data subject in the SHiELD project or parts of it.

This consent should be given in Country A to allow the health care providers to prepare specific data with the intention to make them available in future to other health care providers in the framework of SHiELD. Consent to processing is distinguishable, clear, and is not "bundled" with other written agreements or declarations (according to Article 7(2)). The controller makes clear that there is no obligation to participate and asks for consent to use sensitive data for this (exclusive) purpose (according to Recital 43). The need of a first consent could also provide the side benefit of limiting the number of data subjects available within the system to the strictly necessary and thereby also limiting the potential damage in case of unlawful access to this sensitive data. In order to facilitate making the benefits of project available to data subjects who need treatment in Country B but have not previously given consent in Country A, the SHiELD project could investigate the possibility of allowing patients to give also their first consent for instance in a secure way over the Internet in Country B. The second consent shall be given explicitly for the processing of health data in the case of actual treatment in Country B. As mentioned above, one of the preconditions for a valid consent is that the data subject has received information (from the controller or his representative) which satisfies the requirements of the Articles 12-14 and Recitals 42, 58, 60, 61 and 62 ("informed consent"). In few words,

[17] Article 29 Data Protection Working Party, *Guidelines on Consent under Regulation 2016/679*, WP 259, 28 November 2017. Available in English: http://ec.europa.eu/newsroom/article29/document.cfm?doc_id=50053.

[18] Recital 43 of the GDPR.

[19] Article 29 Data Protection Working Party, 2018, Guidelines on consent under Regulation 2016/679. Available at: http://ec.europa.eu/newsroom/article29/item-detail.cfm?item_id=623051 (Accessed: 28 May 2018).

under the principle of “fair and transparent” processing organizations will need to provide extensive information to individuals about the processing of their personal data, unless the individual already has this information. The GDPR requires more extensive information to be provided than the Data Protection Directive – although much of the additional information is already mandatory in some Member States.

Going back to the use cases, the information regarding the first consent should contain a comprehensive, clear and understandable description of the SHiELD project, mentioning at least the categories of data that would be transferred by which health care providers to which other health care providers and other institutions.

This includes the institutions involved in the further processing of the data in Country B for other purposes than to provide and manage health care. Information must also be provided about the purpose of the transfer, legal basis for processing, identity and contact details of the controller (or its representative, for a non-EU established controller); contact details of the Data Protection Officer, and how long the data would be stored.

Finally, it must be made clear that there is the option of withdrawing consent at any time. The data subject should also be informed about the right of access and rectification of data concerning him/her. The information given before the second consent must at least contain the explanation which health care provider and other institution will process which categories of data and for which purpose. The patient should also have the possibility to give his/her consent only to the transfer and processing of certain categories of health data (modular access rights for health care providers in another country). Finally, it must also be made clear that there is the option of withdrawing consent at any time. The data subject should also be informed about the right of access and rectification of data concerning him/her.

The controller may also have to provide additional information if, in the specific circumstances and context, this is necessary for the processing to be fair and transparent. The information must be provided in a concise, transparent, intelligible and easily accessible way, using clear and plain language (in particular where the data subject is a child).

4.2 Vital Interests of the Data Subject

Article 9 (2) (c) of the GDPR sets out that the processing of sensitive personal data can be justified if it is necessary to protect the vital interests of the data subject or of another person where the data subject is physically or legally incapable of giving his consent. The processing must relate to essential individual interests of the data subject or of another person and it must – in the medical context – be necessary for a life-saving treatment in a situation where the data subject is not able to express his intentions (emergency case).

Accordingly, this exception could be applied only to a small number of cases of treatment and only where the first consent of the two-steps-model has been given. It is of great importance that the scope of this exception should be narrowly defined as to when and how it can be applied. Also, technical measures should be employed in order to prevent misuse of the emergency case (Knoppers B.M., 2014). With regards to the exchange of data in such emergency cases, the data subject should be informed about this possibility in the general information about the SHiELD project, before the first consent is given. Furthermore, considering the importance of the transparency requirement, in this situation it is especially important, that the patient is given access to information about the transmissions that have taken place after the emergency is over. Therefore, information about data transmission have to be properly documented by data controller, allowing possible later checkup of proper data usage. This mechanism is in line with the new accountability principle crystallizedH in Article 5 of the GDPR.

5. DATA SECURITY

Data security plays a prominent role in the GDPR. Compared to the Directive 95/46/EC, the GDPR imposes stricter obligation on data processors and controllers with regard to data security while simultaneously offering more guidance on appropriate security standards (Green M.D. et al., 2000).

The GDPR also adopt for the first time a specific data breach notification requirement. According to the accountability principle controllers and processors have to be compliant with security obligation and able to demonstrate to the authority the compliance. The GDPR also allows them to demonstrate it by either adhere to an approved code of conduct as referred to in Article 40 or an approved certification mechanism as referred to in Article 42. Under Article 32, which delineates the GDPR's "security of processing" standards, controllers and processors are required to "implement appropriate technical and organizational measures" taking into account "the state of the art and the costs of implementation" and "the nature, scope, context, and purposes of the processing as well as the risk of varying likelihood and severity for the rights and freedoms of natural persons". These measures are considered appropriate if they are able to remedy the risk [Article 32 (1)] that are presented by processing, "in particular from accidental or unlawful destruction, loss, alteration, unauthorized disclosure of, or access to personal data transmitted, stored or otherwise processed. Unlike the Directive, however, the GDPR provides specific suggestions for what kinds of security actions might be considered "appropriate to the risk," including: i) pseudonymisation and encryption of personal data (Bolognini L. et al., 2017); ii) ability to ensure the ongoing confidentiality; iii) integrity; iv) availability and resilience of processing systems and services; v) ability to restore the availability and access to personal data in a timely manner in the event of a physical or technical incident; vi) a process for regularly testing, assessing and evaluating the effectiveness of technical and organizational measures for ensuring the security of the processing.

Because of the way SHiELD operates and given the highly sensitive nature of the data processing it is necessary to provide "sufficient guarantees to implement appropriate technical and organizational measures" to meet the GDPR requirements and protect data subject rights.

The GDRP separates responsibilities and duties of data controllers and processors, obligating controllers to engage only those processors that provide "sufficient guarantees to implement appropriate technical and organizational measures" to meet the GDPR's requirements and protect data subjects' rights. Processors must also take all measures required by Article 32, which delineates the GDPR's "security of processing" standards.

Finally, in line with the accountability principle laid down by the GDPR, data controllers and data processors in SHiELD project should develop or update their internal breach notification procedures, including incident identification systems and incident response plans. Such procedures should be regularly tested and re-reviewed. To render the data safe during the data exchanges, secure communication protocols and end-to-end encryption must be adopted based on encryption standards for securing electronic communications. This has to be done in order to prevent acquisition/disclosure of the information in the presence of any intermediaries not controlled by the patient or the authorized healthcare organizations. Endpoints of the encryption must be located within environments either controlled by the patient directly or by the professional healthcare organizations authorized by the patient for processing his/her medical data. Where communications of health data take place by way of applications that rely on publicly available networks (e.g., the Internet) for data exchanges, the system should envisage the use of reliable digital certificates for both the server systems delivering the service and the client devices accessing the data. An unbroken chain of digital evidence must be preserved to safeguard the integrity and authenticity of the health data in an end-to-end manner.

6. CONCLUSION

GDPR aims at is strengthening both individual's rights and the powers of the European authorities, while reinforcing obligations and responsibilities of data controllers - through a directly enforceable hard law-tool, such as EU regulation. The regulation indeed will do away with the huge variation of national law on data protection between member state - due to the different interpretation in transposing the Directive 95/46/EC. As a consequence, the current fragmentation and costly administrative burdens will be overcome and the legal environment will be more harmonized and simple making easier for companies, both domestic and foreign, to conduct their business activities in the EU, including through international data exchanges. The EU will combine openness for international data flows with the highest level of protection for individuals. It has the potential to become a hub for data services, which require both free flows and trust. The Regulation, however, on specific provisions still live some discretion to Member States and this report can't touch the provision the rules on personal data that will be established by national laws with the application of the regulation on May 2018. Many of the aspect linked with the compliance to the GDPR of SHiELD use case, like changes in the data controller and data processor relationship and actual legal bases for processing, that could be fully analyzed only after the Regulation will effectively enter in force.

ACKNOWLEDGEMENT

This work is partially funded by the EU H2020 Frameweork Programme under project SHiELD (grant agreement No. 727301).

REFERENCES

Article 29 Data Protection Working Party, 2007. Working Document on the processing of personal data relating to health in electronic health records (EHR), WP 131. Available at: http://www.dataprotection.ro/servlet/ViewDocument?id=228.

Article 29 Data Protection Working Party, 2017. Guidelines on Consent under Regulation 2016/679, WP 259, 28 November. Available at: http://ec.europa.eu/newsroom/article29/document.cfm?doc_id=50053.

Article 29 Data Protection Working Party, 2017. Guidelines on Data Protection Impact Assessment (DPIA) and determining whether processing is "likely to result in a high risk" for the purposes of Regulation 2016/679. Available at: http://ec.europa.eu/newsroom/document.cfm?doc_id=47711.

Article 29 Data Protection Working Party, 2018, Guidelines on consent under Regulation 2016/679. Available at: http://ec.europa.eu/newsroom/article29/item-detail.cfm?item_id=623051 (Accessed: 28 May 2018).

Bolognini L. et al., 2017. Pseudonymization and impacts of Big (personal/anonymous) Data processing in the transition from the Directive 95/46/EC to the new EU General Data Protection Regulation. In *Computal Law Secure Rev*, Vol. 33, No. 2, pp 171–181. Doi: https://doi.org/http://dx.doi.org/10.1016/j.clsr.2016.11.002.

General Data Protection Regulation (EU) 2016/670. Available in English: http://ec.europa.eu/justice/data-protection/reform/files/regulation_oj_en.pdf

Granieri M., 2017. Il trattamento di categorie particolari di dati personali nel Reg. UE 2016/679. In Le Nuove leggi civili commentate, Vol. 3, No. 1, pp. 165-190. Available at: https://dialnet.unirioja.es/servlet/articulo?codigo=5966213

Green M.D. et al., 2000. Reference guide on epidemiology. Reference manual on scientific evidence. In *Federal Judicial Center*. Available: http://www.fjc.gov/public/pdf.nsf/lookup/sciman06.pdf/file/sciman06.pdf. Accessed 29 June 2012

Knoppers B.M.,' 2014. International ethics harmonization and the global alliance for genomics and health. In *Genome Med*, Vol. 6, N. 2, p. 13. Doi: https://doi.org/10.1186/gm530.

Mittelstadt, B.D., Floridi, L., 2016. The Ethics of Big Data: Current and Foreseeable Issues in Biomedical Contexts. In *Sci Eng Ethics*, Vol. 22, No. 2, pp. 303-41. Doi: http://dx.doi.org/ 10.1007/s11948-015-9652-2.

Mostert, M., et a., 2016, Big Data in medical research and EU data protection law: challenges to the consent or anonymise approach. *In Eur J Hum Genet*, Vol. 24, No. 7, pp. 956-60. Doi: http://dx.doi.org/10.1038/ejhg.2015.239.

Pagallo U., 2017. The Legal Challenges of Big Data: Putting Secondary Rules First in the Field of EU Data Protection". *In European Data Protection Law Review*, Vol. 3 No. 1, pp. 36-46. Available: http://hdl.handle.net/2318/1640445

Piciocchi C.et al., 2017. Legal issues in governing genetic biobanks: the Italian framework as a case study for the implications for citizen's health through public-private initiatives. In *J Community Genet*, pp. 1–14. Doi: https://doi.org/10.1007/s12687-017-0328-2.

HOW PORTUGAL ACHIEVE THE DEMATERIALIZATION OF THE MEDICAL CERTIFICATE FOR DRIVING LICENSE

Rui Pereira, Alfredo Ramalho, Arlete Monteiro, Cristiano Marques and Henrique Martins
SPMS
Av. Da República, N.º 61, 1050-189 Lisboa, Portugal

ABSTRACT

Promoting the easy interaction between the citizens with the multiple sectors of public administration is a tendency in e-government, which have grown hugely in the last year. It is acknowledged as Whole-of-Government.
The Portuguese Government have been endorsing a several of initiatives with this aim and this paper describes the full process of how the dematerialization of the medical certificate for driving license was achieved. From the involvement of the different entities, to the information system architected that promotes the transmission of this information.
One year since its implementation and with more than 780.000 medical certificates issued and transmitted electronically, it was important to review all the process, as well as provide an overview of all the medical certificates issued, the HL7 FHIR messages exchanged and what can be improved to provide a better, simpler and reliable service to all citizens.
This success case taught us the importance of the correct evolvement and articulation of the multiple sponsors of this WoG initiative, of the interoperability between every sponsor, etc. With this process we ensure a greater control, simplification and optimization of the medical certificate for driving license issuing. All these factors can be availed in other dematerialization processes.

KEYWORDS

E-Government, Whole-of-Government, Digital Transformation, Interoperability

1. INTRODUCTION

Making citizens and businesses everyday life easier – by cutting red tape, reducing compliance costs and using ICT to deliver better public services – has become a priority goal for the Portuguese Government (AMA, 2016). There is actually no other way forward, as the availability of electronic governmental information and digital services, frees citizens from space and time restrictions in these new ways of serving called e-government. eGov solutions have a huge potential in promoting transparency, accountability, efficiency and citizen engagement in public service delivery (Palvia and S Sharma, 2018).

In the last years, a new trend in e-government is evident : the evolution towards the provision of a shared response to an issue or life-cycle event integrating public services across several sectors, called whole-of-government (WoG) services (Christensen and Lægreid, 2007) (Ojo et al., 2011). This tendency makes it easier for the public to interact with the public administration and obtain a suitable response to their needs. However, achieving the creation of such services can be very slow due to all the challenges and constraints related to institutional dynamics, regulations, technological difficulties, capacities and resources as well as cultural and developmental dimensions (Ojo et al., 2011) (Seri et al., 2014).

With this aim, in the year 2006, the Portuguese Government launch a scheme called Simplex. Each year, public bodies from all government departments propose simplification initiatives and establish a deadline for its implementation. Simplex also includes ideas directly suggested by the civil servants who participate on the "Simplex Idea Award"; as well as the proposals made by citizens, enterprises and associations who contribute to the Public Consultation.

The main outcomes of this initiative are:

- Simplification of laws and procedures;
- Dematerialization of forms and procedures;

- Elimination of unnecessary certificates;
- Consolidation of existing legal rules;
- Easier access to public services.

Therefore, the current paper pretends to demonstrate all the process that was necessary to achieve a dematerialization of a certificate, with special prominence of the healthcare sector.

2. CONTEXTUALIZATION

The 2016 United Nations e-Government index has showed that the rank of e-government development is led by the United Kingdom of Great Britain and Northern Ireland, followed by Australia and the Republic of Korea. Portugal occupies the 38th place, with an E-Government Development Index (EGDI) of 0.7144 (Nations, 2016). This value is approximated with the average registered in Europe (0.7241) and much higher than the Worlds' EGDI average (0.4922).

In the last years, a lot of good initiatives have appeared, promoting the growth of the WoG. For example, assuring the interoperability in the public administration was one of the great initiatives that have emerged. This was achieved with the creation of an interoperability gateway known as iAP. In the healthcare domain, the Shared Services in Ministry of Health (SPMS) developed a set of components with the aim of promoting the exchange of messages between the various applications at the local (healthcare unit), national or international levels (Ojo et al., 2011).

The Local Interoperability Gateway for Healthcare (light) is responsible for providing an integration mechanism for the internal systems, managing the interfaces in a reliable way and is standardized and highly parameterizable. It is oriented to local integrations, that is, within the institutions themselves, and their communication is based on HL7 v2.5 messages. Communication with the central PNB (Portuguese National Broker) system is done through the HL7 Fast Healthcare Interoperability Resources Specification (FHIR) standard, which in turn can be integrated with iAP, promoting the communication of the Health Ministry with others ministries. Figure 1 pretends to represent the interoperability promoted by these mechanisms.

Figure 1. Interoperability at the Healthcare and WoG Ecosystems

Other initiative that have contributed to the development of the WoG was the SIMPLEX. SIMPLEX is the main ongoing modernization and administrative simplification programme in Portugal. First launched in 2006, SIMPLEX had implemented, by 2011, over 1000 measures of administrative and legislative simplification, including several eGovernment initiatives. In 2016 the programme was launched again, as SIMPLEX+, with the same principals and branding. The overall coordination of SIMPLEX+ lies on the Ministry of the Presidency and Administrative Modernization and the design, management and monitoring of the program are held by the Assistant Secretary of State for Administrative Modernization, with the technical support of AMA. Mechanisms of transversal control through focal points and involvement and evaluation of citizens and companies are in place. The SIMPLEX + 2016 program includes 255 measures, some of them with a WoG perspective. More focused in a health care context, the national strategy for the health information ecosystem (called ENESIS 2020) also provides a common vision for the area of Health Information Technology, based on wide participated governance structures and processes, guiding the action plans and initiatives of the various health information ecosystem (eSIS) stakeholders. It aims to progress as a benchmark of good practices, promoting benefit delivery and risk and resource optimization.

3. MEDICAL CERTIFICATE FOR DRIVING LICENSE DEMATERIALIZATION PROJECT

3.1 Why the Dematerialization Road?

It all started with the conceptualization. With the aim of launching a SIMPLEX programme, for about four months, suggestions for the simplification were collected among citizens, enterprises and also from the public administration. The compilation of all these suggestions are the essence of the SIMPLEX program, which was announced on the 19th of May of 2016.

One of the thirty-eight measures of the SIMPLEX+ 2016 initiatives that the Portuguese Ministry of Health were involved was called "Carta Sobre Rodas" (which can be translated to 'Driving License on wheels'). The responsibility for this initiative was attributed to the Ministry of Planning and Infrastructures, with the collaboration of four other ministries: Ministry of the Presidency and Administrative Modernization; Ministry of Justice; Ministry of Health; Ministry of Internal Affairs.

This initiative aimed at simplifying and reducing the delay in the issuance of Driving Licenses, allowing its issue and revalidation with the usage of the Citizen's Card, without leaving home and it included:

- Automatic sending of a Medical Certificate to the Institute for Mobility and Transport (IMT):
- Removing the address of the title;
- Automatic transmission of photography and signature;
- Sending an SMS or email to let the citizen know that the charter will reach the address chosen by him/her;
- Public publication of quality service indicators (such as the average time to the emission of the license driver).

3.2 How did it Enfolded?

The first step of this process was to understand how was the process at that moment, the existing legislation and what could be improved.

When a citizen pretends to obtain (or renew) his/hers Portuguese driving license, initially he/she must go to a doctor for the assessment of the physical and mental aptitude. This medical examination is intended to evaluate the physical and mental conditions of the candidate or driver, and the Physical and Mental Assessment Report is accomplished by the doctor. Once the evaluation is finalized, the doctor issues the medical certificate.

By the Portuguese law (as well as all the existing regulations), any doctor in the exercise of his professional activity can issue a medical certificate for the driving license. This medical certificate indicates if the citizen is suitable or not to obtain his/her driving license, depending on the evaluation performed before and, if applicable, the respective restrictions and / or adaptations.

The medical certificate is then delivered to the citizen, who must bring this paper to the competent IMT services, to have his/her driving license issued.

In 2015, more than 600.000 paper certificates were registered at the IMT, which means more than 600.000 people literally moving from medical care (public or private context) facilities to the IMT installations with a paper in their hand.

Providing a simpler and quicker service to citizen is very important, to achieve a greater rigor at the medical evaluation, also as a way to improve quality of information as well as to achieve efficiency gains.

Therefore, an inter-sectorial work group was formed, moderated by the Administrative Modernization Agency, with the aim to guarantee that this SIMPLEX measure did succeed. Within this context, several meetings occurred at top management and technical levels to guarantee all milestones required (regulations, requirements, specifications and communications, among others).

3.3 What about the Legal Component?

Once Simplex program was defined and published, work has begun to develop the necessary legislation to enable the transformation process. Decree-Law No. 40/2016, published in July 2016 provides the adequate framework to the WoG process, including roles and responsibilities of the involved organizations.

As stated before, there is Portuguese legislation that requires that drivers and drivers' candidates have the necessary physical, mental and psychological skills to exercise safely the driving of motor vehicles. This new version of the legislation redefines the new minimum requirements for physical, mental and psychological skills required of candidates and drivers.

Moreover, it introduces the electronic issuance and transmission of the medical certificate for the driving license, which was defined to become mandatory from 15 of May of 2017 for health care establishments, regardless of their legal nature, integrated in the National Health Service and for establishments providing health care in the private and social sector.

This legislation also grants to the Shared Services of the Ministry of Health the competence to define and publicize the standards, requirements and specifications of computerized electronic health certificate issuance systems and to promote their implementation by the various operators.

Later, the models and respective contents of the Physical and Mental Assessment Report, Medical Certificate, Psychological Assessment Report and Certificate of Psychological Assessment of the Medical Certificate were defined by IMT and DGS and transported into a Joint Dispatch published in February 2017.

3.4 So, what was the Technical Solution?

The architecture chosen for the system privileges the maintenance of records at the level of the User's Electronic Clinical Process, which must be managed and maintained locally. Each healthcare unit, regardless of whether it is public or private, may use its clinical software. It should make available to health professionals the possibility of electronic registration of the data of the Medical Certificate for the Driving License and its electronic communication to the central health database.

Therefore, a central component was developed, responsible for receiving and validating the requests for certificates from the various software used by physicians in Portugal. The necessary requirements and specifications for the integration between each clinical software and the central health database were published. To help this integration, several meetings were promoted to help the different vendors integrate with the central system, as well as technical sessions were the vendors could make the necessary tests and quickly clarify any doubt/problem that was encounter.

All clinical local software developed new functionality to comply with the Joint Dispatch and technical architecture requirements.

The central component interoperates with the health interoperability platform, the Portuguese Nacional Broker (PNB), that sends them to the Public Administration Interoperability Platform (iAP) to be processed by the IMT systems that acknowledge de operation success. SPMS defined the adoption of the international standard HL7 FHIR for information exchange in the health information ecosystem. Since this standard does not add value to the business of mobility and infrastructures, iAP maps the messages coming from the PNB to XML (which is the format specified by the IMT).

The functional model representation of the technical solution can be seen at figure 2, where we have the doctor issuing the medical certificate for the driving license to a citizen, either on the National Health Service or on a private context, using a public or private software. The information of the medical certificate is integrated with the central system, which is responsible to validate is content. The central system validates the information received from either a technical point (for example, the correct message construction) or from a functional point (for example, the restriction is valid). If any error is detected, the error is communicated to the clinical software. In case of success, there is also a communication to the clinical software (message of success), and it's triggered the transmission of the information of the medical certificated to the IMT, through the interoperability mechanisms (PNB and iAP). The standard HL7 FHIR is used in the integration between the Central Database called ACC and PNB, and also in the integration between PNB and iAP.

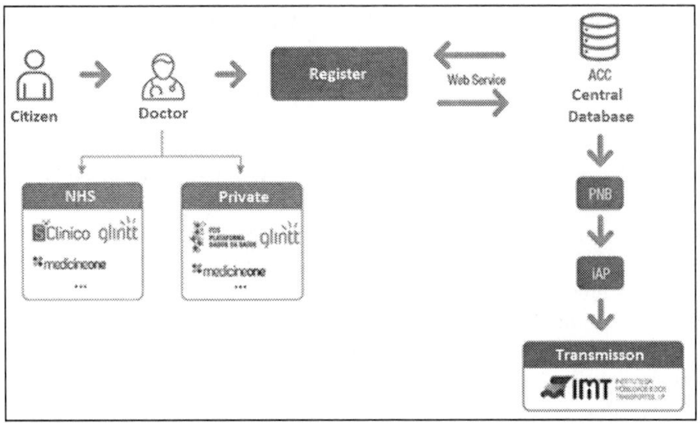

Figure 2. Functional Model Representation

It's important to state that the proper infrastructure, operation processes and auditing mechanism were implemented in the solution.

3.5 What about the Deployment?

So, with the architecture validated and all the entities involved aligned, it was time to make the necessary developments, both on the Electronic Health Record (EHR) provided by SPMS and also on all the central components (central database, interoperability, etc.).

With the objective of validate the new functionality incorporated on SPMS's EHR, a clinical work group was formed and several gatherings were promoted, with the presence of SPMS, DGS and IMT.

This was very important to understand their concerns and suggestions were considered on the development process. Major inputs related to usability and reliability concerns. Special attention was paid to user acceptance tests, communication, training and support. With the outputs obtained of this clinical work group, the functionality was improved in order to respond the user's needs.

After that, SPMS and DGS promoted several clarifying sessions in different geographic locations in the Portuguese territory to inform, present and engage doctors of this new functionality. New suggestions were obtained at this level, which were evaluated.

The main findings collected from the clinical work group and from the session promoted, were shared with the different private clinical software providers, in order to anticipate possible usability and reliability problems. As already stated, SPMS was responsible for defining the standards, requirements and specifications that all software applications must comply with to connect to the central system and issue medical certificates. Awareness communication, technical workshops and several meetings were conducted to engage software providers in the process of transformation. In a later phase, test and verification procedures were performed to ensure the conformity and correct operation of these applications before allowing their interconnection (4).

To engage small prescribers who, individually, wish to issue a medical certificate for the driving license and do not use clinical software, SPMS made available a new functionality in the Health Data Platform (PDS) for that purpose. This measure reduced the digital divide problem.

With the finalization of the development, SPMS promoted the piloting. On the first day of March of 2017, a pilot in a few primary care units were performed. Later, on the 14th of March of 2017 started the first pilots on hospitals units. All the pilot units were chosen in different location of the Portuguese territory and, after the validation of this pilots, this new functionality has been extended throughout all the Portuguese territory.

By the 15th of May of 2018, which was the date that the issuing and transmission of this type of medical certificates was obligatory to be electronic, all the EHR developed by SPMS have the new functionality available, as well as in PDS for doctors in a private context and do not use a clinical software.

As for the different private clinical software providers, some of them where ready before 15th of May, other only have been declare under conformity later that year.

4. ONE YEAR LATER…

So, one year has passed from the obligatory electronic issuing (and transmission) of the medical certificate for the driving license in Portugal. It is time to make a balance.

More than 25 different local software were declared under conformity. But in 2018, SPMS is still evaluating the conformity of new software providers who pretend to add the functionality to issuing this type of medical certificates in their clinical softwares.

As the IMT is not accepting paper based medical certificates, with this SIMPLEX initiative, was possible to full dematerialize this process. And the number of medical certificates electronically issued and transmitted to the IMT have been growing and growing, as can be seen in figures 3 and 4.

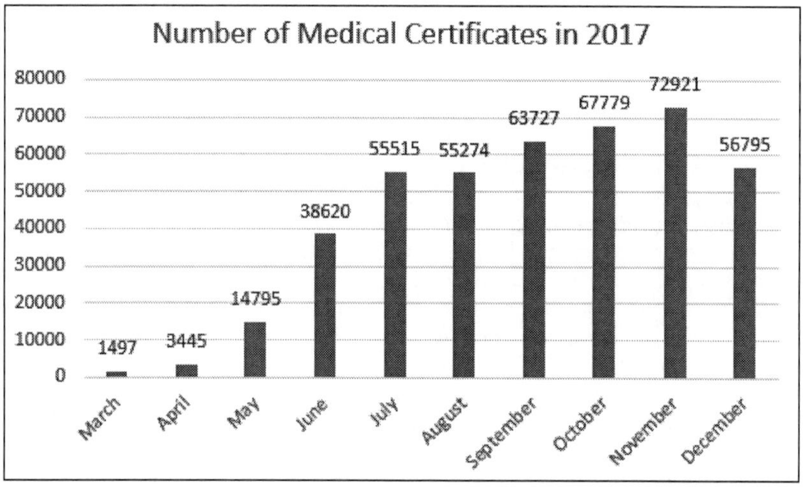

Figure 3. Number of Medical Certificates for Driving License Electronically Issued and Transmitted to IMT in 2017

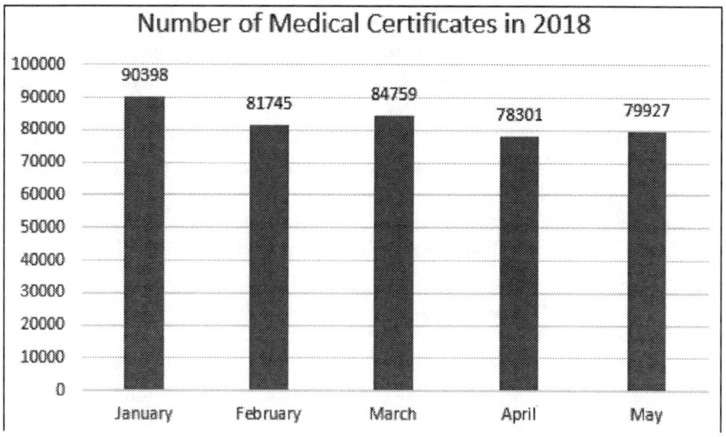

Figure 4. Number of Medical Certificates For Driving License Electronically Issued and Transmitted to IMT in 2018

As it is possible to observe in figure 3, the first two months (March and April) resemble the pilot period, so the low number of electronic issued. May was the month where it became obligatory, more and more softwares were declare under conformity and the number of medical certificates electronically issued have grown and grown, at the point that in 2018 the number of issues are bigger than any month in 2017.

In 2018, 825.745 HL7 FHIR messages have been exchanged in this process and only 107 have produce some type of error (figure5). These error were identified, scrutinized and amended. The main cause of this errors are temporarily connection breaks between the entities.

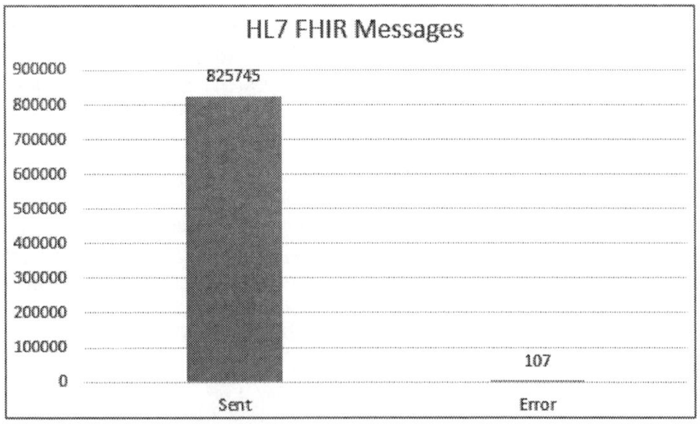

Figure 5. State of the HL7 FHIR Messages Trade in 2018 in the Process of Medical Certificates for Driving License Issuing

If we analyze the messages exchanged in a specific week of 2018 – from May 26 to June 1 – it is possible to perceive the high number of messages that flow. So, in this period, 27905 HL7 FHIR messages were swapped between the players involved in this process and not a single error was identified.

5. FUTURE WORK

It was on May 15 that the electronic issuance of medical certificates for the driving license became mandatory and, since that time, we have seen a high number of emissions. This means that the information is flowing more quickly and in a secure way, the quality of information have improved and efficiency gains have been obtain.

So, what can be done next to improve this project? Logically there is always something that can be improved. In the last year, our focus was on improving the timing of the certificate information processing, in other words, tuning of the system to turn it faster. Our efforts are also involved of the creation/improvement of the automatic mechanisms with the aim of monitoring the messages exchange. More work will be done concerning these aspects.

Moreover, new functionalities can be developed with the intention of promoting a better product. One functionality that can be implemented is the digital signature of the medical certificate for the driving license. This improvement will improve the assuredness of this issuance and achieving legal compliance and global acceptance.

Other major feature that can be introduced is the possibility of the citizen being notified by email or text messages once the medical certificate for the driving license have been delivered to the IMT. During the process of the issuing of this certificate, the physician could ask the citizen if he/her want to be informed of the delivery of the certificate to the final recipient.

After the implementation of the listed features, it is necessary to perceive what can be enhanced in order to improve the interactions of physicians and citizens with this product.

6. CONCLUSION

E-government has huge potential to deliver benefits to both citizens and Public Administrations.

This case showed that WoG e-government may support policy integration and encourage government institutions to work more closely together, revisit existing processes and work flows to simplify and harmonize them, resulting in a shared coherent response to citizen needs. The adopted solution (organizational, legal and technic) has shown fit to purpose and the implementation approach could deliver results in a short time.

The project, in particular, was delivered on time and had good acceptance. It allowed the medical certificated to be electronic transmitted directly by the doctor to the IMT (which results in an exchange of information between two different ministries), ensuing a greater control, simplification and optimization of this process as well as achieving an improvement in road safety and public health.

This could only have been achieved due to the existence of: i) the correct evolvement and articulation of the multiple sponsors of this WoG initiative; ii) interoperability infrastructural IT components with a federative (hierarchical) architecture; iii) formal arrangement of organizational structures for the governance of IT at national and sectorial (health) levels, has shown major critical to achieve the success, demonstrated in a wide adoption at minimum time. All other aspects identified in previous theoretical and empirical studies were also deemed to contribute greatly to the success or constitute a critical success factor.

REFERENCES

Ama. 2016. *SIMPLEX +* [Online]. Available: https://www.simplex.gov.pt/ [Accessed 25/06/2018].

Christensen, T. & Lægreid, P. 2007. The Whole-of-Government Approach to Public Sector Reform. *Public Administration Review,* 67, 1059-1066.

Nations, U. 2016. United Nations e-Government Survey.

Ojo, A., Janowski, T. & Estevez, E. 2011. *Whole-of-government approach to information technology strategy management: Building a sustainable collaborative technology environment in government.*

Palvia, S. & S Sharma, S. 2018. *E-Government and E-Governance: Definitions/Domain Framework and Status around the World.*

Seri, P., Bianchi, A. & Matteucci, N. 2014. Diffusion and usage of public e-services in Europe: An assessment of country level indicators and drivers. *Telecommunications Policy,* 38, 496-513.

SHEDDING LIGHT OVER THE PORTUGUESE HEALTH IT ARCHITECTURE

Joana Cunha and Filipe Mealha
Shared Services of the Portuguese Ministry of Health, Portugal

ABSTRACT

Throughout the past 20 years, interoperability in information systems in the Portuguese National Health Service (Portuguese NHS) has been mostly limited to solutions using *data link layer* in the Open Systems Interconnection (OSI) model. These kind of connection creates a significant and complex problem since their implementation has high variability from sight to sight and used extensively there is a risk of loss of control over who has access to data from a given system. These cibersecurity risks are now even more relevant and linked also to privacy control challenges. This paper aims to inform about the context of the Portuguese Health IT Architecture, being rolled-out at present, the challenges that arise from years of dispersion and heterogeneity in its core information systems, and how the Local Interoperability Gateway for Healthcare (LIGHt) project came to be in providing an event driven approach to system's interoperability, standards adoption and ease of control and monitoring.

KEYWORDS

Interoperability, Monitoring and Prevention, Architecture, Healthcare, Cibersecurity

1. INTRODUCTION

Throughout the past 20 years, interoperability regarding the Portuguese National Health Service (Portuguese NHS) has been limited to *data link layer* in the Open Systems Interconnection (OSI) model. Hospitals and Primary Care Units built and acquired many Information Systems (IS) from several different vendors, without any preexisting normalization and standardization rules. These systems were progressively integrated to the ones provided by the central services of the Ministry of Health (MoH) via end-to-end connections on their databases.

These kind of "*ad-hoc*" connections were customized to address both the specific requirements of each system, and use case, from a technical viewpoint, and the local requirements of each healthcare unit, from a business viewpoint, creating a significant unevenness locally in the information ecosystem.

In Portugal, the Health Sector's Information Systems Infrastructure, i.e. the Ministry of Health's Information System Infrastructure, is largely distributed, mainly because every Hospital or Healthcare Unit has its own local repositories. Currently in the Portuguese Health public sector,, there are fifty-four Hospitals and three-hundred and fifty-nine Primary Care Units, consequently each one of them with its own database In addition to these, there is a set of centralized that must coherent with these local databases in real time.

Over the last twenty years, the number of end-to-end connections grew, in such a way that managing and maintaining them, locally, has become a challenge as well as a burden to the NHS.

This paper sets aim to provide further context on the Portuguese Health IT Architecture, the challenges that arise from years of dispersion and heterogeneity in its core information systems and how the Local Interoperability Gateway for Healthcare (LIGHt) project came to be in providing an event driven approach to system's interoperability, allowing healthcare providers to scale their IT Infrastructure in a way that integrations between systems are made in a standardized, robust and highly customizable way.

It also aims to share lessons learned and challenges that were overcame to assure data's coherency among different databases, right information, at the right place, at the right time.

2. PROJECT OVERVIEW

Throughout the course of recent years, we have seen an exponential increase in "*ad-hoc*" connections between Information Systems at a local level, i.e. an increase in *database links* to account for the sharing and transmission of data among systems.

These kinds of connections, although they apparently "get the job done", they create a significant and complex problem since their implementation varies from sight and consequently their abuse, when they are created in an uncontrolled manner. This will lead to a loss of control over who access and uses data from a given system promoting security issues.

However, this is not a simply technical interoperability problem, but rather a deep architectural problem which undermines Health IT in Portugal.

Since Hospital's and Primary Healthcare Units' administration boards have both financial and administrative autonomy, without a proper set of regulations provided by the government via the central agency for regulating Health IT (SPMS), these problems will only tend to grow bigger.

To address the resolution of these issues in 2014 the project LIGHt (Local Interoperability Gateway for Healthcare) came to be in the Portuguese Hospital's main Hospital Information System (HIS) which is provided by SPMS.

When the development started, the aim of the LIGHt project was to provide the SPMS's HIS with an integration mechanism which would allow to better manage its connections in a standardized, robust and highly configurable way, using an event driven approach LIGHt as a system, i.e., as a software Solution, was designed as a *Message Oriented Middleware* to enable communications via an *Event-Driven Architecture* (EDA), allowing the decoupling among databases and applications from different systems.

As far as system's security is concerned, LIGHt as a *middleware* allows for the removal of the need of one application to directly access the database of another, strongly increasing the level of control and protection of a system, making it less vulnerable and allowing for the monitoring of processed events.

Its aim is to replace all local *dblinks*, with standardized messages exchanged between systems to normalize communications, regardless of the system that originated that message. This goal can only be accomplished if international communication protocols, known to all vendors, are used to exchange messages. The *messaging standard* adopted by LIGHt is the Health Level 7 (HL7) V2 for local (intra-institution) communications and HL7 FHIR for central (inter-intuition) communications.

LIGHt is currently deployed at 48% (26) of Portugal's public Hospitals, and we expect that by the end of 2018 it will be deployed at 100% (49) of all National Public Hospitals, not considering Public-Private Partnerships.

2.1 Approach

2.1.1 Architecture

To replace *dblinks* from local infrastructures, there must be a representation of the health IT infrastructure of Hospitals and other healthcare units in Portugal that maintains a record of how many there are, what purpose they fulfil and which systems they connect.

As previously mentioned the Shared Services for the Portuguese Ministry of Health is the government agency in charge of regulating all matters concerning Health IT. Aside from producing and distributing its own IT solutions, SPMS is also responsible in the regulation of Health IT and its infrastructure.

In late 2017, anticipating the new European Global Data Protection Regulation (GDPR), an opportunity arose setting the right conjecture and allowing for legislation to entrust SPMS with the obligation to construct a nation-wide information repository containing details of every single system and infrastructure in the NHS, as well as the *data* treated by them - thus, a process of mapping local IT ecosystems begun.

This record is crucial for a solid, national IT architecture and will allow for the normalization and regulation of health information systems in Portugal.

2.1.2 LIGHt – Local Interoperability Gateway for Healthcare

LIGHt or **Local Interoperability Gateway for Healthcare**, as a project came to be due to the necessary changes that needed to be made in the way that systems communicated with each other. Its primary purpose is to provide applications with a way to communicate with one-another, without resorting to non-standardized communications, filling the gap for the need of a "*plug and play*" interoperability mechanism.

As a *message oriented middleware*, it allows for a traceable, secure and standards oriented way of sending health data between systems, whilst guarantying data consistency. When in place, it allows for the real-time monitoring of a hospital's or healthcare unit's system's integrations, which, more than allowing system administrators to see if all messages were dully processed, allows managers to check the *real* status of production. With the proper set of *Business Intelligence* mechanisms, it allows decision makers to see how many admission events there were in any given period, how many referrals, surgeries, appointments, discharges, ..., any type of event that triggers a standardized clinical or administrative message.

To make this possible, LIGHt resorts to the use of two healthcare messaging standards: HL7 v2 and HL7 FHIR, being that the translation from one type to another is possible. HL7 or Health Level 7 refers to a set of international standards meant for the transfer of clinical and administrative messages between health information systems. The choosing of this standard was due to its world-wide spread adoption across many different health IT providers. As for the particularity of v2 and FHIR, one has to do with its long-time presence in the market and the fact that it is supported by Integrating the Healthcare Enterprises (IHE) profiles, while the other accounts for a more modern, easier to implement solution that is now seeing adoption by big IT players, like Apple.

Bearing in mind that LIGHt's development comes from government, i.e. taxpayers, money; the entire solution was developed resorting to open-source technology.

From a more technical viewpoint LIGHt is, in its essence, an enterprise message broker designed to use an event-driven architecture and implement a publish – subscribe pattern.

As a Solution, it is composed of 3 applicational components:

- SEC – SPMS Event Controller
- MCS – Mirth Connect Services
- BO – Backoffice

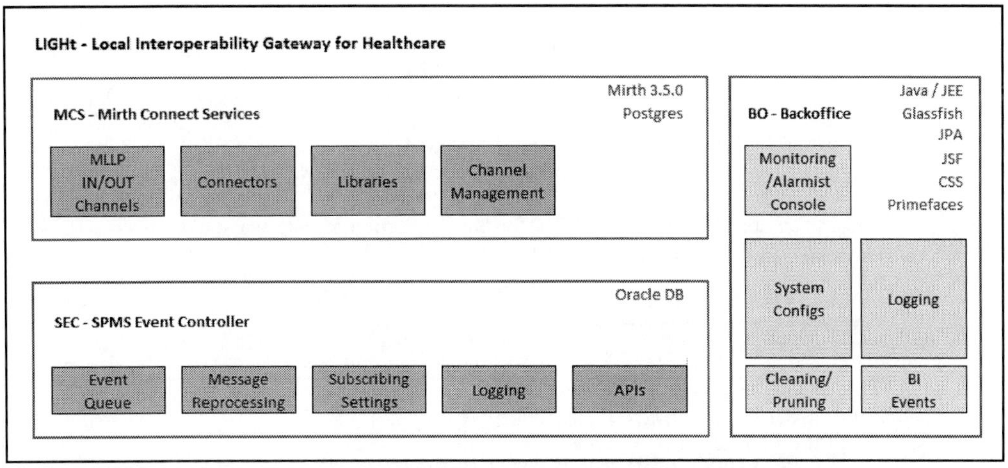

Figure 1. Schematic of LIGHt's Architecture

The SEC module is the LIGHt architectural component that allows the implementation of an event driven architecture (EDA). It is responsible in the creation of events, changing of message's status and the management of logs, allowing a wide and centralized view of integration's status. All the reprocessing events logic, error processing and logging are implemented in this module.

The MCS module is used to exchange messages according to the HL7 standard.

The Backoffice allows for the management, configuration and monitoring of LIGHt in an intuitive and user-friendly interface. It allows system administrators to trace every change that is made at a configurations level and, at the same time, allows support teams to use LIGHt to monitor, in real time, the exchange of messages that is happening in the hospital. Another functionality of this module is the capability to create, register and subscribe new integrations without the need of *bdlinks*. Users can easily validate workflows, access if they are working and act accordingly.

2.1.3 Providing Real-World Scalability

After identifying the need for a mechanism to facilitate the exchange of data between two or more information systems, LIGHt begun its development in 2014, being that 11 February 2016 marks the year of its first production deploy, at a 545 bed Hospital.

The first installment aimed to solve a very specific problem – integrating the HIS with the Laboratory Information System (LIS) at a Portuguese Public Hospital. At that time, physician orders were made in a Clinical Information System and then passed on directly to the LIS, without being registered in the HIS. Once the laboratory results were ready, a message would then be triggered to the HIS, and the only record maintained in the hospital's administrative system was the realization date of a given examination.

This posed a problem, as managers could not easily tell how many exams were prescribed, how many were left unattended, how much time took place between the prescription and the realization, as well as other kinds of problems that naturally arise from this scenario.

After LIGHt's deployment, the communication flow drastically changed. Messages began to flow between the LIS, HIS and the prescription software with LIGHt as an intermediate. This allowed for the Hospital to monitor, in real time, the number of events processed, how many were successful, errors or warnings, identify information load spikes, and see how long each transaction took. Also, every single transaction was traceable and auditable. The following table shows the number of messages concerning laboratory (LIS) events received by the HIS up until 2016 (before LIGHt) and after 2016 (after LIGHt), on the Hospital that served as the proof of concept institution.

Table 1. Number of events transmitted from the LIS to the HIS before 2016 (before LIGHt) and after 2016 (after LIGHt)

Year	Nº Analysis
2009	19 019
2010	19 675
2011	20 400
2012	20 309
2013	19 874
2014	19 261
2015	19 517
2016	1 406 117
2017	2 595 679

LIGHt has since been updated and now allows for the transaction of data via standardized (HL7v2.5 and FHIR DSTU3) messages in the following contexts:

Table 2. Number of standardized HL7 messages supported by LIGHt

Module	Nº of standardized HL7 message transactions supported
LABORATORY	16
RADIOLOGY	16
INPATIENT_MANAGEMENT	10
OUTPATIENT_MANAGEMENT	19
ER MANAGEMENT	10
DAY HOSPITAL MANAGEMENT	8
CLINICAL DATA	9
MEDICATION	2
MASTERFILES	51

For every supported transaction[1] LIGHt has the capability to push data to its subscribers, i.e. the capability to notify them whenever a change occurs within a given information source.

Moreover, filters can be added on request from the consumer so that only specific kinds of notifications are received. This allows for a Master Patient Index (MPI) to notify all systems that a change was made to a patient's demographic data and, for the subscribing system to choose whether it wants to receive said notification or not. For example, a system may want to receive an update concerning a change in a patient's address, but not a notification in a change in his marital status. Another great example would be a LIS receiving appointment notifications concerning only clinical pathology examinations. This kind of configuration contributes heavily in matters of information security and data confidentiality.

The following sections of this chapter will emphasize three case-studies that will provide knowledge of the benefits that came from the implementation of this project, as well as an insight over the real-world challenges that arose and had to be overcome for LIGHt to now cover 50% of all public hospitals.

2.1.4 Freeing Agendas

The Portuguese NHS tends to approximately 11 000 000 Portuguese's[2], so medical agendas are a precious asset. Late 2017 a new message was implemented in LIGHt that allowed for the signaling of a patient's passing away. In Portugal, every citizen has his/her own NHS unique identifier. That information is held in a Master Patient Index (MPI) alongside the patient's demographic information. Upon the implementation of the "passing away" transaction, hospital HISs started to receive information whenever a died. With that information, 6 public hospitals now have the capability to free their agendas whenever a patient is pronounced dead in our national information infrastructure. The freeing of agendas can have a tremendous impact in reducing the running costs of the Portuguese NHS. The following table shows the number of cancelled appointments due to a patient's passing away and the potential money saved from making that agenda available for another patient to take.

Table 3. Number of cancelled appointments after a patient is pronounced dead, estimated savings and LIGHt's "passing away" message activation date

Institution	Nº of cancelled appointments after the patient being pronounced dead	Estimated potential savings	Message activation date
Hospital A	182	9 797.06 €	21-11-2017
Hospital B	276	14 857.08 €	28-11-2017
Hospital C	223	12 004.09 €	12-11-2017
Hospital D	60	3 229.80 €	23-11-2017
Hospital E	121	6 513.43 €	30-01-2018
Hospital F	1 251	67 341.33 €	12-09-2017
Total:	**2 113**	**113 742.79 €**	

The numbers of cancelled appointments presented in table 3 were taken on May 16, 2018. These savings were calculated considering that the average medical appointment in Portugal has an approximate cost of 53,83 €[3]. If every single one of those appointments were to represent a patient "no show", then the NHS would still assume that cost per consultation, which would translate in a loss of 113 742.79 €, since no patient was being tended to at that time. However, if that slot is made available for another patient to take, then no monetary loss is accounted for, and the costs translate into savings.

This process is still permeable to errors, specifically, human errors. Since the death certification is always declared through human intervention, i.e. a medical act, if a mistake is made, it could potentially be propagated. This could turn into a "worst-case scenario" in which the re-scheduling of appointments could cause missed appointments or potentially the loss of a patient's turn on a waiting list. This, however is very unlikely because Portugal as a unique electronic death certification system, SICO, and that system involves doctors, police and other ministries. Only after the full verification, is the information given to MPI about a patient's passing away. This mechanism avoids cases were a patient would be wrongly pronounced dead in

1 Link to all LIGHt supported HL7 transactions: https://spmspt.atlassian.net/wiki/spaces/PD/pages/42958990/2.18.+Opera+es+Mensagens

2 Data obtained from the Portuguese MPI at 06/06/2018.

3 This number reflects the actual average cost of a medical consultation to the NHS. The price a patient would pay for this appointment would be a small fraction of the presented number, since in Portugal healthcare should be, tendentially, free-of-charge.

our national information infrastructure, and that error indivertibly canceling every appointment that patient had. In the eventuality that an error is made, strategies from the business side can help mitigate these unwanted events.

2.1.5 Monitoring and Alarmistic

The ability to reprocess events, i.e. messages, and the ability to trigger alarms are two core features in LIGHt that currently play a vital role in any hospital where it is deployed.

Since LIGHt assumes the role of the primary hospital service bus it processes, on an average daily basis, around 85 000 messages, each carrying valuable data and information that must be delivered for the patient to get the care he/she needs. In 27 January of 2017 the scheduling system of a Portuguese hospital had a catastrophic failure, losing all its records from 12-01-2017 to 27-01-2017. That information was said to be irretrievable by the vendor, however, due to LIGHt's message reprocessing capability the hospital managed to regain that history once more. At that same hospital, over the course of one year, 12 493 423 messages were sent via LIGHt, with a success rate of 99.72%.

LIGHt allows for the real-time monitoring of all events processed, through a Backoffice available to the hospital. System administrators can see the evolution of all transactions being made, identify and sort out events from 3 different categories: success, errors and warnings; and identify load spikes or abnormal message processing times from a *queued* state.

Complementary to this monitoring, an Alarmistic component set in place during the first quarter of 2018 allows users to be alerted, in real time and 24*7, in the presence of errors or any abnormal events that may occur, thus allowing for a faster, more efficient response. This feature was first put to test in 03 May of 2018 when a human intervention caused a glitch on the Portuguese MPI infrastructure, at 18h37. The MPI provides data to both central and local applications, so any downtime on this infrastructure can spell disaster to care providers. In this case, the glitch made it so that general practitioners couldn't prescribe medication and patient admissions were compromised. However, since one hospital was subscribing a LIGHt message to query the MPI, an email was set off at 18h39 (2 minutes after the glitch) that notified both the hospital information systems administration team, as well as LIGHt's support team.

Upon receiving said email, it was possible to notify the MPI support team and have the problem solved at 19h54. Without this kind of alarm mechanism in place, this problem could have extended over a larger period, potentially and severely disrupting the way services should be provided by care givers.

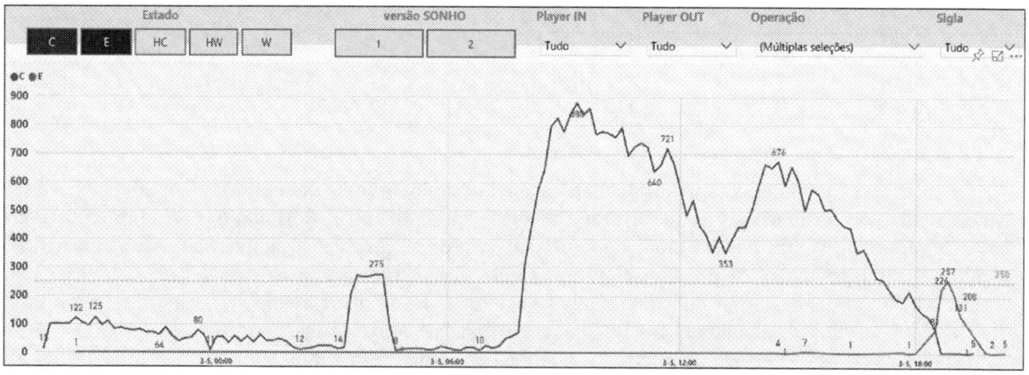

Figure 2. Real time monitoring of events in 03 May 2018

From: light@xxxxxxx.min-saude.pt
Sent: 3 of may de 2018 18:39:10 (UTC+00:00) Dublin, Edimburgo, Lisboa, Londres
To: Backoffice Light | SPMS
Subject: [IPOLX1] - LIGHt Alert (320)

Dear Mr/Ms,

An alert was registered in the LIGHt of your institution, xxxxxxx, at 18:39:09 on 03-05-2018 with the ID 290 and the following information:

Alert: **[320] – Abnormal Variation in the number of errors (Criticality: Medium)**

Frequency:[10/3] – **Number of event with errors superior to 1**
4 error events detected.

Subscription: **[1] - 10110-PATIENT_RNU_QUERY | SONHO->RNU**

Best regards,
LIGHt support team
E-mail: light@spms.min-saude.pt

Figure 3. Translation of the alarm email set off to notify the downtime at the MPI

LIGHt's alarmistic capabilities allow for the signaling of errors and contribute to reduce system downtimes. Alarms can be configured to, as an example, set off emails if no messages are received from a certain system during a set period of time. For instance, if the LIS stays more than one hour without sending a message to the HIS, a notification can be set off to signal that something might be wrong. Emails (and in the near future SMS) alarm notifications are subscribed in the same manner as message subscriptions. There is a pre-determined alarm subscription catalog from which users can choose which alarms they want to set in place, being that each institution can then customize the attributes of each message in LIGHt's backoffice[4].

3. FUTURE WORK

In line with the work presented in this paper, we strongly believe that the next big step is to conclude the mapping of our national infrastructure that begun late 2017. This map will allow us to normalize and regulate of system integrations in Portugal. Additionally, a set of semantic catalogs needs to be defined and adopted, so that normalization in the messages exchanged can be achieved. Alongside these processes, LIGHt will play out an important role in the unification of primary care and hospital systems'. Since most HIS and EMR in Portuguese hospitals and primary care units are provided by SPMS, and since those systems are now beginning to show their age, a restructuring is overdue. LIGHt will be one of the *middlewares* that will integrate these systems. Also, most of LIGHt's messages are ADT related. We will now start working in making more clinical messages available to all players. Another big area for us is in clinical decision support, where the driver is the patient. Finally, we will keep on developing our FHIR standard messages.

4. CONCLUSION

As it happens in any structural change, even if it is for a better end, people can oppose to change. To solve the current interoperability challenges in our national health information ecosystem just addressing the technical aspect of it isn't enough. Several non-technical factors must be taken into consideration, being that ultimately, the most prominent one are the people behind systems. Systems will communicate if the people behind them collaborate. What tends to happen is ambiguity in their communication – sometimes the teams involved in a transformation process aren't well aligned. This ambiguity is carried on to the very semantic that defines messages.

4 List of LIGHt's supported alarm messages: https://spmspt.atlassian.net/wiki/spaces/PD/pages/444071981/M+dulo+de+Alarm+stica

In order to achieve true interoperability, a semantic agreement must be achieved. Otherwise, no sense can be made from the information exchanged. To do so, a strong set of rules and legislation must be in place, that regulate communication patterns and that contribute for the universal adoption of those standards. Our work demonstrates how an event-oriented middleware approach brings added value to system's security, data integrity, information quality, business management, resource optimization and costs minimization. As we demonstrated by the use-cases presented, health information technologies can certainly have a positive impact on health reforms. The power of data provided by the coupling of BI solutions to a message-oriented *middlewares* brings enormous benefits to all of our healthcare professionals, be it practitioners, IT specialists and patients.

System's interoperability ultimately allows doctors the time to treat our patients instead of our diseases.

ACKNOWLEDGEMENT

This work was supported by the Shared Services of the Portuguese Ministry of Health (SPMS).

REFERENCES

Anil Jain, 2016, Interoperability: A Common Challenge for Population Health.

In Keun Lee, 2011, The Development of a Graphical User Interface Engine for the Convenient Use of the HL7 Version 2.x Interface.

Jatinder Singh and Jean M Bacon, 2014, On middleware for emerging health services.

Olaronke Iroju, Abimbola Soriyan, Ishaya Gambo, and Janet Olaleke, 2013, Interoperability in Healthcare: Benefits, Challenges and Resolutions.

MODELING OF THE GEMA: A NOVEL WEB SYSTEM FOR MANAGEMENT OF COLLABORATIVE CARE IN MENTAL HEALTH

Hugo Granjeiro Soares Fontes, Cefas Rodrigues Freire, Emerson Arcoverde Nunes,
José Arilton Pereira Filho, Hugo Sailly Moura Bezerra
and Ricardo Alexsandro de Medeiros Valentim
Federal University of Rio Grande do Norte, Brazil

ABSTRACT

The Collaborative Care (CC) process, in Brazil also known as Matrix Support (MS), has as main purpose integrate Primary Care (PC) and Psychiatry Care in Mental Health (MH) in a shared and collaborative care construction of pedagogical-therapeutic interventions. This model, created in the 21st century, approaches a new structural concept between PC and MH in several cities of Brazil. Therefore, this paper aims to propose the modeling a novel web system that is able to improve the CC process by systematization, computerization, standardization and documentation of the practices involved in the process. The system should be able to promote the communication between the health team professionals and enhance the networking. Hence, the Unified Modeling Language (UML) was used to build the modeling based in the specified requirements from collaborative care workshops with several mental health and primary care professionals in the city of Natal, organized by the Onofre Lopes University Hospital (HUOL) in partnership with the Laboratory for Technological Innovation in Health (LAIS) of Rio Grande do Norte Federal University (UFRN) and Natal Health Department. Thus, to compose the proposed modeling, the UML diagrams of Use Cases, Classes and Activities were developed, modeling the GEMA system, which intends to facilitate and potentialize the integration between Primary Care and Mental Health Teams.

KEYWORDS

Mental Health, Collaborative Care, System Modeling, ICT for Health

1. INTRODUCTION

The mental illnesses prevalence, as well as its costs, keeps increasing. European Union data shows that the overall financial costs, including direct medical costs, as well as indirect costs related to care and loss of productivity, account for more than 450 million euros per year (György P. et. al. 2016).

In Brazil, it remains the third main cause of costs for social security, with an average annual impact of 186 million reais to the pension system. A study with Primary Care (PC) users showed a rate of mental disorders ranging from 51.9% to 64.3%. Mental health problems were especially high in women, the unemployed, and people with low education and/or low income.

Other studies show that mental health actions in primary care are still insufficient in the face of rising demand. Among the reasons mentioned are: the lack of specialized mental health professionals; the lack of preparation of primary care physicians (PCPs) for mental health cases; and the need to work the mental health network as intersectoral, which transcends the field of health. (Gonçalves DA et al, 2014). Online mental health interventions are substantially less costly to implement than conventional interventions, and access to care for these people remains a key challenge in the field of mental health and wellbeing (György P. et. al. 2016).

The gap in mental health care is a worldwide problem that can find a possible solution in the integration between the Primary Care and the Psychiatry (Fortes, 2014). In this scenario, Collaborative Care (CC), most known in Brazil as matrix support (MS), emerges as an essential process in order to achieve this integration based on collaboration and sharing of care in partnerships between generalists and specialists in Psychiatry.

A Collaborative Care intervention is a multifaceted approach to patient care provided by a PCP and a Psychiatrist, in the context of the primary care setting (Katon W et al, 1995). This is a new way of producing health wherein two or more teams develop a proposal of pedagogical-therapeutics intervention in a process of shared construction (Chiaverini DH et al, 2011).

In this work we propose the modelling of a web system to facilitate the communication and the targeting between the primary care, the mental health team and other resources of the Psychosocial Attention Network (PAN), so that they can work integrated as one collaborative care network. The system must be able to enhance the process through its systematization, computerization, standardization and documentation practices.

Thereby, the modelling is one of the most important activities in the software development context. According to Jacobson and Booch (2004), the model provides the system visualization, behavior and structural specification, acting as a documentation guide of all the decisions taken in regard to the project (Rumbaugh, Jacobson and Booch, 2004).

For the design of these specifications we used the Unified Modeling Language (UML), which is a general purpose modeling language that graphically represents the requirements of a system (Adamu and Zainoon, 2016). The UML is widely used because of its standardization, where it is possible to have a macro view of the system by using different diagrams.

2. GEMA: MANAGEMENT OF THE COLLABORATIVE CARE IN MENTAL HEALTH

In the Brazilian Public Health System specific situation, the primary care team, also known as Family Health Strategy teams, is the first contact of the patient within the Health System, who work in Units with interdisciplinary professionals specialized in primary care. When in the need of mental health consultation, today they can only count on the referral system, calling the Mental Health Teams (MHT) and depending on the availability of these specialists; usually, it takes weeks or months to get this consultation, with low efficiency. One of the reasons is because this process is centered mainly on the availability of the professionals, rather than the needs of the patient. Concerned about this issue, we intend to facilitate and accelerate this process with the GEMA system.

In the collaborative care process approached by GEMA, we have three team profiles: Reference Team (RT), Matrix Support Team (MST) and External Team (ET). RT is a team formed by professionals of a Primary Care Unit. MST is a team formed by professionals of one or more Mental Health Units. And ET involves all of the devices not necessarily from health care that are available at the district (such as Social Assistance Centers, and some NGOs). All professionals (RT professionals, MST professionals and ET professionals) involved in a shared case are known as Collaborative Care Team (CCT).

With GEMA, the RT Practitioner can send a summary of any case selected to the MST in order to decrease the time required for the referral specialist response. The MST has a referral professional who can or cannot accept the case (usually the head of the mental health team). Once the case is accepted, the flow continues with the professionals of both teams (RT and MST) composing the CCT. Through the proposed system, they can discuss the case and establish the Singular Therapeutic Project (STP), which is a way to focus the treatment on the patient rather than on the disease, respecting their particularities. Thus, goals and targets are traced and can be tracked, with an easier identification of the local health system potentialities as well as some flaws.

For instance, an autistic child may have severe communication problems, needing the help of several specialists. By producing the STP, the CC team can enumerate several problems and the best professionals to solve them, like a phonoaudiologist to language problems. But in many cities, those professionals are hard to find in any primary setting. The system facilitates this process by allowing all professionals to communicate with each other about the needs of each patient and accompanying how the therapy is working. For example, where this child can find this specialist, and how to call him or her to collaborate. Therefore, respecting the local availability of resources, and trying to overcome various possible difficulties (such as lack of professionals, long waits, general medical illnesses and social problems), each professional involved collaborates and most of the patient's needs can be attended.

3. REQUIREMENTS ENGINEERING AND MODELING OF GEMA

In order to accomplish a good specification and modelling to the system proposed here, it is essential to identify which needs this system should supply and the potentialities that it must promote. So, it is necessary to understand how it improves and systematizes the Collaborative Care in Mental Health..

The first requirement elicitation was performed from an open interview with the HUOL psychiatrics. This interview was guided in two baselines: a psychiatric interview guide found at the Guia de Matriciamento em Saúde Mental (Chiaverini DH et al, 2011), Applied Matrix Support Guide, from the Brazil Department of Health, and software requirements described by Sommerville (2008). Then, these requirements were specified and the result guided the creation of the first prototype. A new interview with HUOL stakeholders who worked on this prototype allowed a new requirement elicitation.

The next steps follow this workflow. In total, six interviews and two workshops, with simulated scenarios on system's prototypes, were made, as the example shown in the Frame 1. About 40 professionals attended the workshops. This public was composed of community health agents, physicians, nurses, psychiatrists, psychologists, etc. All of them from Natal Department of Health, primary care units (PCU) or mental health units (MHU). Through this process the requirements have been elicited and the UML diagrams such as the Use Case Diagram, Class Diagram and Activity Diagram are presented in the next sections.

Initial assumption:
A mother takes her son with autism to a Primary Care Unity.
Normal:
The primary care physician from Reference Team requests a collaborative care of this case by a Matrix Support Team. A professional from the MST accepts the request. So, a case discussion area is created where the professionals from both teams fill the fields referring to disease history, family and personal histories and clinic hypothesis.
A MST administrator ends the discussion. Thereafter, the systems create the patient STP workspace.
A physician fills the patient diagnosis, with organic and psychosocial evaluation. The professionals add goals and tasks to the STP.
The boy shows speech problems and none of the teams have a phonoaudiologist. A MST professional searches within the system for the required specialist from his or hers district and ask one of them to collaborate. If he or she accepts, automatically enters the STP workspace and has access to the discussion case report.
After the STP is finished a patient review must be scheduled.
Other activities:
The system generate the case discussion and STP reports.

Frame 1. Example of a Scenario

3.1 Requirements

The Table 1 presents some user functional requirements which were extracted by the requirements engineering process presented before. First, the system organization division is perceived: cities divided into districts, each district have MHUs and each one of these is responsible to support the PCUs. Moreover, the professional can own more than one profile (a professional can work in more than one health unit).

It is important to note in UFR005 that when a professional from RT (RTP) makes a CC Request, he or she must be able to choose which of his/hers team members will act on the case. Only those professionals will have access to Case Discussion and the Singular Therapeutic Project (STP) and will be part of the Collaborative Care Team of this patient. UFR006 shows that a professional from MST (MSTP) can accept or disagree that his team share the care of this case. If he accepts the request, he must specify which MST professionals will act in that case.

UFR013 indicates that all actions on the system that are of interest to the user should generate notifications for the user. E.g., the practitioner should receive an email and a notification in their browse if such user is added to a Collaborative Care Request or receives an invitation to contribute to a patient's STP.

Table 1. Summarizes Some of the User Requirements that meet the Needs of the Main Flow of Collaborative Care

Identificação	Descrição do requisito
UFR001	The software must allow the District maintenance by city.
UFR002	The software must allow the Matricial Support Team maintenance related to a District. It should be done in a way that a MST must be an abstraction in the system to a professional grouping from one or more mental health specialized units.It does not mean that all professionals in these unity will be system users.
UFR003	The software must allow the maintenance of the Reference Team related to MST. It should be done in a way that a RT must be an abstraction in the system to a professional grouping from one primary care unit. It does not mean that all professionals in this unity will be system users.
UFR004	The software must allow that a professional have more than one profile related to a same system user. The user must choose the profile that will be used at each time after the system login.
UFR005	The software must allow that a professional from a basic health unity could request the matrix support to a specialized team and select the people who will work on this case.
UFR006	The software must allow that a specialized team, upon receiving a request of matrix support, could accept or refuse the request. If it is accepted, the team can select the people who will work on the case.
UFR007	The software must allow that a matrician case could have its documented discussion.
UFR008	The software must create a dashboard to a Singular Therapeutic Project of the treatment for each patient after the case discussion finalization.
UFR009	The software must allow that goals can be assigned, on the dashboard of a STP, to selected professionals.
UFR010	The software must allow the referred patient diagnosis maintenance on dashboard of a STP.
UFR011	The software must allow to request the inclusion of new professionals, on dashboard of a STP, in the referred patient treatment.
UFR012	The software must allow to each user a dashboard with the cases in which he/she is working and his/hers opened goals, according the chosen profile after the login (if the user has more than one profile).
UFR013	The software must exhibit notifications on the browse every time that a goal is attributed to the user.
UFR014	The software must generate reports/documents of each matrix process step.

3.2 UML Diagrams

The UML diagrams are used to connect the system elements through different ways. This aims to represent static, structural and behavioral software aspects (Rumbaugh, Jacobson and Booch, 2004). In the present work, the used diagrams were the Use case diagram, activity diagram and the class diagram. Hereinafter, we present the summarized diagrams of the main flow of the proposed web system.

3.2.1 Use Case Diagram

The use case diagram models the system requirements and its actors. It is a great tool for the early development of the system, since it can graphically summarize how actions each actor will engage (Rumbaugh, Jacobson and Booch, 2004). Because of its intuitiveness, it can also be understood by the customer and helps validating the specification requirements. In Figure 1, only the main use cases of the suggested system are represented.

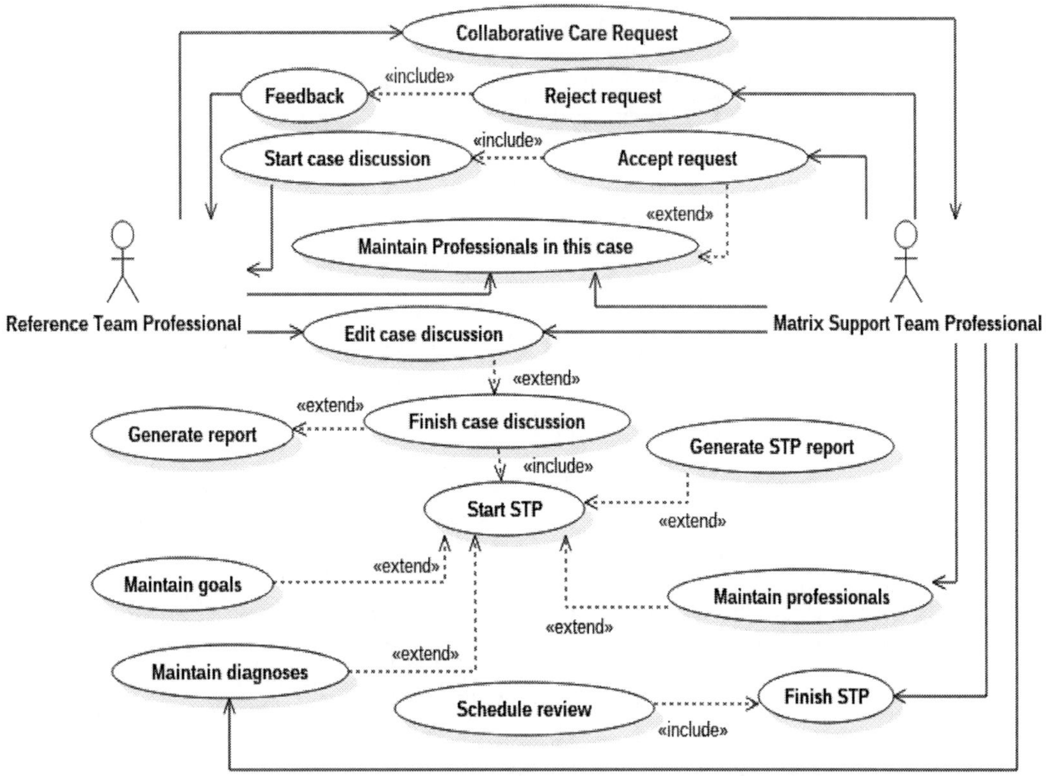

Figure 1. Use Case Diagram

Through this diagram it is notable that the STP starts after the case discussion finishes. Afterward, the MSTPs and RTPs can manage the tasks and goals related to the patient treatment. Additionally, a MSTP can make the patient diagnosis and include new professionals to the STP, according to the case. As shown in Frame 1 if a STP needs a phonoaudiologist and the CCT does not have one of them, a MSTP can invite a phonodiologist from any RT or MST of your district to contribute.

3.2.2 Class Diagram

The main goal of the classes diagram is to give a static view of the developed model, showing some elements, as class and types, its relationships and content (Górski and Sowa, 2016). The Figure 2 shows the main flow of the system proposed in a summarized classes diagram. The *Specialized Team* and the *Primary Care Team* are the abstractions of the health teams in the collaborative care process.

Furthermore, the *Matrix Support Team* and *Reference Team* classes are extensions of the *Network Team* class. This has been concepted to allow the join of another team types in system, such as social service institutions, non-governmental organizations, etc.

Trough this Class Diagram it is possible to see that a Collaborative Care Request may have two indexed formularies: the PHQ-2 and PHQ-9 questionnaires. They serve as a screening and diagnostic tools for Major Depression Disorder, and the CC Team can have access to them through the system. Other tools may be added according to the demand of the professionals involved in each case.

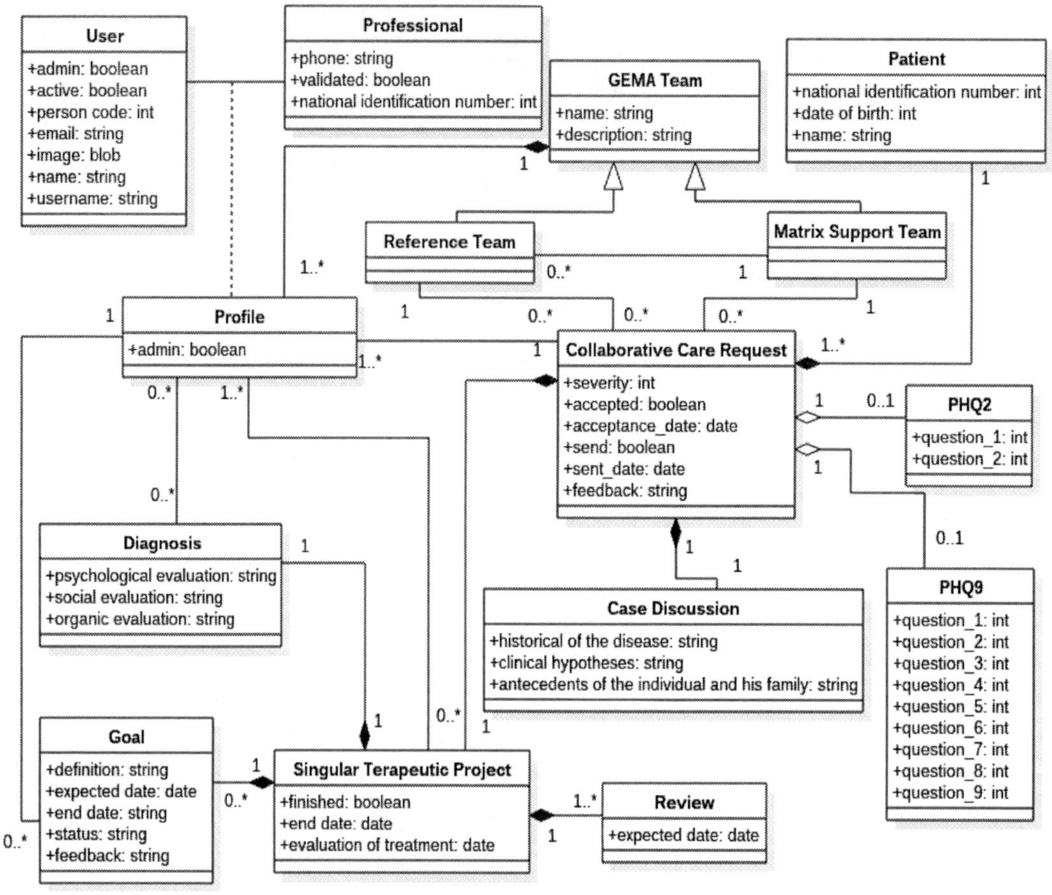

Figure 2. Simplified Class Diagram

3.2.3 Activity Diagram

The activity diagram is capable to represent the software functional behaviour of a particular process. Thusly, after requirements elicitation, it is quite useful to show the system information flow, in accordance with certain conditions.

Analyzing the diagram in Figure 3, succeeding the dispatch of a collaborative care request by the RTP, a profissional from MST can accept or not the request. In case of rejection, the MSTP must send a feedback with some clinical directions for the RT to handle the case, even without the collaborative care. For example, a case with drug dosing doubt which may be easily responded. If the request was accepted, a feedback is also send to establish a initial communication. Hereafter, the case discussion begins. Likewise, when the STP ends, a review must be scheduled to the patient.

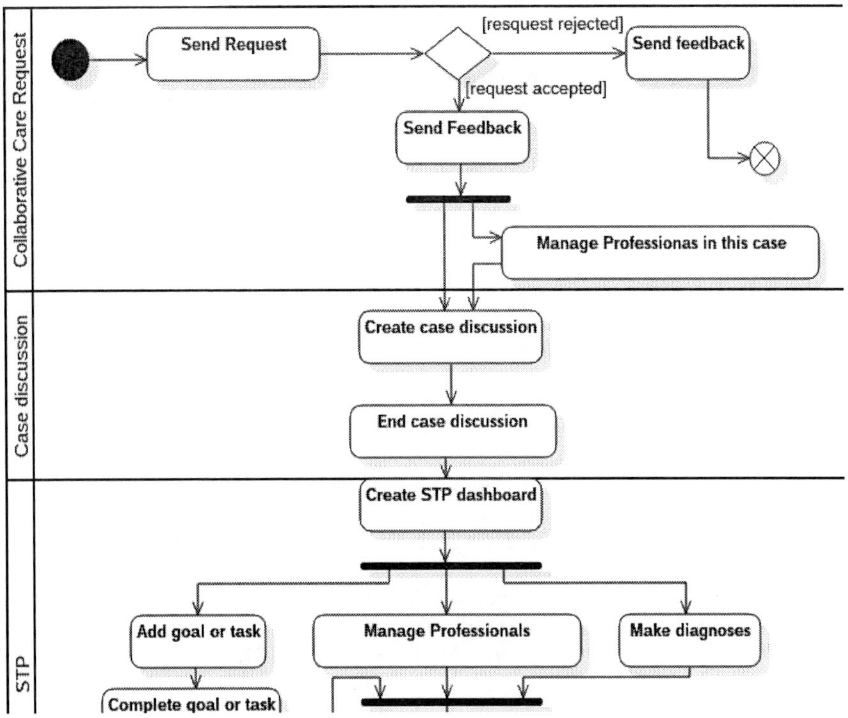

Figure 3. Activity Diagram of Collaborative Care Request process

4. CONCLUSION

After a series of workshops made in partnership with the HUOL/LAIS and the Health Department of our city, through the years of 2016 and 2017, with over 40 professionals of MT and PC, the architectural modeling of the GEMA web system was accomplished.

This work presents our proposal of a web system capable of promoting the management of collaborative care and potentiating this process in mental health. Thus, the proposed system can broaden the integration between the primary care network, mental health teams and other devices of the psychosocial care network, identifying problems in an agile way and favoring the management of services in the territory, with access to information of the situation of each patient in real time, allowing integration among health teams, shared, multidisciplinary and continuous care.

It may also provide a more effective and efficient resources usage and increase the professionals' abilities and satisfaction for them as for the patients, being a way to overcome known difficulties in our territory.

ACKNOWLEDGEMENT

The authors would like to acknowledge the University Hospital Onofre Lopes (HUOL), as much as the Laboratory for Technological Innovation in Health (LAIS) and the Federal University of Rio Grande do Norte (UFRN) for all the infrastructure and support given during the development of this work.

REFERENCES

Adamu and W. M. N. W. Zainoon, "A Framework for Enhancing the Retrieval of UML Diagrams," Software Reuse: Bridging with Social Awareness, G. Kapitsaki and E. Santana de Almeida (Eds.), ICSR 2016, LNCS 9679, 2016, pp. 384-390.

Bower P, Gilbody S. *Managing common mental health disorders in primary care: conceptual models and evidence base*. BMJ. 2005; 330(7495): 839–842.

Brasil. Ministério da Saúde. *Diretrizes do Nasf: Núcleo de Apoio à Saúde da Família*. Série A. Normas e Manuais Técnicos. Cadernos de Atenção Básica, n. 27; 2009. Ministério da Saúde, Brasília, Brasil.

Chiaverini DH et al. *Guia prático de matriciamento em saúde mental.* Ministério da Saúde: Centro de Estudo e Pesquisa em Saúde Coletiva, 2011. Ministério da saúde, Brasília, Brasil.

Fortes, Sandra et al. Psychiatry in the 21th century: changes from the integration with Primary Health Care through matrix support. Physis [online]. 2014, vol.24, n.4, pp.1079-1102. ISSN 0103-7331. http://dx.doi.org/10.1590/S0103-73312014000400006.

Gonçalves DA et al. *Brazilian multicentre study of common mental disorders in primary care: rates and rela*ted social and demographic factors. Cad. Saúde Pública. 2014; 30(3):623-632.

Górski, T., & Sowa, M. (2016). Construction of UML class diagram with Model-Driven Development. *Biuletyn Wojskowej Akademii Technicznej*, *65*(1), 111-129.

Grote, N. K., Katon, W. J., Russo, J. E., Lohr, M. J., Curran, M., Galvin, E., & Carson, K. (2015). Collaborative care for perinatal d*epression in socioeconomically disadvantaged women: a randomized trial.* Depression and anxiety, *32*(11), 821-834.

György P. et. al. EU JJoint Action on Mental Health and Well-being . European Union, 2016.

Katon W, Von Korff M, Lin E, Walker E, Simon GE, Bush T, et al. Collaborative managem*ent to achieve treatment guidelines. Impact on depression in primary care*. JAMA 1995;273:1026–31.

Kawai, S., & Matsuura, S. (2015, July). Model driven development by separating concerns in UML requirements specification. In *Computer Software and Applications Conference (COMPSAC), 2015 IEEE 39th Annual* (Vol. 3, pp. 668-669). IEEE.

Li, X., Liu, Z., & Jifeng, H. (2004). A formal semantics of UML sequence diagram. In *Software Engineering Conference, 2004. Proceedings. 2004 Australian* (pp. 168-177). IEEE.

Rumbaugh, James, Ivar Jacobson, and Grady Booch. *Unified modeling language reference manual, the*. Pearson Higher Education, 2004.

Sabharwal, Sangeeta; KAUR, Preeti; SIBAL, Ritu. Theoretical and Empirical Validation of Use Case Diagram Complexity Metric. International Journal of Mathematical and Computational Methods, v. 2, 2017.

Silva Junior JS, Fischer FM. *Adoecimento mental incapacitante: benefícios previdenciários no Brasil entre 2008-2011.* Revista de Saúde Pública. 2014; 48(1), 186-190.

Singh, M., Sharma, A. K., & Saxena, R. (2016). Formal Transformation of UML Diagram: Use Case, Class, Sequence Diagram with Z Notation for Representing the Static and Dynamic Perspectives of System. In *Proceedings of International Conference on ICT for Sustainable Development*(pp. 25-38). Springer, Singapore.

Torchiano, M., Scanniello, G., Ricca, F., Reggio, G., & Leotta, M. (2017). Do UML object diagrams affect design comprehensibility? Results from a family of four controlled experiments. *Journal of Visual Languages & Computing*, *41*, 10-21.

WHO. *WHO Report 2001: Mental health: new understanding, new hope*. World Health Organization, Geneva; 2001.

Sommerville, Ian. Software engineering. New York: Addison-Wesley, 2010.

FEATURES OF PARENTS' DIFFICULTIES IN USING TECHNOLOGIES TO MANAGE DIABETES IN CHILDREN AND ADOLESCENTS

Wala'a Barifah[1,2], Bill Davey[1] and Vince Bruno[1]
[1]*School of Business IT & Logistics, RMIT University, Melbourne, Australia*
[2]*Management Information Systems, TAIF University, Taif, Saudi Arabia*

ABSTRACT

Type 1diabetes (T1D) is a chronic disease (CD) affecting people of different ages including children. The growing number of diabetics has placed a burden not only on healthcare providers, but also on caregivers at home including parents, close relatives and friends. Information and communication technologies (ICTs) have been employed to support the self-management of CDs. However, little is known about how these ICTs help patients' parents to manage their children's disease. This study aimed to identify the challenges experienced by parents, which may inhibit the use and effectiveness of ICTs. Employing a qualitative approach, 58 semi-structured in-depth interviews were conducted with 37 parents of T1D children and 20 allied health professionals. Data was analysed using Atlas software. Initial coding identified two major themes namely that ICTs are ineffective in facilitate the acquisition of T1D information and not suitable for parents' preferences and culture. Results from this study can help various organisations such as ministries, and ICT companies, individuals and groups including nurses, doctors and web designers to provide recommendations for the diverse ICT tools that could empower parents' use of ICT and facilitate the adoption of ICT tools for other stakeholders.

KEYWORDS

ICT, T1D, Difficulty, Parents, Management, Health Information

1. INTRODUCTION

Enhancing parents' understanding of their children's T1D is one of the ways to ensure that T1D patients receive quality home care and management of the condition. T1D is more prevalent among children and adolescents (C&A henceforth coming) than ever, affecting children globally (Newell & Hahessy, 2013; Streisand & Monaghan, 2014). As Mayer-Davis et al (2017) state, diagnosis of T1D in C&A presents a considerable medical and public health issue, mainly because of the challenges related to the condition's management, coupled with the threat of severe and chronic complications. Parents of the children with T1D are responsible for the daily management of the disease (Carter, 2015; Xiros, 2016), including frequent glucose monitoring, diet regulation, insulin administration, and physical activity regulation (American Diabetes Association, 2013). Therefore, T1D management can be a challenging task for parents for various reasons. Not only the associated physiological factors and heightened sensitivity to insulin, but also the children's behavioural, cognitive and socio-emotional development.

To effectively provide support to children and ensure effective T1D management, parents must apply novel strategies, such as education and support concerning their coping skills. These strategies should be embedded in the healthcare services to empower parents and reduce their stress in providing care to their children (Grey et al, 2011). To cope with the difficulty of helping their children to conform to the rigorous and frequent monitoring of blood glucose, parents of children with T1D need access to information that can help them effectively manage their children's disease. In a multinational survey conducted by Peyrot (2009), it was revealed that having access to multiple sources of information was linked to better outcomes for parents of children with T1D. A common approach through which parents can gain access to valuable information is using information and communication technology (ICT).

ICT platforms allow caregivers to access valuable information on the management of CDs. The internet is an important source of information for both patients and caregivers through online forums and networks. Application of ICT in accessing information encompasses what is commonly known as e-Health, which could become a strategy for ensuring that children with T1D receive proper support and care from their parents (Lindberg et al, 2013; Wittenberg, 2015).

Despite the importance of ICT and related technologies in the management of CDs, parents of children with T1D experience numerous challenges in the use of ICT for managing their children's disease. However, little has been documented concerning the challenging aspects of understanding the use of ICT among parents of children with T1D. This study aims to particularise the difficulties experienced by parents of children with T1D when using ICT and related technologies for managing the disease and providing support to their children.

2. LITERATURE REVIEW

The need to obtain a deeper understanding of how parents perceive T1D treatment technology and the use of ICTs was identified by Nordfeldt et al (2013). Multiple studies have documented various factors that make it challenging for consumers to use ICT and related technologies. Via scenario-based tests and in-depth interviews of 14 patients and 14 caregivers, Nijland et al (2008) reported that it was challenging for patients and caregivers to search for self-care information using a digital media encyclopaedia and the navigation structures of the digital medical encyclopaedia website were hard for caregivers to understand when trying to obtain the desired information. While many caregivers reported the need to retrieve health information immediately from a website, but it was found that the websites were not equipped with search options to help them find the information quickly (Nijland et al, 2008).

Some researchers have reported that the use of ICT has profound effects on users' health (Heinrich et al, 2012; Boels et al, 2017), while various studies have demonstrated that the attrition problem threatens many e-Health interventions (Kelders et al, 2011; Brown et al, 2007). Nijland et al (2008) attributed the decline of web application usage among patients and caregivers to their lack of user friendliness and the ceiling effect. This was determined from email interviews, with the results indicating that patients and caregivers were likely to forget how to use web applications due to the lack of a reminder feature. In contrast, previously tested web-based self-management programmes for diabetes reported to have enhanced efficacy, feasibility and user acceptability (Pouwer, 2009; McKay et al, 2001; Heinrich et al, 2010). A meta-analysis by Angeles et al (2011), aiming to evaluate the effectiveness of web-based tools for diabetes patients, demonstrated that the difficulties participants experienced in using web-based interventions caused higher dropout rates. In addition, a study evaluating the use of web-based intervention programs to support diabetes self-management via elements obtained from social cognitive theory concluded that web-based trials ought to plan for an approximately 50% dropout rate as early as the first month following an intervention. Further, van Vugt et al (2016) reported that nearly every second patient failed to log on to their personal health record more than once, where the record also provided self-management support and personal feedback for patients with T2D.

Concerning the reduced use of web-based interventions, Lie et al (2017) demonstrated that participants were discouraged by the frustrating technology used in a Guided Self-Determination (GSD) eHealth intervention. Although the participants initially reported being interested in the intervention to save time and other resources when receiving information from their nurse, they were concerned about the difficulty of navigating the webpage as a result of errors they experienced with the website portal. Thus, they regarded the webpage as tiring and time consuming (Lie et al, 2017). Some web portals are considered cumbersome, as participants must download and save PDFs and then fill out reflections.

According to Sayakhot et al (2016), most websites that patients can rely on as sources of valuable information for managing diabetes offer static content that is hard to understand. Lie et al (2017) also found that the content of most websites addressing diabetes management and intervention was confusing for most patients and web users. Many websites have been reported to offer caregivers and patients too many options, thereby causing confusion about what to select to obtain useful information. In this respect, a study by Nolan et al (2015), based on patients' reactions to a web-based cardiovascular risk calculator for T2D, found that users had difficulty in understanding the information displayed by the calculator, and they did not realise that the different formats presented the same information in varying ways. While web-based interventions have been demonstrated to improve self-management (Nes et al, 2013), in a study seeking to compare diabetes-specific Android smartphone applications, Demidowich et al (2012) concluded that, out of 42 studied smartphone applications, only 4 had a composite usability score exceeding 20, indicating that only a few applications had the potential to offer a comprehensive method of effective diabetes management. Many diabetes webpages, including the Portal da Diabetes and Learning About Diabetes, Inc., have been described as providing static information about diabetes management and lacking interactive tools. This is in line with the assertions of Yu et al (2012) that existing diabetes websites are characterised by instructive information of varying eminence, newspaper format and high reading levels, with little or no interactive technology, problem-solving assistance or social support.

Although the challenges experienced with web solutions have sometimes been described as minor, Lie et al (2017) reported that a significant number of participants in their study perceived the GSD content as not being designed to meet their needs and expectations for effective T1D management. Electronic-based diabetes tools and applications must be comprehensible, interactive and promote social support among patients and caregivers for continued use of websites and electronic-based tools, resulting in increased clinical improvement.

3. METHOD

3.1 Participants and Population

The research question for this study is as follows: *What do parents experience difficulties using ICT for seeking information to manage children's and adolescents' T1D*? A qualitative approach was adopted, as gaining insight into the features of parents' difficulties requires a deep understanding of the relevant phenomena (Trochim, & Donnelly, 2001). Semi-structured questions allowed a wide range of issues to be introduced and permitted the parents to guide the interviews. As primary care providers, thirty-six parents of C&A with T1D were included in the study. The sample of participants was drawn from two countries, Australia and Saudi Arabia, to broaden the characteristics of included parents.

Pilot interviews were held with eight participants from the two countries, and 58 interviews were conducted. Several categories of professionals were interviewed with a view to capturing their understanding of parents' problems and provide triangulation to the study. These included paediatric endocrinologists, diabetes educators, dieticians and company representatives for diabetes devices. The interviews used prompting techniques, as proposed by Browne and Rogich (2001), which provide a context for testing the relative effectiveness of new approaches. The questioning scheme entailed asking questions based on a theoretical model of knowledge structures (Browne & Rogich, 2001). Use of the prompting techniques in interviews elicits more requirements from the participants, and thus, more information.

3.2 Analysis

The interview transcripts were uploaded to Microsoft Word and then Atlas software for management, coding, visualisation, retrieval of queries and reporting data (Bazeley, 2007). Two external researchers reviewed the coding and themes obtained from Atlas for validity checking. The transcripts were analysed starting with the assumption that parents' difficulties when interacting with the ICTs to seek information about T1D and its management.

'Actor–network theory' (ANT), which determines the nature and views of actors in health innovation via interviews, was used. ANT provides insight into the ICT adoption framework by indicating the diversity of the actors (Awa et al, 2016). In this paper, two major actors were studied – parents and ICT. Emerging themes were tentatively grouped and presented with a statement for identifying the barriers in the two actors' relationship following the ANT approach. This paper examines a key translation between these actors, describing parents' interaction in using relevant ICT, thereby incorporating the two themes discussed in the next section.

4. RESULTS

In this qualitative study, many participants identified key elements of their difficulties using ICT, as they relate to the management of T1D in children. These difficulties aggravate the need for new innovations in the context of ICT for CD management and awareness.

4.1 Theme one- ICT is not Facilitating the Acquisition of T1D's Information

This theme incorporates the ICT's tools and information is not assisting parents for acquiring information about T1D. The theme consists of one major category as described below.

4.1.1 ICT Tools and Information is not Available

It was evident from the analysis that most parents faced difficulties due to the lack of availability of ICT tools related to diabetes management. Many parents explained that ICT tools were non-existent or not effective for their children's condition. Despite the attention of T1D, parents' need for information was ignored by the stakeholders, thus it was less available.

[There are so many types of ICT that we are missing it. If we search the Internet and social media looking for information, it is not available. Even when we find some information, apps, or websites, it is insufficient and ineffective.] ***Par-k-1***

Researcher*: Why do you think this happens?*

[Because no attention is being paid to us or our kids' condition.] ***Par-k-1***

Another mother was concerned about the lack of information:

[It's not really common! If there is ICT training about diabetes, it is not available.] ***Par-a-1***

All interviewees also mentioned that T1D is complex in its nature as it requires extensive information and shared responsibilities between different ICT designers, developers, and other medical stakeholders. One parent saw the need for collaboration between official authorities and experts in the field acting as ICT facilitators due to the difficulties in understanding the nature of the diseases and information available about them.

[..... there is the need for a partnership between different fields: medicine, medication, experts in diabetes treatment technologies, diabetes health informatics, and technicians. That's why it is not in here.] ***Par-k-8***

This study showed that lacking T1D sources contributed to ICT's difficulties. An endocrinologist stated that little information on ICT usage for T1D management and for parents was available in English and practically none in any other languages.

[I am sure there is no website about insulin administration for parents in Arabic, Spanish, or Chinese, but there is a little in English.] ***Endo-k-4***

Most parents noted the same issues about difficulties which aligned with the endocrinologist's opinion. Further, one of the causes of the scarcity of information in regards to ICT and diabetes and treatment technologies, according to parents, is that information providers, such as professionals, diabetes care companies, and technologists don't seem to want to publish this information in ICT. A mother illustrated this issue when she lost a book given to her by the hospital and had difficulty finding the same information about diabetes management in ICT.

[It is hard to find information about these medical devices in all ICTs. They [professionals and providers] said to me to keep it [the book], do not lose it, because you will not find this on the Internet. Obviously I don't think they want this to get out there and I misuse it.] ***Par-a-7***

The analysis also shows that the use of ICT in diabetes management has financial boundaries for both; ICT designers and parents. For designers, their effective designs should be affordable for parents who find it difficult to spend additional money on ICT over the money they already spend on their medical condition. This is important because it seems, as our analysis shows, that the companies for diabetes treatment and technology follow profits-based approach ignoring users' needs. An endocrinologist deduced:

[The aim of companies A and B [companies for diabetes treatment and technology] is first and foremost the doctor.

***Researcher**: Why?*

[For their advertising and profitability, but unfortunately they are not volunteering to do something special for the parents and patients. Unfortunately it is impossible!]. ***Endo-k-1***

The next section describes the second theme emerged from the preliminarly analysis.

4.2 Theme two- Current ICT is not Suitable for Parents Preferences and their Children

The content and design of ICT has emerged as one of the features hampering the use of ICT and had an adverse influence toward parents' information gathering. This theme consists of two categories:

4.2.1 Current ICT is not User-Friendly

The focal meaning of this category is ICT is not easy to use or understand in finding T1D's information for the end users. One interesting finding was that the need to document every single detail in ICT made the use of ICT more difficult. In the past, caregivers were just read information from a book or pamphlet without being told more facts about the management of diabetes and the problems of medical technologies. However, currently parents prefer using friendly ICT that makes accessibility easier. The following diabetes educator **(Dia-edu-a-1)** explains the difficulties of parents using ICT applications and having to document all the details.

[People don't want to document every single detail in an app to try to determine things. So try to make the technology as user friendly as possible and make the information sources as quickly and easily accessible as possible.] ***Dia-edu-a-1***

4.2.2 Creates Information Overload, is not User Oriented and can be Culturally Inappropriate

The focal meaning of this category is that, parents finding difficulties in understanding the T1D's information and unable to making a decision because of having too much information through using of ICTs. Also, it means the content of ICT does not align with the norm of the local culture of the parents. The need of parents in using ICT information for diabetes management means they are susceptible to finding information from both trusted and untrusted sources. When parents search through ICT, they often experience information overload, such as finding algorithms for insulin doses from unknown authors that make the situation more complex and could end in negative health outcomes.

[I think for people to trust who you know rather than trying to keep on top of all this information overload.] ***Dia-edu-a-1***

Furthermore, parents need comprehensive information about their children's condition in an organised and customised structure to help improve their care management. Accurate and timely data are often not available, and some ICT tools do not support parents' needs.

[I started on Facebook and looked at some different sites a bit and all the people talking on it. I found it to be annoying and confusing because it doesn't suit my child's case. I would use an ICT tool if it said it suits my case!] ***Par-a-7***

Another interesting finding of this study was that parents face cultural difficulties using ICT tools and resources. It appears that what is missing is a unique pattern of ICT that can fit into both contexual contexs. Taking the example of finding the nutirtion information using ICTs, parents revealed their concerns regarding the lack of sufficent information and the inaccuarte content available for carbohydrate counting. **Par-k-18** used a mobile application at home in which the search ended with no results found for particular foods caused the parent to lose the trust of using ICT and even affected his next experience.

5. DISCUSSION

Constant interaction between healthcare institutions and patients has been an important aspect of improving quality of care. ICT approaches can increase the effectiveness of patient care by enabling structured and accessible valuable information for managing CDs (Lisiecka-Biełanowicz & Wawrzyniak, 2016). The findings show that the effectiveness of the current utilisation of ICT by parents of T1D children confined to finding emotional support rather than a primary source of information. This is in line with Greene et al's (2011) findings, which showed that diabetes communities on Facebook serve as support communities, promotional spaces and places where individuals can solicit various forms of disease management information not found elsewhere. Many of the participated caregivers stated the social media platforms like Facebook and Twitter to share personal experiences on diabetes self-management. Whiles a few participants indicated the minimal use of mobile applications to receive updates on insulin management tips. A practical implication for this finding pertains to the need for effective design of ICT to better manage the condition.

Despite the benefits of ICT and related technologies, the participants stated that they experienced difficulties while using or interacting with the various platforms. Like the findings in prior studies, this research demonstrated that parents of children with diabetes experienced problems in trying to understand the message or information on most diabetes webpages, which usually lacked interactive tools. Moreover, the information on webpages and other ICT platforms were complex; parents found it difficult to understand and employ such information in managing their children's diabetic conditions. They stated that they would prefer intuitive, simple and user-friendly websites with only a few options to ensure effective diabetes self-management. Unlike previous studies, the study found that the parents viewed the information on most websites as indistinct and generalised, and thus, unsuitable for their children's conditions.

6. RECOMMENDATIONS

The following recommendations seek to address the challenges of using ICT tools to support the management of CDs:

a. ICT tools must provide an interactive component to bring the parents, health community and healthcare services together to enable better dissemination of diabetes management information;

b. The designers' and programmers' skills and knowledge should be improved for producing ICT tools that assist parents in providing care for children with a CD. This can be done through university courses providing a combined approach, which includes both the context (i.e. diabetes information) and technologies (i.e. ICT);

c. Technicians could adopt a machine learning approach, designing ICT tools based on previous interactions in order to improve the recommendations and personalisation. This information needed for such tools will require ethical consideration. This will require collaboration between healthcare organisations and the IT community to provide effective designs for managing CD with ICT;

d. Educational providers can design specialised courses focussing on the combination of the key areas of health-informatics, medical practitioners, CD, ICT, and users' needs. The current specialisations available in he filed of health informatics and diabetes education are designed with a minor-technology focus.

e. The design of emerging ICT tools is recommended to be inclusive and culturally responsive to meet the demands of its particular users.

f. Financial barriers should be addressed in a collaborative approach including medical professionals, ICT providers, and communal support.

7. CONCLUSION

Interviews with parents of T1D children, and ancillary medical staff has shown that there are two main themes are that parents finding difficulties: 1: ICT is not Facilitating the Acquisition of T1D's Information and 2: Current ICT is not Suitable for Parents Preferences and their Children. The first theme has one aspect: 1- ICT Tools and Information is not Available. The second theme has two aspects: 1- Current ICT is not User-Friendly and 2- Creates Information Overload, is not User Oriented and can be Culturally Inappropriate.

It is clear from the data that, parents use various ICT tools, including diabetes webpages, emails, social media sites, smartphone apps and messaging apps. These tools enable parents to access and share information meant to improve self-management of CDs. The ICT tools have also eased the sharing of critical information by physicians and nurses with patients and caregivers and helped patients and caregivers to share their experiences on social sites like Facebook and Twitter. This study's findings suggest that the use of diabetes ICT tools can be difficult, which may discourage their use. Some such difficulties include technical problems in trying to log in to health websites, webpage errors and complex and invalid information provided for queries. This information on managing diabetes and other conditions is hard to read, hard to understand, ambiguous and generalised. In addition, users of ICT tools experience difficulties in navigating health websites, since they often lack appropriate navigational tools.

REFERENCES

American Diabetes Association, 2013. Standards of Medical Care in Diabetes – 2013. *In Diabetes Care*, Vol. 36, No. 1, pp. S11-S66. http://doi.org/10.2337/dc13-S011

Angeles, R. et al, 2011. The Effectiveness of Web-based Tools for Improving Blood Glucose Control in Patients with Diabetes Mellitus: A Meta-analysis. *In Canadian Journal of Diabetes*, Vol., 35, No. 4, pp. 344-352. doi: 10.1016/S1499-2671(11)54011-0.

Awa, H.O., Ukoha, O. and Eke, B.C., 2016. Adoption of emerging ICTs: The role of actors in a social network. *In Cogent Business & Management*, Vol., 3 No.1, pp.1259879. https://doi.org/10.1080/23311975.2016.1259879

Bazeley, P., 2007. Qualitative Data Analysis with NVIVO. *Thousand Oaks*, CA: Sage.

Boels, A. M. et al, 2017. Diabetes Self-management Education and Support Delivered by Mobile Health (m-Health) Interventions for Adults with Type 2 Diabetes Mellitus. *In Cochrane Database of Systematic Reviews*, No. 11 (Art. No.: CD012869). DOI: 10.1002/14651858.CD012869.

Browne, G.J. and Rogich, M.B., 2001. An empirical investigation of user requirements elicitation: Comparing the effectiveness of prompting techniques. In Journal of Management Information Systems, Vol., 17 No. 4, pp.223-249.

Brown, L. L. et al, 2007. A Review of Web-assisted Interventions for Diabetes Management: Maximizing the Potential for Improving Health Outcomes. *In Journal of Diabetes Science and Technology (Online)*, Vol. 1, No. 6, pp. 892-902.

Carter, M. B., 2015. Stress in Parents of Children with Type 1 Diabetes. In DNP Projects, 57. https://uknowledge.uky.edu/dnp_etds/57

Demidowich, A. P. et al, 2012. An Evaluation of Diabetes Self-management Applications for Android Smartphones. *In Journal of Telemedicine and Telecare,* Vol. 18, No. 4, pp. 235-238. Doi: 10.1258/jtt.2012.111002.

Greene, J. A. et al, 2011. Online Social Networking by Patients with Diabetes: A Qualitative Evaluation of Communication with Facebook. *In Journal of General Internal Medicine,* Vol. 26, No. 3, pp. 287-292. http://doi.org/10.1007/s11606-010-1526-3

Grey, M. et al, 2011. Coping Skills Training for Parents of Children with Type 1 Diabetes: 12-Month Outcomes. *In Nursing Research,* Vol. 60, pp. 173-181.

Heinrich, E. et al, 2012. Evaluation of the Web-based Diabetes Interactive Education Programme (DIEP) for Patients with Type 2 Diabetes. *In Patient Education and Counseling,* Vol. 86, No. 2, pp. 172-178.

Heinrich, E. et al, 2010. Self-management Interventions for Type 2 Diabetes: A Systematic Review. *In European Diabetes Nursing*, Vol. 7, No. 2, pp. 71-76. doi: 10.1002/edn.160.

Kelders, S. M. et al, 2011. Effectiveness of a Web-based Intervention Aimed at Healthy Dietary and Physical Activity Behavior: A Randomized Controlled Trial about Users and Usage. *In Journal of Medical Internet Research,* Vol. 13, No. 2, p. e32. http://doi.org/10.2196/jmir.1624

Lie, S. S. et al, 2017. Dropout from an eHealth Intervention for Adults with Type 2 Diabetes: A Qualitative Study. *In Journal of Medical Internet Research,* Vol. 19, No. 5, p. e187. http://doi.org/10.2196/jmir.7479

Lindberg, B. et al, 2013. Using Information and Communication Technology in Home Care for Communication between Patients, Family Members, and Healthcare Professionals: A Systematic Review. *In International Journal of Telemedicine and Applications,* Vol. 2013, p. 461829. http://doi.org/10.1155/2013/461829

Lisiecka-Biełanowicz, M. and Wawrzyniak, Z., 2016. Healthcare Model with Use of Information and Communication Technology for Patients with Chronic Disease. *In Annals of Agricultural and Environmental Medicine*, Vol. 23, No. 3, pp. 462-467. Doi: 10.5604/12321966.1219188.

Mayer-Davis, E. J. et al, 2017. Incidence Trends of Type 1 and Type 2 Diabetes among Youths, 2002–2012. *In The New England Journal of Medicine,* Vol. 376, No. .15, pp. 1419-1429.

McKay, H. G. et al, 2001. The Diabetes Network Internet-based Physical Activity Intervention: A Randomized Pilot Study. *In Diabetes Care,* Vol. 24, No. 8, pp. 1328-1334.

Nes, A. A. G. et al, 2013. Web-based, Self-management Enhancing Interventions with e-Diaries and Personalized Feedback for Persons with Chronic Illness: A Tale of Three Studies. In Patient Education and Counseling, Vol. 93, No. 3, pp. 451-458. https://doi.org/10.1016/j.pec.2013.01.022

Newell, E. and Hahessy, S., 2013. Social Support for Parents of Children with Type 1 Diabetes: An Ethnographic Study. *In Journal of Diabetes Nursing,* Vol. 17, No. 1, pp. 27-31.

Nijland, N. et al, 2008. Evaluation of Internet-based Technology for Supporting Self-care: Problems Encountered by Patients and Caregivers when Using Self-care Applications. *In Journal of Medical Internet Research*, Vol. 10, No. 2, pp. e13. http://doi.org/10.2196/jmir.957

Nolan, T. et al, 2015. Patient Reactions to a Web-based Cardiovascular Risk Calculator in Type 2 Diabetes: A Qualitative Study in Primary Care. *In British Journal of General Practice*, Vol. 65, No. 632, pp. e152-e160. Doi: 10.3399/bjgp15X683953

Nordfeldt, S., Ängarne-Lindberg, T., Nordwall, M. and Krevers, B., 2013. Parents of adolescents with type 1 diabetes-their views on information and communication needs and internet use. A qualitative study. In PloS one, Vol. 8 No.4, pp.e62096. https://doi.org/10.1371/journal.pone.0062096

Peyrot, M., 2009. The DAWN Youth WebTalk Study: Methods, Findings, and Implications. *In Pediatric Diabetes*, Vol. 10, No. 13, pp. 37-45.

Pouwer, F., 2009. Should We Screen for Emotional Distress in Type 2 Diabetes Mellitus? *In Nature Reviews Endocrinology*, Vol. 5, No. 12, pp. 665-671. Doi: 10.1038/nrendo.2009.214.

Sayakhot, P. et al, 2016. Use of a Web-based Educational Intervention to Improve Knowledge of Healthy Diet and Lifestyle in Women with Gestational Diabetes Mellitus Compared to Standard Clinic-based Education. *In BMC Pregnancy and Childbirth*, Vol. 16, p. 208. http://doi.org/10.1186/s12884-016-0996-7

Streisand, R. and Monaghan, M., 2014. Young Children with Type 1 Diabetes: Challenges, Research, and Future Directions. *In Current Diabetes Reports*, Vol. 14, No. 9, p. 520. http://doi.org/10.1007/s11892-014-0520-2

Trochim, W. M. and Donnelly, J. P., 2001. The Research Methods Knowledge Base. [Internet-based textbook]. Retrieved from http://www.socialresearchmethods.net/kb/

Van Vugt, M. et al, 2016. Uptake and Effects of the e-Vita Personal Health Record with Self-management Support and Coaching, for Type 2 Diabetes Patients Treated in Primary Care. *In Journal of Diabetes Research*, Vol. 2016, p. 5027356. Doi: 10.1155/2016/5027356.

Wittenberg, E., 2015. Textbook of Palliative Care Communication. Oxford University Press, Oxford, UK.

Xiros, E., 2016. The Quality of Life of Parents of Children with Diabetes Type 1: A Qualitative Study of the Quality of Life of Parents with Diabetes Type 1 and Their Coping Experiences. Department of Health Management and Health Economics, University of Oslo.

Yu, C. H. et al, 2012. Designing and Evaluating a Web-based Self-management Site for Patients with Type 2 Diabetes – Systematic Website Development and Study Protocol. *In BMC Medical Informatics and Decision Making,* Vol. 12, No. 57, pp. 1-10. https://doi.org/10.1186/1472-6947-12-57

SICO MOBILE – THE APP THAT SUPPORTS MOBILE DEATH CERTIFICATION IN PORTUGAL

Rui Pereira, Paulo Nogueira, Arlete Monteiro, Cristiano Marques and Henrique Martins
SPMS , Av. Da República, N.º 61, 1050-189 Lisboa, Portugal

ABSTRACT

The use of electronic death certificate is a worldwide emerging issue. In Portugal, since 2014, all death certificates are issued electronically, thanks to existence of the system called SICO (Sistema de Informação dos Certificados de Óbito). With this, the Portuguese Ministry of Health became a pioneer in the electronic death certificate area with the launch of an information system for this type of records. It is well known that the information and communication technology (ICT) in health are quickly developing. Taking into account the evolution of technology, new tendencies and needs, it was necessary to expand the options given to doctors to use the system. This was the driver to further innovate and create the mobile e-death certificate. With this pioneering project, we intend to address the issue of electronic certification through a solution architecture that is close to the needs of users.

Afterwards the study of the Portuguese mobile device market share, the defining of the strategy of aim for, in the first instance, limit users of SICO and all the necessary development, we obtained a mobile app that allow the issue of death certificates. In order to validate the app, a Beta phase took place before it was made available to the iOS and Android's stores.

Nowadays, a physician can issue death certifications in Portugal from a mobile app called SICO Mobile. This new app, which already counts with 121 downloads, provides greater convenience and mobility to physicians. With this, death certificates with better accuracy and quality of data are achieved. New features are already being prepared with the objective of improving even more its usability, the physicians' convenience and the data's quality.

KEYWORDS

Mobile, App, Dematerialization, Death Certification, eHealth

1. INTRODUCTION

In most industrialized countries, mortality data is the only medically relevant wide-ranging statistics for report on health and disease in a population. Even though mortality data reflects just the opposite of the state of health of a population, age and sex-specific mortality rates are important indicators of the health status of a population and are relevant factors for the effectiveness of the healthcare subsystems (Dorn and Moriyama, 1964). Validation studies show that the diagnoses on death certificates are much more reliable than generally thought (Frentzel-Beyme et al., 1980).

The Mortality Information System (MIS) exists in Portugal, as a paper-form death certification model, since 1950. Recently, this paradigm has changed, after the Directorate-General of Health (DGS) had coordinated a joint working group involving different ministries, resulting in the implementation, by the Shared Services of the Ministry of Health (SPMS), of an information system for electronic death certification: the Death Certificate Information System (SICO). A strategic multistep approach defined by geographic areas of the country was planned for the implementation of the system. The national implementation was completed in December of 2013 and 100% e-death certification was achieved beginning January 2014.

After SICO's national implementation with mandatory registration (publication of a law and four decree-laws in the Portuguese Official Journal (República, 2012a) (República, 2012b) (República, 2012c) (República, 2012d) (República, 2012e) and the approval of the Portuguese Data Protection Authority), it was time to continue the grow towards a multi-platform system. In this aim, SICO Mobile, as it was entitled, takes form as a mobile app where physician can certify case, in much the same way as on the existing web-based platform, but regaining the mobility that papers forms had, and that desktop solutions had removed from the work flow.

2. BACKGROUND

A huge milestone was achieved, in Portugal, when at the 1st of January of 2014 all the death certificates had to be issued in SICO. With this new information system, the certifications of death were issued immediately, safer and with better quality. Also citizen's data are now under more protection, as each institution only receives the information that it legally needs. The fully dematerialized process was preceded by a long maturation period and by an experimental phase, since autumn 2012 in Coimbra, which throughout 2013 has gradually been extended to the whole country, including autonomous regions.

SICO is based on the registry of an adult or a fetal and neonatal death certificate through an electronic form that, in its inherent form, has verifications to fill all the relevant fields, and the circumstances of each death type are considered. At the same time, this system also collects the Clinical Information Bulletins issued in accordance with the law, and reports of clinical and medical autopsies performed. These are the main operations that are available for the physicians. SICO, however, is much more than that, as it can be used by other actors and persons involved in the processes per their respective roles in the flow of the communication of a death certificate.

3. WHY A MOBILE APP?

We live in an era where billions of information records are created every day, so there is a need to use this information in an effective way. In the case of governments, for good governance, information must be used strategically for the better good. Good governance requires, among other things, decisions based on "information sets" or knowledge. Digitizing this information and making it available on a network that is accessible to all individuals opens up possibilities for the use of this information by everyone - paving the way for Digital Governance or e-governance (Deloitte, 2013).

The digitization of information minimizes the need of human intervention, making the administrative process more efficient and transparent. The greatest benefits of e-Governance are the elimination of delays in providing services and the provision of fair and equal access to government services / products to its citizens, making the process more transparent and inclusive (CDW, 2012) (Corbin, 2014).

It is a known fact that the use of mobile phones can support healthcare and public health interventions. Moreover, in the recent years, the mobile phones are being used as platforms for delivery of health interventions, as to provide diabetes education (Klasnja and Pratt, 2012) or to encourage attendance of primary care appointments (Fairhurst and Sheikh, 2008), for instance. Mobile phone applications permit physicians to monitor patients with chronic heart failure (D. Scherr, 2006) and to detect early signs of arrhythmia or ischemia that can indicate an imminent heart attack (Rubel et al., 2005). In addition, patients themselves can use phone applications and sensing and measurement devices to keep track of their physical activities (Klasnja and Pratt, 2012) and to monitor physiological markers relevant to their health status (Klasnja and Pratt, 2012).

With the mandatory registration of death certificates in SICO, it becomes necessary to provide SICO software on different platforms. This requirement arises as in many real situations physicians don't have access to a computer to register the respective death certificate (for instance, deaths that occur in the street, deaths that occur at home, and deaths that occur in nursing homes or similar institutions).

This need was confirmed when we analyzed the different type of devices that were used to access SICO. The great majority of the access to SICO were made on a desktop or a laptop, but more than 10.000 access where made using either a mobile phone or a tablet. Therefore, it was clear the need to either adapt the web-based application to be responsive to mobile or develop an app to fulfil the needs of these physicians who already are using mobile devices to access SICO.

Either way, physicians gain the capacity to issue the death certificate at the moment they are verifying the death, which can lead to a decrease the loss of data quality. This challenge is even more important when we look at the trend of penetration of smartphones in Portugal that increases monthly, and to the fact that the users prefer to use mobile devices - 2h42m of mobile application use per day – than browser applications - about 22 minutes of browser application use per day (Gabe Zichermann, 2011).

4. MARKET RESEARCH

In order to define the strategy (adapt the web-based application to be responsive to mobile or develop an app), we evaluate the market share of the usage of the desktop, the mobile phone and the tablet. In Portugal, nearly 75% use a desktop and 22% a mobile device (while tablet only has 4% of usage). But when we analyze this data worldwide, people are already using mobile devices over desktop (51% vs 45% respectively) (StatCounter, 2018). Based on the number of the usage of mobile devices in Portugal (figure 1) and supported by the worldwide tendency, the option for the developing of a mobile application was taken.

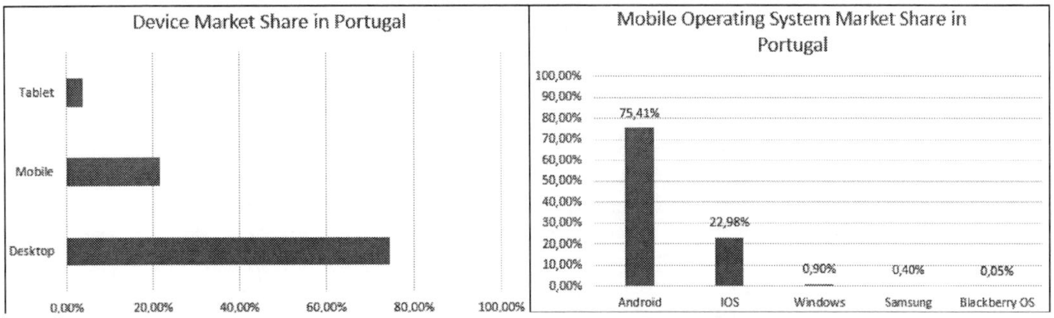

Figure 1. Device and Mobile Operating System Market Share in Portugal

After this decision, the research continued with the different operating systems existing for the mobile devices. SPMS' strategy regarding this question is to develop mobile app's that can be used either in iOS or Android operating system. This strategy is also aligned with the usage in Portugal which can be seen in figure 1 (StatCounter, 2018). The operating systems from Windows, Samsung (called Tizen) and Blackberry are of little usage.

By analyzing each operating system at a version context (figures 2), it is obviously that they are very different. For instance, more than 70% of iOS users have the 11th iOS version, which was launched in the 3rd quarter of 2017. However, in Android, there are a greater range of versions in use. For instance, the last version which is called Oreo, launched also in the 3rd quarter of 2017, only is use in 6% of the Portuguese Android devices. 39% use the version Nougat, which was launched at 2016 (StatCounter, 2018).

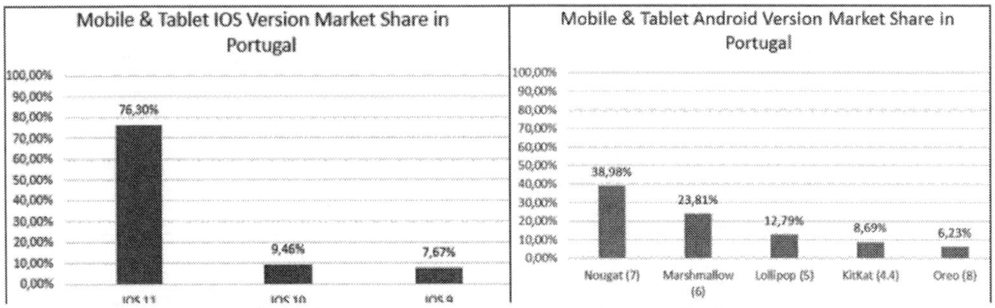

Figure 2. Mobile & Tablet OS Version Market Share in Portugal

Another major concern is the mobile vendor market share in Portugal (figure 3). If for iOS operating system, there is only one vendor (Apple), for Android there are very different vendors, like Samsung, Huawei, Xiaomi, Asus and others (StatCounter, 2018).

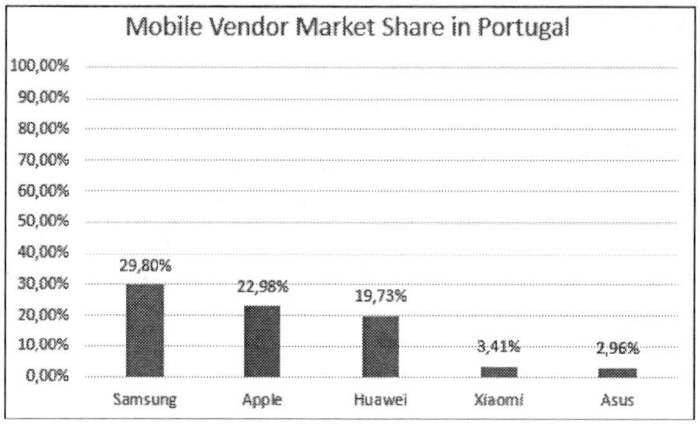

Figure 3. Mobile Vendor Market Share in Portugal

To sum, with these researches, it was obvious the strategy to develop a mobile app (for Android and iOS), and after the development of the app, a strong component of testing had to be applied in order to ensure its operation regardless the vendors, the version of the operating system, the screen resolutions and others facts that could interfere with the quality of the product.

5. THE DEVELOPMENT

For the first version of this innovating mobile app, the strategy was to develop an app directional to the physicians and only with some of the functionalities that are available in the web version, as can be seen in table 1. With this approach, we can validate the usage of the app, its acceptance and, in case of success, evolve with the addition of other functionalities for the physicians and providing access to users with different roles (Deloitte, 2013).

Table 1. Different Features on SICO (Web Version) and SICO Mobile

SICO	SICO Mobile
✓ Register/query death certificates (data protection issues apply to queries);	✓ Register/query death certificates (data protection issues apply to queries);
✓ Register/query Clinical/Circumstantial Bulletins	✓ Register/query Clinical/Circumstantial Bulletins;
✓ Query National Emergency Institute clinical form	✓ Print Transport Permit (to Funeral Homes)
✓ Register/query autopsy report	
✓ Amend and correct death certificates	
✓ Print Transport Permit (to Funeral Homes)	

Hence, the intention was to develop a mobile app to integrate the existence architecture of this information system. The propose can be seen in figure 4, where several types of roles can access the web-based SICO, but the mobile app SICO is only available, at least for the first version, to physicians to, for instance, issue death certificates.

The death certificate goes through all the validations (NHS users' registry and National Health Professional Register) and flows through the respective circuit, independently if it was issued on the SICO's web version or mobile app. The validation from the Public Prosecution Service or from the Police Authorities can be involved, if necessary. So, the disposal of the mobile app only acts as another tool to the physician and act along with the web version (and certificates issued in the mobile app can be seen on the web version and the other way around).

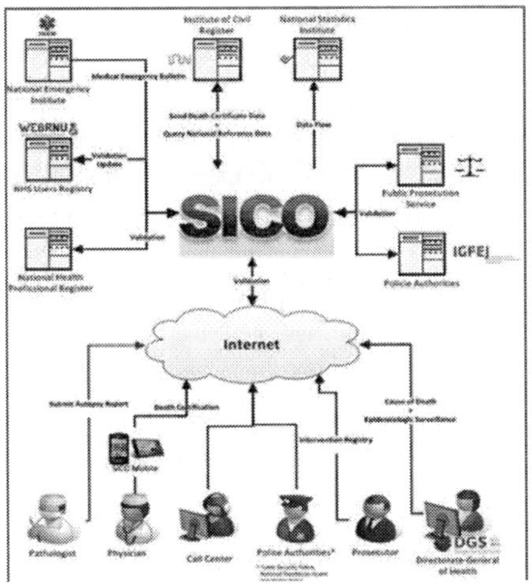

Figure 4. Overview of the Actual Mortality Information System Architecture

After the definition of the functional requirements, it was time to define the technical component to fulfill it. The option went with the usage of an open-source Software Development Kit (SDK) known as Ionic, which is built on top of AngularJS and Apache Cordova. Using tools and services to develop hybrid applications, this app takes part of Web technologies such as CSS, HTML5 and SASS.

The option for the usage of Ionic is related to the fact that we wanted a mobile application that would work both on iOS and Android operating systems. Another relevant aspect is that Ionic uses Cordova plugins to gain access to different features, such as GPS, camera, push notifications, automated builds and much more.

In the progress of the development of this application, some challenges were faced. The first struggle experience was how to display the information to the user. For instance, how to turn the extend form to issue a death certificate in a mobile device? No field could be excluded (as they are mandatory by law) and the fulfill of the form should be simple, usable, friendly and intuitive. And if errors are detected in the fill in of the form, how can they be shown in a simple way? It took several tests and usage options, some of them customized, to make sure that the experience was decipherable from the web version and, at the same time, innate to use. Furthermore, some technology available on the mobile devices were used to bring a better experience to the user: the possibility to click on the error message and automatically navigated to the field where the error was detected (Figure 6); usage of other apps installed in the mobile device to allow the visualization of the pdf generated or to email the pdf; among others.

Guaranteeing the compliance of the business logic was another major issue that we faced. The mobile version must have had to work equally as the web version. This was a lengthy process, which forced many tests with several people who had knowledge of the business, as well as a strict quality assurance.

Another important aspect was to ensure the data was being collected properly as well as it was submitted to the server. With this propose, some web services had to been developed to allow the communication between the client (the mobile app) and the existing server of SICO. With these new web services, we guarantee the communication between the two when registering and consulting a death certificate.

The other major issue was to ensure it was working properly in the different mobile devices (at least the devices of higher percentage of usage in Portugal). Several tests were performed in various iOS and Android devices, with different versions of the operating system. Based on the results of these tests, the design was changed and improved according to the needs of each platform as well as taking care of the performance and impact it would have on it.

Exceeded the different encounters faced, we end up with a mobile application to issue or query death certifications, that works both in Android and iOS, is intuitive, clear and simple. Another relevant aspect of the solution developed, is that we took part of the technology available on these mobile devices (smartphones and tablets) with the aim of providing a better experience in the usage of the mobile application.

6. THE RELEASE OF THE APP

As stated before, we knew the necessity of testing the app in different versions of Android, in devices from different vendors and not only on mobile phones but also on tablets. Testing the business rules and assuring that the behavior of one functionality on the mobile device must be the same on the web version. So, no release plan was put in action until we have the assurance of this assumptions.

With the testing component finalized successfully, it was time to organize the Beta version. The physicians who certified more deaths in 2016 and 2017 where invited to become the first physicians to issue death certificates with more mobility. After this period of beta testing, and with not one error reported, the official launch was made on the 22th of March of 2018, at the 2nd Portugal eHealth Summit, in Lisbon, Portugal.

The app is, since that date, available to be downloaded at the App Store (iOS) and Play Store (Android).

7. RESULTS

After all the developing process, we end up a mobile app directed to the physicians issue the death certificates. The aspect of the opening screen of this mobile app can be seen at figure 5.

Figure 5. Illustration of the Opening Screen of SICO Mobile

The login of this app is made with the same username and password that it is used to login into the SICO's web application. After the successfully login at the app, the menu appears where the physician can select the operation that he/she intends.

As exposed previously, in the process of the development of the app, the technology available on each mobile devices was used. For example, if in the process of issuing a death certificate the physician forget to fulfill one or more fields that are obligatory, an error is presented and clicking on it, the physician is automatically guide to that specific field, avoiding the unnecessary scrolls (figure 6).

Figure 6. Illustration of the Mobile Technologies Applied in SICO Mobile

When the Beta phase was materialized, we had some physicians that were excited with the mobile application approach in this context. At the official launch of the app, the physician that had certified more death by the mobile app were invited to talk a bit of his perspective of the usage of the app.

Having the perspective of the users it's very relevant since they are the users of the product. This physician was indeed a legist doctor that appraised the capability that have gained to issue de death certificated with a lot more mobility, comfortably and release him of having to go to his desk in order to issue this certificate. Also said that, although the death certificate is an extensive document, the app was very intuitive, clean and flows with the native functions of his operating system.

Since the release of the app, this mobile application have been downloaded 121 times (85 on iOS and 36 on Android). More than 160 successful accesses were registered from 50 different physicians in order to issue/query death certificates or Clinical/Circumstantial Bulletins or to print transport permit. In the last case, a pdf file is generated that can be visualized in the mobile phone and send to a printer or automatically integrated with an email client to email it.

More than 35 death certificates were issued by the mobile app and the process has occurred as it was expected and not a single inconsistence has been reported yet. Full disclosure of these values can be observed at table 2, as well as the average time that the system processes each operation.

Table 2. Resume of Different Operations Performed at SICO Mobile

Operations:	Number of times	Average time operation (milliseconds)
Login	161	5060
Death Certificated issued	36	609
Print Transport Permit	74	329
Print Death Certificate	32	325
Query Death Certificate	23	419

8. FUTURE WORK

With the data from Table 2 it was identified the need in improving the login system. Five seconds is not a long time as it has to verify the username and password, the role(s) associated and load the page with the different options that his/hers role allows to access.

The addition of more functionalities to this app are on the verge too. The possibility of rectify information of a death certificate that was incorrectly introduced previously is one of them. This functionality will increase the adhesion of this app by the physician as it has been the major suggestion that we have received concerning SICO Mobile. Evidently, when a mistake is identified by the Institute of Civil Register, for instance, and the physician has to correct it, he/she has to access the computer, them to a web-browser to open SICO and emend the identified mistake. The possibility of revise the information directly on the smartphone, will allow a better convenience for the doctor and improved data quality.

Now that the app has been validated by physicians, it can be made available for other users (apart from the physicians). This will allow, for instance, the pathologist to register the autopsy reporter or the police authority to validate death situations in the public road, for instance.

Geographical coordinates could be another interesting feature to be added to this mobile app. Once the place of death is an essential field in fulfilling of a death certificate, the geographical coordinates obtained by the smartphones' GPS could be integrated if the physician is issuing the death certificate at the exact point where the death occurred.

Evidently, there are many others features that can be developed or integrated in this app in order to expand its effectiveness. The aim is to innovate more and more, in a way to simplify the flow of this process. However, it is imperative to assure the relevance of each feature

9. CONCLUSION

The operationalization of an information system that allows the electronic certification of deaths (as well as the transmission of the appropriate information to the various stakeholders involved in this process) was a great step in the dematerialization of the National Health Service.

A mobile version of this information system would open the range of possibilities for health professionals, who acquire an adequate tool to certify a death wherever they may be, regardless of the surrounding circumstances. Therefore, the development of the mobile version of SICO were put into action.

The development of this mobile application was not easy since we had to map the fields of the death certificate form that physicians are accustomed to filling in the desktop with a generous screen to a mobile device, where the screen is noticeably smaller. In addition to this constraint, the usability of the application could not have been forgotten. In the end, everything has been overcome thanks to the use of the existence technologies of the devices themselves and the performance of various quality assurance tests.

This app, and its success, can open new opportunities as we validate the methodology, architecture and the technology used. Why not use the knowledge gain with this process and expand and replicate it on other information systems? The possibilities are endless.

REFERENCES

CDW 2012. Mobile Strategies for Government. *Taking advantage of smartphones and tablets to boost efficiency and improve services.*

Corbin, K. 2014. Mobile Apps Becoming Default for Government Digital Content. Available: https://www.cio.com/article/2376399/government-use-of-it/mobile-apps-becoming-default-for-government-digital-content.html.

D. Scherr, R. Z., A. Kollmann, P. Kastner, G. Schreier, F.M. Fruhwald 2006. Mobile phone-based surveillance of cardiac patients at home. *Journal of Telemedicine and Telecare,* 12, 255-261.

Deloitte 2013. Government meets its mobile future: Seven essential strategies for state leaders.

Dorn, H. F. & Moriyama, I. M. 1964. Uses and Significance of Multiple Cause Tabulations for Mortality Statistics. *Am J Public Health Nations Health,* 54, 400-6.

Fairhurst, K. & Sheikh, A. 2008. Texting appointment reminders to repeated non-attenders in primary care: randomised controlled study. *Quality and Safety in Health Care,* 17, 373.

Frentzel-Beyme, R., Keil, U., Pflanz, M., Struba, R. & Wagner, G. 1980. [Mortality data and statistics, importance for health services and epidemiological research (author's transl)]. *MMW Munch Med Wochenschr,* 122, 901-6.

Gabe Zichermann, C. C. 2011. *Gamification by Design: Implementing Game Mechanics in Web and Mobile Apps*, O'Reilly Media, Inc.

Klasnja, P. & Pratt, W. 2012. Healthcare in the pocket: Mapping the space of mobile-phone health interventions. *Journal of Biomedical Informatics,* 45, 184-198.

República, D. D. 2012a. Lei nº 15/2012 de 3 de Abril de 2012. Diário da República.

República, D. D. 2012b. Portaria nº 329/2012 de 22 de Outubro de 2012. *In:* SAÚDE, M. D. A. I. D. J. E. D. (ed.). Diário da República.

República, D. D. 2012c. Portaria nº 330/2012 de 22 de Outubro de 2012. *In:* SAÚDE, M. D. A. I. D. J. E. D. (ed.). Diário da República

República, D. D. 2012d. Portaria nº 331/2012 de22 de Outubro de 2012. *In:* SAÚDE, M. D. J. E. D. (ed.). Diário da República.

República, D. D. 2012e. Portaria nº 334/2012 de 23 de Outubro de 2012. *In:* SAÚDE, M. D. J. E. D. (ed.). Diário da República.

Rubel, P., Fayn, J., Nollo, G., Assanelli, D., Li, B., Restier, L., Adami, S., Arod, S., Atoui, H., Ohlsson, M., Simon-Chautemps, L., Télisson, D., Malossi, C., Ziliani, G.-L., Galassi, A., Edenbrandt, L. & Chevalier, P. 2005. Toward personal eHealth in cardiology. Results from the EPI-MEDICS telemedicine project. *Journal of Electrocardiology,* 38, 100-106.

Statcounter. 2018. *StatCounter Global Stats* [Online]. Available: http://gs.statcounter.com/ [Accessed 25/06/2018].

DESCRIPTIVE MODELLING OF TEAM TROUBLESHOOTING IN NUCLEAR DOMAIN

Jari Laarni
VTT Technical Research Centre of Finland Ltd, Vuorimiehentie 3, Espoo, P.O.Box 1000, FI-02044 VTT, Finland

ABSTRACT

Even though modern engineering systems are becoming more reliable, there is evidence that main control room (MCR) operators and maintenance technicians have problems in diagnosing complicated events and multiple simultaneous events in nuclear power plants (NPPs). There is a lot of research on finding faults in complex incident situations, but there is quite little knowledge of the cognitive strategies, states and activities of MCR teams in troubleshooting situations. This knowledge is, however, important in modelling collaborative diagnostic reasoning and in developing training interventions and decision support systems to support reasoning and problem solving in NPPs. The present paper will firstly critically review existing methods to model collaborative cognition. Our results suggest that some modelling tools, such as social network analysis and information network analysis are valuable in analysing information sharing and team situation awareness in collaborative troubleshooting, but they do not provide detailed information about the sequential evolution of a team's knowledge state throughout the diagnostic process. On the other hand, the tools representing an individual troubleshooter's successive knowledge states, transformed by information processing activities, are ignorant of the distributed nature of complex troubleshooting. Therefore, these approaches have to be tailored to suit better the analysis of team collaboration and co-operation in incident and accident situations, i.e., we have to better understand the collaborative aspects of cognition and team-level collaboration in fault finding and in diagnosing faults. We propose a modelling approach suitable for analysing collaborative diagnostic reasoning and troubleshooting of a NPP MCR crew which is based on existing methods and tools. The approach describes the progress and evolution of a MCR operator crew's knowledge states throughout the critical sections of a simulator run.

KEYWORDS

Troubleshooting, Operator, Process Control, Collaboration

1. INTRODUCTION

Even though modern engineering systems are becoming more reliable, it is estimated that even 70% of maintenance technicians' work hours are spent in troubleshooting system failures (Morrison et al. 1997). There is also evidence that main control room operators have problems in diagnosing complicated events and multiple simultaneous events in process industry (Kim et al. 1999, Rouse 1978, Toms and Patrick 1987, Wohl 1982). Since troubleshooting is often quite challenging, cognitive processes involved in troubleshooting have been quite extensively studied. However, even though troubleshooting complex failures is more or less a collaborative activity, there is quite little knowledge of the problem solving team's cognitive strategies, states and activities in troubleshooting situations (see, however, Hoc and Carlier 2000, Patrick 1999). This knowledge would be important in modelling diagnostic reasoning and in developing team-level training interventions and decision support systems to support reasoning and problem solving in industrial settings.

Diagnostic reasoning and troubleshooting consists of various kinds of tasks and subtasks. Different authors have identified different sets of stages in troubleshooting, but all these sets of stages include some key elementary activities: Problems and malfunctions have to be found, interpreted and conceptualized; causes have to tracked and confirmed; and remedies and repair activities have to be initialized and evaluated (De Mast, 2013). Problem finding means problem recovery; in problem interpretation the problem is categorized, and a mental representation of the problem is constructed. Action retrieval means retrieval of appropriate strategies that are needed to solve the problem; in decision making one action is selected from the array of options; and finally, during outcome evaluation, it is assessed the outcome and progress of problem solving (Nokes et al. 2010).

The above-mentioned steps do not, however, follow each other in a linear fashion in every troubleshooting task. Troubleshooting is more or less a strategy-based activity, in which one important strategic task is to fluently orchestrate troubleshooting activities. Moreover, in troubleshooting conditions, diagnostic reasoning typically involves frequent shifts among diverse strategies and perspectives, and the troubleshooter may shift his/her strategies many times during a troubleshooting activity (Rasmussen 1993).

This paper is divided into three parts: Firstly, we shortly review previous research in modelling troubleshooting and in collaborative process control and troubleshooting; secondly, we present a critical evaluation of some modelling techniques specifically designed for the analysis of troubleshooting at the individual and collaborative level; and thirdly, we outline a modelling approach suitable for analysing collaborative troubleshooting of a NPP MCR crew.

2. TROUBLESHOOTING OF A MAIN CONTROL ROOM CREW

2.1 Modelling Troubleshooting

Human troubleshooting has been modelled by various means, and there are two main clusters of models that can be used: predictive and descriptive modelling. Predictive modelling includes techniques used to predict future outcomes, whereas descriptive modelling describes how the thing actually works summarizing the main features of data. Accurate descriptive models are a good starting point for the development of predictive models. It is characteristic to descriptive models that they simplify complexities of the reality, identify constraints, and compile results (Stanton et al. 2008). In this paper, we will focus on descriptive modelling of team troubleshooting.

Diagnostic reasoning has been considered as a search process through a problem space (Newell & Simon 1972). The problem space is the mental model of the task environment that the troubleshooter constructs in hypothesis generation phase (Newell and Simon 1972). After constructing the problem space, he/she begins the diagnosis process by scrutinizing and investigating the faulty system and comparing the system states to similar problems he/she has encountered earlier (Newell and Simon 1972). If he/she is not able to recall a previous problem, the troubleshooter must generate novel hypotheses by analysing the initial information gathered in order to identify discrepancies between the system's states and by interpreting these discrepancies based on the problem space he/she has constructed (De Mast 2013).

Throughout the hypothesis generation process the troubleshooter tries to narrow the problem space and to identify the potential faults (Newell and Simon 1972). When all reasonable hypotheses have been generated, they are tested and evaluated through an iterative approach (Newell and Simon 1972). Next, the troubleshooter generates possible solutions for fixing the problem based on the results of the tests he/she has executed. During the evaluation of the solutions, the troubleshooter may notice that additional information has to be gathered in order to be able to make a final decision of the selected solution. Finally, each troubleshooting experience is remembered and fixed in his/her memory (De Mast 2013).

One possible approach to illustrate the troubleshooter's mental states and activities is based on the problem behaviour graph described by Newell and Simon (1972), which represents a troubleshooter's successive knowledge states, which are transformed by information processing activities. Patrick (1999) has further elaborated this notation system and generated a process tracing approach called "Mental States and Activities in the Problem Space" (MAPS) to investigate the nature of problems the MCR operators have in diagnostic reasoning in nuclear domain and in other process industries (Bereiter and Miller 1989). In this method verbalizations are transcribed onto a timeline, transcripts are encoded into behaviourally-related categories, and a behaviour graph is developed to illustrate the progression through knowledge states and activities during troubleshooting (Patrick 1999).

States of knowledge reflect the operators' current understanding of the plant, and they include hypotheses related to the nature and location of the fault: Hypotheses are specified and resolved, and knowledge is gathered from symptoms, plant states and alarms (Patrick 1999). Activities in MAPS refer to means of altering and connecting states of knowledge. Six activities have been identified in MAPS: information gathering, symptom identification, hypothesis generation and testing, reviewing and reflection, and system control: information is searched for and gathered, possible signs of the fault are identified, hypotheses are constructed and tested, test results are evaluated and actions are executed (Figure 1; Patrick 1999).

The main limitation with the MAPS is that it is only applicable to the analysis of troubleshooting at the individual level. The approach has to be tailored to suit better the analysis of team collaboration and co-operation in incident and accident situations. Our aim is to extend the MAPS notation system in such way that it can also be applied in analysing of collaborative team troubleshooting.

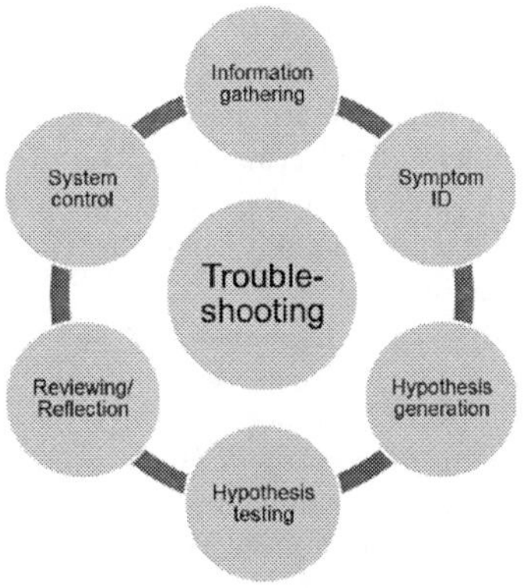

Figure 1. Main Stages of Problem Solving/Troubleshooting (Patrick 1999). ID: Identification

2.2 Collaborative Process Control and Troubleshooting

As suggested above, process control can be considered as collaborative command and control practice characterized by processing of shared information through several stages (Figure 2). During this practice, information is distributed and changed to knowledge; situation awareness is built at the team level; actions of crew members are coordinated, and possible troubleshooting is performed. Collaborative process control is also an iterative process in which this series of activities is repeated several times in succession.

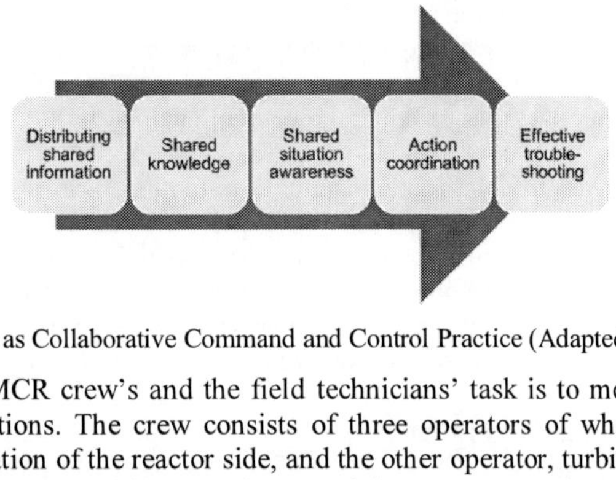

Figure 2. Process Control as Collaborative Command and Control Practice (Adapted from Stanton et al. 2008)

For example, the NPP MCR crew's and the field technicians' task is to monitor and control the power process and repair malfunctions. The crew consists of three operators of whom one, reactor operator, is responsible of the safe operation of the reactor side, and the other operator, turbine operator, is responsible of the turbine and electrical systems. A shift supervisor's task is to manage and coordinate the crew's behaviour in all plant conditions. In incident situations, the group process involves specific communication and collaboration activities that are well described in operational procedures. The final group product in incidents and accidents is a safe final state of the plant.

Typically, people work collaboratively using different technologies to produce diagnoses and troubleshooting systems (Büscher et al 2009). Sasou and Reason (1999), among others, have shown that deficiencies in crew communication is one of the most critical factor which is responsible for a MCR crew's failure to detect a problem in nuclear power plants. Patrick et al. (2006), in turn, found in their study that supervisors failed to monitor the operators to ensure that they performed adequately in troubleshooting situations, and they did not get involved in the crew's activities in order to check the status of failure management and reorient the crew members' work activities.

Apparently, also, regular communication between control room operators and maintenance personnel will improve the ability of the operators to coordinate their actions with those of the maintenance technician (Rhodes et al., 1995). Aaltonen et al. (2015) showed that the automation mechanics collaborate extensively with various other work sectors at the plant. The control room operators are kept aware of the maintenance activity, and the shift supervisor has to approve the automation maintenance work orders. This is important for upholding the general situation awareness on the plant's functioning and preventing surprising or misleading the operators and maintenance technicians.

At the crew level, collective problem identification is important, that is, all the members of the crew should realize whether other members of the crew have identified the problem (Fiore and Schooler 2004). Problem conceptualization, i.e., understanding and diagnosing of the malfunction, has also to be shared. Without a shared understanding of the problem, the crew may be solving the wrong problem.

Fiore and Schooler (2004) propose that process mapping, i.e. representation of the various stages that are involved in a particular work activity, can be considered as an effective problem-solving tool, because it helps the crew to construct a shared mental model of the problem. It also helps information sharing among the crew by guiding the transfer of information occurring in group discussions. The knowledge must be made explicit in order to be able to support the construction of a shared mental model (Fiore and Schooler 2004). Process mapping guides the crew to build up collaboratively their shared understanding of the process, and it forces the crew to identify deficiencies in their problem understanding, before they begin to address solutions to the problem (Fiore and Schooler 2004).

To summarize, the crew must have a shared mental model, including a shared understanding of the critical phases in the work activity, a shared problem structure, roles and skills required, and shared understanding that each crew member actually has all the critical knowledge that is needed in order to solve the problem. For example, a shared problem structure is important for the generation of an effective problem solution. The crew should also actively share views with each other on their understanding of the problem. (Fiore and Schooler 2004).

2.3 Descriptive Models of Team Troubleshooting

According to (Laughlin 2011), team troubleshooting can be analysed from different perspectives: We can analyse it from the group task, structure, process and product point of view. That is, we can focus on what the group is doing; how the group is organized; how the group members interact and influence each other; and what the collective response of the group is, respectively (Laughlin 2011). Different modelling tools put emphasis on different perspectives, but an optimal tool would consider all the perspectives or at least as many of them as possible.

Descriptive models of team troubleshooting can be classified in various ways (Stanton et al. 2008). Some examples of this kind of models are structural models that describe the functional elements performing signal transmission and adjustment, network models that describe links and interrelations between elements, dynamic models that describe dynamic interactions between units over time, and socio-technical models that try to identify factors that constrain or enable behaviour at the individual or organizational level (Stanton et al. 2008). In this paper, we will focus on the following models that have shown to be useful in modelling collaborative cognition:

- Task Network Analysis (TNA);
- Social Network Analysis (SNA);
- Information/Knowledge/Propositional Network Analysis (KNA);
- Co-ordination Demands Analysis (CDA);
- Time-Window Sequential Analysis (TWSA);
- Functional Situation Modelling (FSM).

Task Network Analysis and Social Network Analysis are included in the Event Analysis of Systemic Teamwork (EAST) toolbox developed by Salmon et al. (2005). TNA illustrates the linkages between non-sequential tasks, as they are performed. A task network can be compared to the "ideal case" based on a task analysis of operator activity, and the analysis would then focus on the discrepancies between the ideal and actual case. SNA analyses the communicative and other acts among the crew and between the workers and user interfaces. It is useful in characterizing the communication networks that promote or hinder human performance, but it does not say anything about the content of the communication. Figure 3 shows a visualization of the results of SNA data of MCR operator communications in a simulated accident scenario. The line thickness indicates the strength of the connection, i.e., the number of communicative acts between a pair of operators. As can be seen, the number of communicative acts is higher between the reactor and turbine operator than between other pairs of operators (e.g., shift supervisor and turbine operator). In order to get information about the causes of these differences, a more detailed analysis of crew communication is needed.

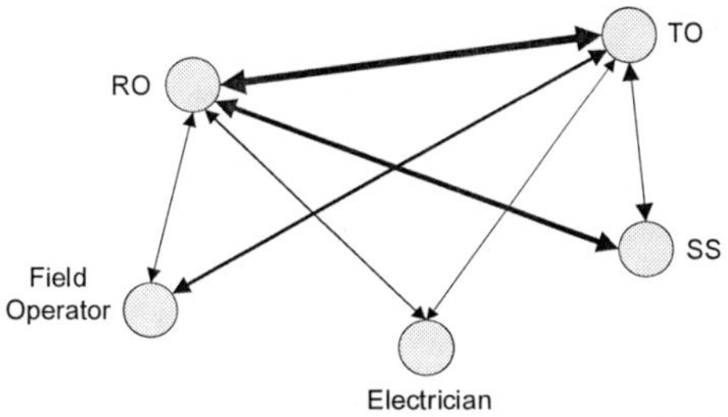

Figure 3. Depiction of SNA data of operator communications in a simulated accident run. RO: Reactor Operator; TO: Turbine Operator; SS: Shift Supervisor; FO: Field Operator. The line thickness indicates the strength of the connection between a pair of operators

Knowledge Network Analysis describes and analyses information concepts in the main control room (Stanton 2013). It is useful in illustrating the information that is readily available to operators in a particular situation, but it does not say anything about the reasoning behind these concepts or the reasoning in which these concepts are deployed. One example of a KNA is shown in Figure 4 illustrating the information discussed and processed by operators in a simulated accident scenario. A leakage in the primary side leads first to the reactor and turbine scram, and finally through several steps to the detection of the cause of the failure.

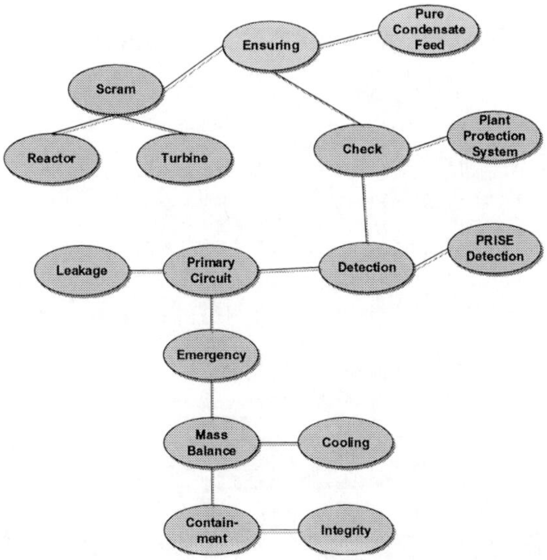

Figure 4. Depiction of KNA Data of a Simulated Accident Run

Sequential analysis refers to a set of methods that shows patterns of behaviours over time and investigate their contingencies. Time-Window Sequential Analysis answers to the question whether the presence of a particular activity increases the probability of an occurrence of another activity. According to Salmon et al. (2005), Co-ordination Demands Analysis depicts the activity and the interaction between actors participating in an activity over time using a standardized symbology. The main aim is to describe the extent to which the operators are working together to accomplish a particular mission. There are also other methods suitable for sequential analysis (Chorney et al. 2010). For example, we have described MCR crew communication patterns over time in a simulated accident scenario by using the notation system illustrated in Figure 5 (Laarni et al. 2016). It shows the communicative acts/interruptions encountered by operators in a crew from the beginning of the simulated accident to the end of the session. The arrows refer to interruptions or communicative acts encountered in a scenario run, and the numerals indicate the durations of these events. This kind of sequential analysis is useful in identifying those parts of a simulator run in which task load is the highest.

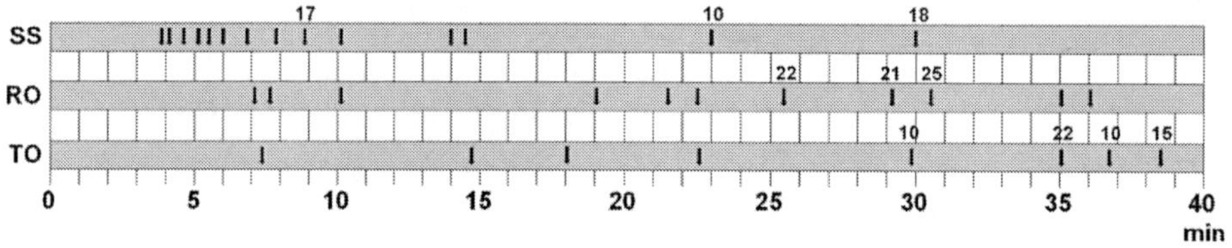

Figure 5. Depiction of sequential patterns of MCR crew communication/interruptions over time in a simulated accident scenario (adapted from Laarni et al. 2016). Each arrow refers to one interruption/communicative act. SS: Shift Supervisor; RO: Reactor Operator; TO: Turbine Operator

Functional situation models have been constructed for the analysis of simulated incident and accident scenarios. Functional Situation Model combines both chronological and functional views (Savioja 2015). These two dimensions define a two-dimensional space in which the most important operator actions and process phenomena are mapped. Actions are depicted under the functions they are related to. The critical function of the domain provides each action a contextual meaning. In the chronological view, the scenario is divided into different phases based on the goals of the operating activities; in the functional view, the critical functions of the process that are endangered in a specific situation are depicted. In Figure 6, three goals (safety, production and health) and four stages (detection, mitigation, diagnosis and stabilization) are depicted. Frame type indicates the kind of function that is displayed.

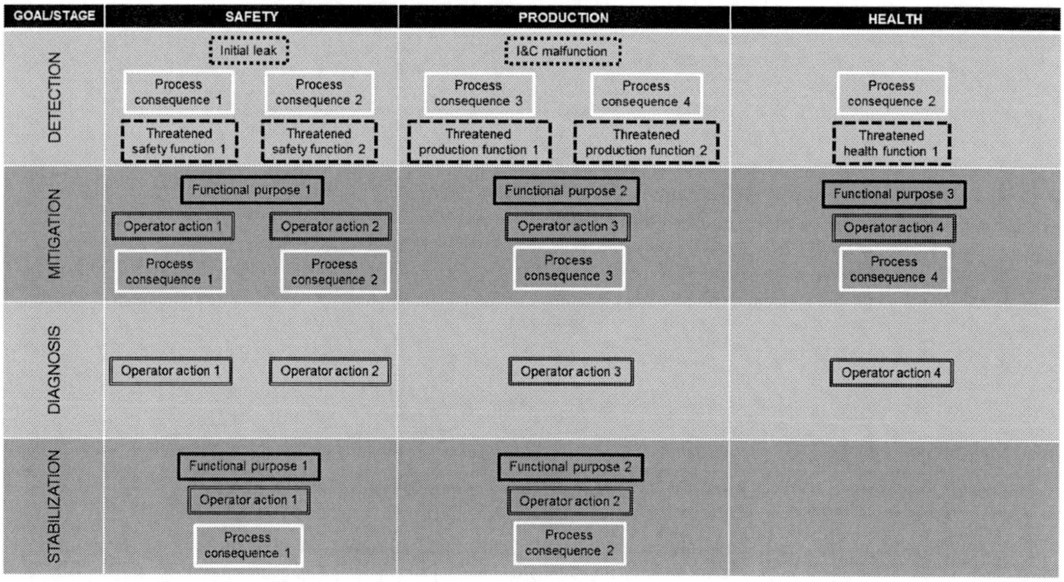

Figure 6. Schematic depiction of a Functional Situation Model of operator crew behaviours in a simulated accident scenario (modified from Savioja 2015). Frame type indicates the kind of function that is displayed (dotted: process failure; white: process consequence; broken: threatened function; black: functional purpose; double: operator action)

2.4 Evaluation of Modelling Tools

It is important to try to assess the suitability of the above-mentioned models to the analysis of team troubleshooting. In modelling team troubleshooting and cognition in general, we are interested in, e.g., whether the model is theoretically sound, whether the model is based on reasonable assumptions and how the model makes a distinction between good or bad performance (Stanton et al. 2008). To that aim, we have evaluated the tools according to the following criteria:

- Soundness of theoretical basis;
- Usefulness to the analysis of operator work in nuclear domain;
- Availability of suitable software;
- Labour and resource intensity;
- Training requirements;
- Ability to differentiate good and bad performance;
- Presentation of temporal order of events;
- Presentation of structure of network relations;
- Presentation of critical process functions;
- Presentation of internal state of actors.

One expert classified the models with their applicability to model team troubleshooting on the above-mentioned criteria on the scale from 1 to 3 (see Table 1). Of those existing methods, FSM seems to be the most suitable and the CDA the least suitable for the evaluation of team troubleshooting. FSM was considered suitable for the analysis of operator work; it is quite easy to conduct, and it is capable of differentiating good and bad performance and present critical process functions. CDA, in turn, has not much practical value in describing team troubleshooting.

However, as can be seen, none of these methods is a definitive technique for the analysis of team troubleshooting, and none of them combines all the critical features of an optimal technique, i.e., presentation of temporal order, network relations, critical functions and internal states of critical stakeholders. One of the main limitations of most of the tools is that they do not provide information about the troubleshooters' internal states, whereas the only one (i.e., MAPS) which illustrates the troubleshooter's internal states is not suitable for the analysis of team troubleshooting.

Table 1. Scores of the 7 methods for the evaluation of team troubleshooting on ten criteria on a scale from 1 to 3 (1: little value; 2: some value; 3: high value). Some of the criteria are not applicable to some of the methods (marked by N/A)

Tool/Theme	TNA	SNA	KNA	CDA	TWSA	FSM	MAPS
Theoretical basis	N/A	N/A	N/A	N/A	N/A	2	2
Usefulness	2	2	2	3	2	3	2
Availability of software	3	3	3	N/A	3	1	1
Labour intensity	2	2	3	2	3	3	3
Training required	2	2	2	2	2	3	2
Differentiation of good and bad performance	1	1	1	2	2	3	2
Presentation of temporal order	1	1	1	1	3	2	3
Presentation of network relations	3	3	3	1	1	1	1
Presentation of critical functions	1	1	1	1	1	3	1
Presentation of internal states	1	1	2	1	1	1	3
SUM	16	16	18	13	18	22	20

2.5 Integration of Existing Approaches

Since there is a clear need for a better tool, we propose a modelling approach suitable for analysing collaborative diagnostic reasoning and troubleshooting of a NPP MCR crew which is based on existing methods and tools. The approach describes the progress and evolution of a MCR operator crew's knowledge states throughout critical sections of a simulator run by combining all the above-mentioned elements into a single presentation format. The notations are based on observations of a simulator run and on retrospective process tracing interviews. Figure 7 outlines the main features of this kind of format in which internal states of individual operators are combined into a crew-level description of problem solving communications and presented along a timeline.

In Figure 7, mental states and knowledge acts are depicted by a layer of shaded bars; one set of bars referring to a problem space of one MCR operator. Six mental, behavioural or communicative acts are considered: information gathering, symptom identification, hypothesis generation and testing, reviewing and reflection, and system control. Each bar refers to one kind of mental/behavioural/communicative act, and question and exclamation marks following numerals indicate the type of a communicative act (i.e., whether it is a question of a request). All the knowledge states and acts are located on a timeline. Initiation of main acts is indicated by numerals big in size; brief replies and comments are indicated by small numerals. If more than one of the operators have participated in an act, the numeral has been repeated on all the bars of relevant acts. There is a separate layer of bars at the bottom depicting the crew-level "problem space". The black arrow on the right hand side indicates that the crew-level problem space is produced by summing up all the mental states/acts of individual operators. This arrow is only shown to illustrate the latter operation, and it is not visible in a real application.

Data of two MCR crews in the same simulator accident scenario is depicted in Figure 7. As can be seen in Figure 7A, there is a great deal of information gathering, symptom identification and system control acts in the beginning of the simulator run, whereas the number of hypothesis generation/testing is smaller. And even though some hypotheses were generated during the period of five minutes, none of them was tested (ie., there is no single numeral on the corresponding bar at the bottom of the figure). In Figure B, the number of information gathering, symptom identification and system control acts in the beginning of the simulator run is smaller, and it seems to be that the crew got off to a good start quite slowly. On the other hand, only the latter crew could generate a plausible hypothesis of the cause of the failure and test it during the five-minute period of time. In conclusion, it can be said that performance of the latter crew was somewhat smoother and more efficient during this time period.

A diagram illustrating the collaborative problem space may be useful in demonstrating collaborative problem identification and conceptualization that have shown to be important in shared understanding of the problem (Fiore and Schooler 2004). It may also be useful in identifying the most mentally taxing sections of a simulator run.

Some of the main challenges are that the diagram is quite laborious and time-consuming to create, and a lot of information is needed for its development. In addition, information about the troubleshooter's mental states is based only on his/her observable behaviour and verbal comments during the simulator run. A more reliable description of the mental states should be based on transcriptions of their think-aloud reports.

3. CONCLUSIONS

Our results suggest that some modelling tools, such as social network analysis and functional situation modelling are valuable in analysing information sharing and team situation awareness in team troubleshooting, but they do not provide detailed information about the sequential evolution of a team's knowledge state throughout the diagnostic process. On the other hand, the tools representing an individual troubleshooter's successive knowledge states, transformed by information processing activities, are ignorant of the distributed nature of complex troubleshooting. Therefore, these approaches have to be tailored and their most promising features have to be combined to suit better the analysis of team collaboration and co-operation in incident and accident situations, i.e., we have to better understand the collaborative aspects of cognition and team-level collaboration in fault finding and in diagnosing faults.

Since there is a great need for a method suitable for analysing collaborative aspects of cognition and team-level collaboration in fault finding and in diagnosing faults, we have proposed a modelling approach which is based on existing methods and tools. The approach describes the progress and evolution of a MCR operator crew's knowledge states and communication throughout the critical sections of a simulator run. The model helps us to identify the distribution of knowledge states and the most demanding sections of the scenario run. It would be useful, for example, in the analysis of operator collaborative fault-finding behaviour in the Human Factors validation of MCR systems. We are going to apply it to the analysis of challenging simulated accident scenarios in which there are clear differences in performance between the operator crews.

Since the generation of modelling diagrams is a laborious task when working with a pencil and paper, one of our more distant goals would be to develop a software tool for assisting researchers in the analysis and systematization of observational, think-aloud and interview-based data on troubleshooting activities and in its presentation in a graphical form.

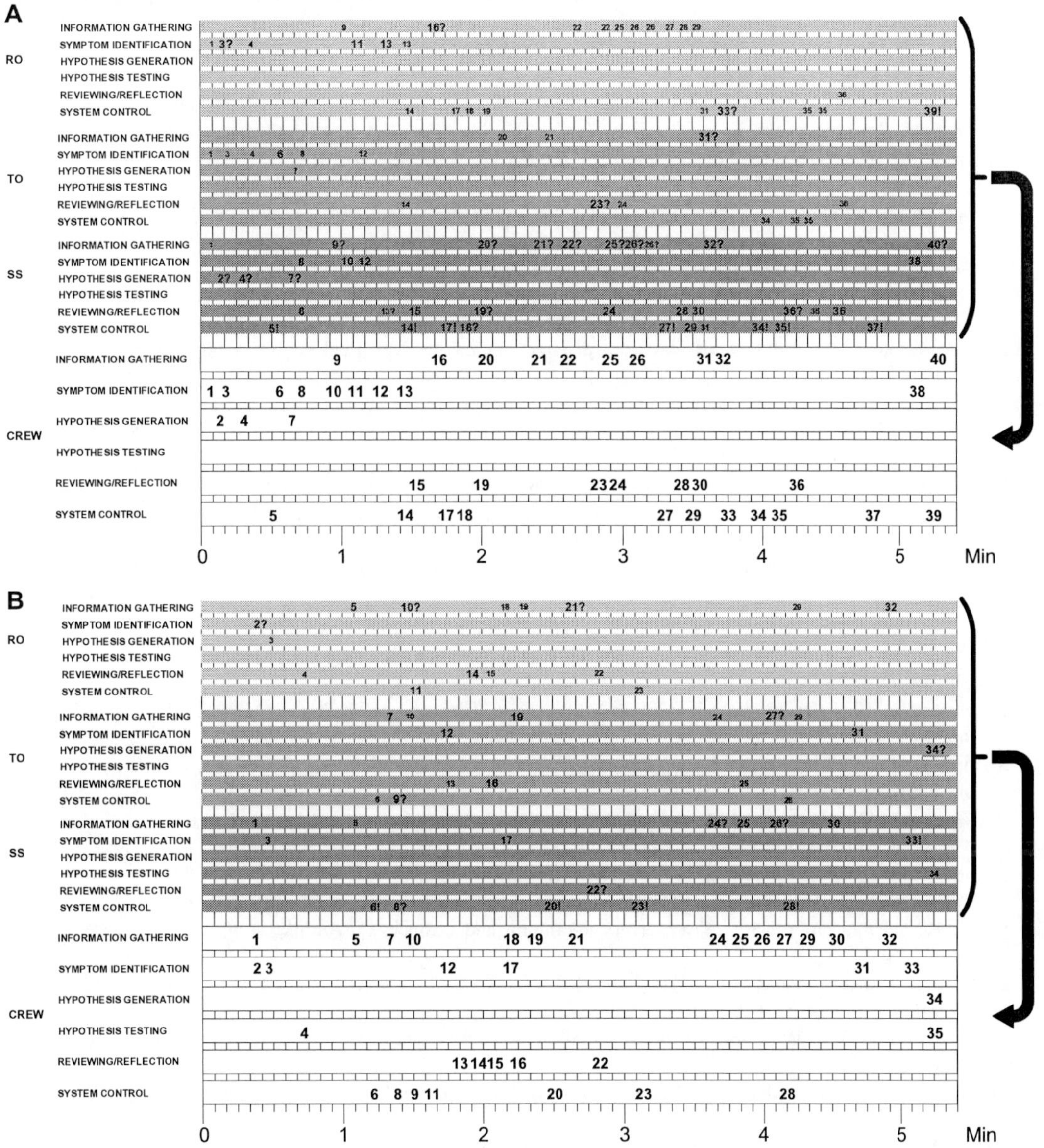

Figure 7A and B. Two diagrams of a sequential analysis of the evolution of two MCR crew's problem space in terms of critical functions in a five-minute period of time in a simulated accident scenario. The numerals refer to individual troubleshooting acts. Main acts initiating a new phase in troubleshooting are indicated by big numerals, whereas small numerals refer to answers to other operators' questions and reflections. The question and exclamation marks indicate a question or a request directed to another operator. The black arrow on the right hand side indicates that the crew-level depiction of troubleshooting performance can be derived by summing up the operator-level behaviours. It is only shown to illustrate the latter operation, i.e., it is not visible in a real application. SS: Shift Supervisor; RO: Reactor Operator; TO: Turbine Operator

REFERENCES

Aaltonen, I. et al., 2015. Challenges and training needs in NPP automatic maintenance. *Proceedings of ANS NPIC&HMIT 2015*, American Nuclear Society, Charlotte, USA.

Bereiter, S.R. and Miller, S.M., 1989. A field based study of troubleshooting in computer-controlled manufacturing systems. *IEEE Transactions on Systems, Man, and Cybernetics*, Vol. 19, pp. 205-219.

Büscher, M. et al., 2009. Designing for diagnosing: Introduction to the special issue on diagnostic work. *Computer Supportive Cooperative Work*, Vol. 18, pp. 109-128.

Chorney, J.M. et al., 2010. Time-window sequential analysis: an introduction for pediatric psychologists. *Journal of Pediatric Psychology*, Vol. 35, pp. 1061–1070.

De Mast, J., 2013. Diagnostic quality problem solving: A conceptual framework and six strategies. *The Quality Management Journal*, Vol. 20, pp. 21-36.

Fiore, S.M. and Schooler, J.W., 2004. Process mapping and shared cognition: Teamwork and the development of shared problem models. In E. Salas and S. M. Fiore (Eds.), *Team Cognition. Understanding the Factors That Drive Process and Performance*. Washington, DC, American Psychological Association (2004).

Hoc, J.-M. and Carlier, X., 2000. A method to describe human diagnostic strategies in relation to the design of human-machine cooperation. *International Journal of Cognitive Ergonomics*, Vol. 4, pp. 297-309.

Kim, H. et al., 1999. Aiding fault diagnosis under symptom masking in dynamic systems. *Ergonomics*, Vol. 42, pp. 1472-1481.

Laarni, J. et al., 2016. Multitasking and interruption management in control room operator work during simulated accidents. *Proceedings of the 13th International Conference on Engineering Psychology and Cognitive Ergonomics, held as Part of HCI International 2016*. Toronto, Canada, pp. 301–310 (2016).

Laughlin, P.R., 2011. *Group Problem Solving*. Princeton University Press, Princeton and Oxford.

Morrison, D.L. et al., 1997. Predictors of fault-finding skill. *Australian Psychologist*, Vol. 32, pp. 146-152.

Newell, A. and Simon, H.A., 1972. *Human Problem Solving*. Englewood Cliffs, NJ, Prentice-Hall.

Nokes, T.J. et al., 2010. Problem solving and human expertise. *International Encyclopedia of Education*, Vol. 5, pp. 265-272.

Patrick, J., 1999. Analysing operators' diagnostic reasoning during multiple events. *Ergonomics*, Vol. 42, pp. 493-515.

Patrick, J. et al., 2006. Human processes of control: tracing the goals and strategies of control room teams. *Ergonomics*, Vol. 49, pp. 1395-1414.

Rasmussen, J., 1993. Diagnostic reasoning in action. *IEEE Transactions on Systems, Man, and Cybernetics*, Vol. 23, pp. 981-992.

Rhodes, W. et al., 1995. *Training Troubleshooting Skills in Nuclear Power Plants, Report*. Atomic Energy Control Board in Canada.

Rouse, W.B., 1978. Human problem solving performance in a fault diagnosis task. *IEEE Transactions on Systems, Man, and Cybernetics*, Vol. SMC-8, pp. 258-271.

Salmon, P et al., 2005. *Analysing C4i Activity: A Review of the Event Analysis of Systemic Teamwork (EAST) Methodology*. HFIDTC/WP1.1.3/10. Human Factors Integration Defence Technology Center.

Sasou, K. and Reason, J., 1999. Team errors: definition and taxonomy. *Reliability Engineering and System Safety*, Vol. 65, pp. 1-9.

Savioja, P., 2015. *Evaluating systems usability in complex work Development of a systemic usability concept to benefit control room design*. VTT, Espoo.

Stanton, N.A., 2013. Representing distributed cognition in complex systems: how a submarine returns to periscope depth. *Ergonomics*, Vol. 57, pp. 403-418.

Stanton, N.A. et al., 2008. *Modelling Command and Control: Event Analysis of Systemic Teamwork*. Ashgate, Abingdon, UK.

Toms, M. and Patrick, J., 1987. Some components of fault-finding. *Human Factors*, Vol. 29, pp. 587-597.

Wohl, J.G., 1982. Maintainability prediction revisited: Diagnostic behaviour, system complexity, and repair time. *IEEE Transactions on Systems, Man, and Cybernetics*, Vol. SMC-12, pp. 241-250.

POLITICAL ANALYTICS FOR PUBLIC ELECTRONIC PROPOSAL DELIBERATION AT MUNICIPAL LEVEL

Georgios Drakopoulos[1], Vasilis Triantafyllou[2] and John Christodoulopoulos[2]
[1]*Department of Informatics, Ionian University, Tsirigoti Square, Kerkyra 49100, Hellas*
[2]*Computer Informatics and Engineering Department,*
Technological Educational Institute of Western Greece, GR-30020, Antirio, Greece

ABSTRACT

The development of e-government is a major technological achievement allowing broader citizen participation to political representative systems, enhancing thus the flexibility and the openness of these systems. This electronic participation can take place in various scales ranging from the national to local. This report describes a series of fundamental and advanced political analytics which can form the basis of an electronic system for addressing the need of municipalities for public deliberation of proposals. Specifically, social factors such as citizen digital reputation, participation frequency and innovation, vote relevance, and the number of neutral votes are taken into consideration.

KEYWORDS

Public Deliberation, Political Analytics, Keyword Set, Keyword Variability, Reputation System

1. INTRODUCTION

With the advent of reliable, secure, and scalable information systems e-government has become a reality in Western world, with a relevant culture already beginning to form. A crucial part of e-government is the deliberation of proposals at the municipal level, as it directly influences the quality of everyday lives of citizens.

The primary objective of this conference paper is to present a diverse number of basic and advanced political analytics for evaluating proposals submitted either by the municipality or by citizens. These analytics form the backbone of a public electronic deliberation system where proposals are constantly evaluated under a wide array of criteria ranging from geolocation filters and custom keywords to citizen participation frequency.

The remainder of this work is structured as follows. The scientific literature regarding the taxonomy of political systems and the political analytics is briefly reviewed on section 2. Section 3 enumerates the principles behind any realistic implementation of political analytics. Vote handling is covered in section 4. The role of geolocation information to political analytics is highlighted in section 5. Section 6 focuses on citizen defined keywords and comments. The social factor of participation frequency is analyzed in section 7, which can be coupled with the discussion of section 8 concerning digital citizen reputation. Finally, section 9 recapitulates the main points of this report and indicates possible future research directions.

Table 1 summarizes report notation. Throughout this text, q_i stands for the i -th available proposal in the system, while c_k denotes the k -th citizen enrolled in the system. Also, $v_{i,k}$ is the vote of c_k regarding q_i , while v^i_j is simply the j -th vote for q_i . Finally, matrices are represented as boldface uppercase, vectors as boldface lower case, and scalars as plain lower case.

Table 1. Paper Notation

Symbol	Meaning
$\triangleq$	Definition or equality by definition
$\{s_1,\ldots,s_n\}$	Set with elements $s_1,\ldots,s_n$
$\lvert S\rvert$	Set cardinality
$\langle w\rangle$	String length
$\lVert x_1-x_2\rVert_2$	Euclidean distance between points x_1 and x_2

2. RELATED WORK

The analysis of political systems dates back to the days of Aristotle, who classified them based on the size of the ruling class compared to that of the underlying society in *Politika*. Earlier Plato had expressed his views on the perfect political system in *Politeia*. Along a similar line of reasoning, although from a qualitative point of view, political systems are categorized according to the relationships between the ruling class and the society by (de Montesquieu, 1867).

In the post-war scholar world, a comprehensive and analytical framework for understanding modern political systems has been proposed in (Easton, 1957). The political actions of a citizen in a democracy based on purely economic motives are studied in (Downs, 1957). Similarly, the relationship between wealth acquisition, political party competition, and political stability is explored in (Ferguson, 1995). From a geopolitical perspective, the importance of political systems and their proper management through selected analytical tools are examined in (Brzezinski, 1997a) and (Brzezinski, 1997b). For a critical view of modern Western democracies regarding, among others, the overlook of analytics see (Huntington, 1997). Finally, for comments on the diffusion of ideas and the multitude of ways a complex idea can be decomposed to elementary memes see (Blackmore, 2000). Political analytics are instrumental in the way governments manage mid- and long-term strategic objectives (Braibanti, 1968). Finally, the deep and multi-way connection between social media and the modern political analytics is explored in (Zeng, 2010).

3. GENERAL PRINCIPLES

This section describes the basic guidelines for the analytics starting from the most general to the most specific. These principles cover a broad spectrum of topics and contribute to the longevity, reliability, and significance to the local community of the deliberation system.

Principle 1. *The analytics must be designed in a way such that citizens using it are engaged. Citizen participation is paramount, especially for a municipal system which addresses small communities. Thus, it is imperative that citizens should capitalize on the opportunity offered by such a system to voice their opinions. A related requirement is the following.*

Principle 2. *A portion of the analytics must be open ended to allow citizen input. By allowing citizens to freely express their opinion in open format besides the traditional multiple choice, radio, and number boxes not only expands the freedom of choice but also harvests important information which may not have been anticipated by the system and the analytics designers. Moreover, it reinforces the public sentiment reflection to the voting results, which is useful in a multitude of levels.*

Principle 3. *Questions concerning citizens should address objective aspects of their lives and personalities. In order to obtain information which is as objective as possible about the citizens, it is vital that open ended questions must be formulated in a way that address objective and tangible aspects of both the personality and the activities of a citizen.*

Therefore, questions such as

- How will this proposal influence your fortune?
- Name three of your best achievements as a professional

are preferable to questions such as

• How much do you know about this topic?
• Are you the best person to decide on this topic?

as the former can find in most cases objective information, even partial, inside a subjective judgment.

Principle 4. *Citizens should be able to be categorized by various criteria. This is a fundamental requirement since the analysis of the votes depends on a number of socioeconomic criteria beyond the age, gender, and income. Today factors such as education, work experience, religious views, political alignment, exposure to social media, and family size play an important yet unconscious role in daily decision making. Thus, the system should easily categorize votes depending on various citizen traits.*

Principle 5. *The system should be able to fuse heterogenous and possibly multimodal data. Since there is a number of data contributors to the system such as administrators and citizens who upload data ranging from proposals, votes, and geolocation data to keywords, comments, and metadata, it makes sense that these data can be fused into a functional schema, not necessarily in the sense of a relational database but rather in that of a data model.*

Principle 6. *The system should display the decision criteria to citizens. In order to reinforce a culture of openness and accountability, the system should display at least the number of votes, the criteria used to rank proposals, and the way proposals were merged or rejected.*

Principle 7. *There should be a human in the decision loop. Since any computerized system, even an open and self-adaptive one, is bound to be imperfect, it is vital that every decision should be reviewed by human administrators. This is especially important given that the decisions voted in such as system will influence the everyday lives of citizens.*

4. VOTES

The most basic action a citizen is called to perform is to vote on a proposal. Although in certain electoral systems citizens may be allowed to cast multiple votes, in this report a citizen is supposed to cast only a single vote. Thus, the total number of votes corresponds to the number of citizens who evaluated a given proposal.

To any given proposal q_i corresponds an ordered triplet (p_i, n_i, e_i) with nonnegative elements corresponding to the number of positive, negative, and neutral votes respectively. The fraction π_1 expresses the popularity of q_i where

$$\pi_1 \triangleq \frac{p_i}{p_i + n_i} \tag{1}$$

since the neutral votes e_i carry no popularity. Thus, the number of citizens actively involved in q_i is $p_i + n_i$.

A similar fraction π_2 measures the acceptance of q_i. Specifically

$$\pi_2 \triangleq \frac{p_i + e_i}{p_i + e_i + n_i} \tag{2}$$

While π_1 indicates the percentage of voters who actively welcome q_i among the voters having an opinion on that matter, π_2 shows the acceptance of q_i among all voters.

A proposal may be considered as having insufficient traction when e_i exceeds at least half the total number of votes, namely when

$$e_i \geq \frac{p_i + e_i + n_i}{2} \Leftrightarrow e_i \geq p_i + n_i \tag{3}$$

In general, it should be noted that the number of neutral votes is an open topic in voting systems theory. For instance, in ancient Athens neutrality was connected to objectivity in the sense that a voter with nothing to lose or gain from a given matter was more likely to vote with a clear mind. However, this has been disputed in the modern world, as neutrality is often associated with lack of information about the matter at hand or with indifference.

5. GEOLOCATION INFORMATION

One way to determine both the applicability width of a proposal qi and the relevancy of a vote v_j^i to q_i is to attach geolocation information to both. That is, in the proposal case, the area A_i influenced by it. In most cases, A_i is or at least should be the spatial union of some predetermined rudimentary geographical areas S_m, which in turn usually correspond to districts or other local administrative partitions or certain points of local interest such as a port, a power plant, a square, or a monument. Therefore A_i can be written as

$$A_i = \bigcup_{m=1}^{M_i} S_m \tag{4}$$

where M_i is the number of rudimentary areas spanning A_i. This form facilitates the implementation using a spatial data structure such as an R tree, an R* tree, or a Quadtree. For a comprehensive list of spatial data structures see (Samet, 1990). Note that there may be different administrative partitions of the same area for different proposals. For instance, there may be a separate partitioning based on the power network, the school locations, and the health-related infrastructure.

Once A_i is set, then the spatial cover of q_i can be computed as follows. If S is the set of all possible districts, chosen appropriately, and its cardinality $|S|$ equals M, then a metric μ_1 for the impact q_i can have on the municipality is

$$\mu_1 \triangleq \frac{|A_i|}{|S|} = \frac{M_i}{M} \tag{5}$$

In the vote case, the residency r_j and the work address r'_j of the voter are recorded and attached to $v_{i,j,k}$, becoming thus an inherited attribute of the vote. Naturally, a special consent from the voter must be granted according to the local and national laws pertaining to personal data protection. One way to measure the relevancy of v_j^i to q_i if Ai consists of a single S_0 is to

$$y \triangleq \left[\frac{1}{1+\|s_0^* - r_j\|_2^2} \; \frac{1}{1+\|s_0^* - r'_j\|_2^2}\right]^T \tag{6}$$

where s_0^* is the center of S_0. Thus y contains the modified inverse Euclidean distances of the home and work address from the "center" of the effect of q_i. The two components $y[1]$ and $y[2]$ of vector y can be combined in many ways, depending on q_i and the topic it corresponds to. For instance, the maximum of these two values can be kept

$$\max\{y[1], y[2]\} \tag{7}$$

Another option is to use the weighted $F1$ metric to derive a robust metric

$$\frac{1+\rho_0}{\frac{1}{y[1]} + \frac{\rho_0}{y[2]}} \tag{8}$$

If A_i consists of M_i S_m whose corresponding epicenters are s_m^*, then there are several options. One is to redefine y as

$$[\frac{1}{M_i}\sum_{m=1}^{M_i}\frac{1}{1+\|s_m^* - r_j\|_2^2}\frac{1}{M_i}\sum_{m=1}^{M_i}\frac{1}{1+\|s_m^* - r'_j\|_2^2}]^T \tag{9}$$

6. KEYWORDS

Another way to motivate engagement is to allow citizens to attach their own keywords to any proposal. Thus, proposals can be enriched with additional information from any citizens who may have immediate knowledge of a local issue. Besides being a strong participation motive, keywords can indicate the degree of reliability and the level of online activity for c_k.

One way to quantify the level of innovation owed to the keywords appended by c_k is the following. Let $w_{k,i,j}$ denote the j-th keyword that c_k appends to q_i and also let $W_{i,k} \triangleq \{w_{k,i,j}\}$ be the set of these keywords. Moreover, let W_i be the set of keywords already associated with q_i. Then, the Tanimoto coefficient measures the degree of novelty of $W_{i,k}$ compared to W_i

$$T_{W_i,W_{i,k}} \triangleq \frac{|W_i \cap W_{i,k}|}{|W_i \cup W_{i,k}|} = \frac{|W_i \cap W_{i,k}|}{|W_i| + |W_{i,k}| - |W_i \cap W_{i,k}|} \tag{10}$$

The second form of (10) is better for large sets with small intersection. Their cardinalities can be estimated with techniques like the one proposed in (Drakopoulos, 2016d). Besides the cardinality of $W_{i,k}$, the number of keywords contributed by c_k, the average keyword length for q_i

$$\bar{l}_{i,k} \triangleq \frac{1}{|W_{i,k}|}\sum_{j=1}^{|W_{i,k}|}\langle w_{k,i,j}\rangle \tag{11}$$

as well as the overall average keyword length

$$\bar{l}_k \triangleq \frac{1}{I_k}\sum_{i=1}^{I_k}\bar{l}_{i,k} \tag{12}$$

indicate the level of engagement for and potentially education level of c_k. I_k denotes the number of proposals c_k has submitted keywords for.

Moreover, the variability of the values for $\bar{l}_{i,k}$ defined as

$$\sigma_k^2 \triangleq \frac{1}{I_k - 1}\sum_{i=1}^{I_k}(\bar{l}_{i,k} - \bar{l}_k)^2 \tag{13}$$

may serve as an indication of the interest of c_k. A low variability indicates an almost constant level of interest and, thus, of engagement, while higher values reflect a fluctuating interest.

Finally, by using standard online vocabularies such as WordNet and SentiWordNet the part of speech for each keyword and its sentimental charge, if any, can be found. Therefore, an estimate of public sentiment of a given topic can be formulated. More verbs or shorter and repeated nouns may be a call for immediate action, possibly on a highly charged topic. On the other hand, a wider array of nouns and less charged words may come from a local community which has a more restrained and calm approach on a topic.

7. PARTICIPATION FREQUENCY

Recording the voting frequency f_k of citizen c_k may reveal significant information at the community level. First, by assigning f_k to five levels of participation will indicate the level of citizen engagement. Those levels correspond to very low, low, moderate, high, and very high participation. For a healthy system, there should be a considerable segment of citizens to the top two levels.

Also, by sorting f_k in descending order and plotting these values may reveal interesting patterns. The most important of them is probably the decay rate of f_k, which can be used for citizen clustering. Two popular models for this decay rate are the exponential law and the power law which are expressed respectively as

$$f_k = \alpha_0 e^{-\beta_0 (k-1)}, k \geq 1$$
$$f_k = \eta_0 (k-1)^{-\gamma_0} = \eta_0 e^{-\gamma_0 \ln(k-1)}, k \geq 1 \quad (14)$$

The exponential law is a line in a log-linear scale, while the power law is a line in a log-log scale. The parameters α_0, β_0, η_0, and γ_0 are all positive. Their actual values can be computed from the available data using nonlinear least squares, linearized least squares, or a total least squares approach, although maximum likelihood methods are becoming increasingly popular. Based on experience from various large, real world, social graphs a power law is expected to emerge, however this remains to be seen in an actual political electronic system.

The number of citizens enrolled in any public electronic deliberation system is an important Key Performance Indicator (KPI) as it reflects the acceptance of the system from the local community. Provided that registration is optional and that citizen accounts are not enabled by default, user base models can serve as the basis for the viability analysis of the system. Two such models are the Malthus and the Verhulst models.

8. DIGITAL REPUTATION

The reliability of a deliberation system can be enhanced with a digital reputation system which citizens vote for the popularity of other citizens. Although these votes may be biased, when the number of votes is large certain objective patterns may emerge depending on the properties of the reputation system. Therefore, it makes sense to have such a system in order to track which proposals attract citizens with high digital reputation, which is an indication of the expectations of the local community.

One reliable digital reputation system based on PageRank which can fuse information from personality, e_q, and other tests is the following. Assume that each c_i can rate c_j with a nonnegative integer $a_{i,j}$ from 0 to L_0 and that there are Q_0 citizens enrolled in the system. Let A be a matrix which is elementwise defined as

$$A[i,j] \triangleq \alpha_{i,j} A \in [0, L_0]^{Q_0 \times Q_0} \quad (15)$$

Now assume that each c_k takes up to K_0 tests. Furthermore assume that each test score is normalized such that it lies in $[0,1]$. Note that there is no limitation to the type of tests, so the latter can range from personality tests like Myers-Briggs, the Five Factor, the Voight-Kampff, or the Rorschach to EQ tests and IQ tests. Let b be a vector with the normalized test results and w be a system defined distribution vector which weighs the relative importance of the elements of b. Then, let D be a diagonal matrix whose i-th element contains the weighted sum of the tests administered to c_i and it is common for all citizens. In other words, D is elementwise defined as

$$D[i,j] \triangleq \begin{cases} w^T b_i, i = j \\ 0, i \neq j \end{cases} \quad (16)$$

The final step of the reputation subsystem lies in normalizing A such that each column adds up to one and finding the eigenvector g^*

$$(\zeta_0 A + (1-\zeta_0)D)g^* = g^* \quad (17)$$

The power method can efficiently compute g^*. The parameter ζ_0 indicates the relative importance of both factors.

9. CONCLUSION

This report examines the design principles behind a public electronic deliberation system. Moreover, it presents a series of fundamental and advanced political analytics which are based on significant social factors. Specifically, the roles of positive and neutral votes, geolocation information, citizen proposed keywords and comments, participation frequency, and digital reputation are explored.

Proposal merging can be done in a number of ways. One is to use a similarity metric based on a combination of the geographic information attached to the proposals and to the keywords appended to them by the citizens. In this manner, a great degree of relevance between two proposals can be ensured. Finally, for advanced applications of tensors to social media see (Drakopoulos, 2017a) and for applications to information retrieval see (Drakopoulos, 2017b).

REFERENCES

R. Albert and A.-L. Barabasi. Statistical mechanics of complex networks. *Reviews of modern physics*, 74(1):47, 2002.

A.-L. Barabasi and H. E. Stanley. Fractal concepts in surface growth. Cambridge University Press, 1995.

S. Blackmore. *The meme machine*, volume 25. Oxford Paperbacks, 2000.

R. D. Bock and M. Aitkin. Marginal maximum likelihood estimation of item parameters: Application of an EM algorithm. *Psychometrika*, 46(4):443–459, 1981.

R. Braibanti. Comparative political analytics reconsidered. *The Journal of Politics*, 30(1):25–65, 1968.

Z. Brzezinski. The grand chessboard. *Harvard International Review*, 20(1):48, 1997a.

Z. Brzezinski, B. Scowcroft, and R. Murphy. Differentiated containment. *Foreign Affairs*, pp. 20–30, 1997b.

C. Cmu. A numerical method for nonlinear age dependent population models. *Differential and Integral Equations*, 3(4):767–782, 1990.

C.-L. d. S. de Montesquieu. Esprit des lois. Libr. de F. Didot Freres, 1867.

S. Douglas and T.-Y. Meng. Linearized least-squares training of multilayer feedforward neural networks. In *IJCNN*, volume 1, pages 307–312. IEEE, 1991.

A. Downs. An economic theory of political action in a democracy. *Journal of political economy*, 65(2):135–150, 1957.

G. Drakopoulos. Tensor fusion of social structural and functional analytics over Neo4j. In *IISA*, pages 1–6. IEEE, 2016a.

G. Drakopoulos. Knowledge mining with tensor algebra. Technical report, Ionian University, October 2017. doi: 10.13140/RG.2.2.25548.92803.

G. Drakopoulos and A. Baroutiadi. Implementing centrality measures for Neo4j. In *ECESCON*, 2015.

G. Drakopoulos, P. Gourgaris, A. Kanavos, and C. Makris. A fuzzy graph framework for initializing k-Means. *IJAIT*, 25(6):1–21, 2016b.

G. Drakopoulos, A. Kanavos, and A. Tsakalidis. A Neo4j implementation of fuzzy random walkers. In *SETN*, 2016c.

G. Drakopoulos, S. Kontopoulos, and C. Makris. Eventually consistent cardinality estimation with applications in biodata mining. In *SAC*, pp. 941–944. ACM, 2016d.

G. Drakopoulos, A. Kanavos, I. Karydis, S. Sioutas, and A. G. Vrahatis. Tensor-based semantically-aware topic clustering of biomedical documents. *Computation*, 5(3), May 2017a.

G. Drakopoulos, A. Kanavos, P. Mylonas, and S. Sioutas. Defining and evaluating Twitter influence metrics: A higher order approach in Neo4j. *SNAM*, 71(1), 2017b.

G. Drakopoulos, A. Kanavos, and K. Tsakalidis. Fuzzy random walkers with second order bounds: An asymmetric analysis. *Algorithms*, 10(2):40, 2017c.

A. Dubkov and B. Spagnolo. Verhulst model with Levy white noise excitation. *The European Physical Journal B*, 65(3): 361–367, 2008.

D. Easton. An approach to the analysis of political systems. *World politics*, 9(3):383–400, 1957.

T. Ferguson. *Golden rule: The investment theory of party competition and the logic of money-driven political systems*. University of Chicago Press, 1995.

G. H. Golub and C. F. Van Loan. An analysis of the total least squares problem. *SIAM Journal on Numerical Analysis*, 17(6):883–893, 1980.

A. Guttman. R-trees: A dynamic index structure for spatial searching, volume 14. ACM, 1984.

W. He and A. Wu. Estimation of Verhulst model parameter based on linear programming. *Systems Engineering-Theory & Practice*, 8:141–441, 2006.

S. P. Huntington. *The clash of civilizations and the remaking of world order*. Penguin Books India, 1997.

X.-W. Jiang, L. Ding, Z.-H. Guan, and F.-S. Yuan. Bifurcation and chaotic behavior of a discrete-time Ricardo–Malthus model. *Nonlinear Dynamics*, 71(3):437–446, 2013.

A. Kanavos, G. Drakopoulos, and A. Tsakalidis. Graph community discovery algorithms in Neo4j with a regularizationbased evaluation metric. In *WEBIST*, 2017.

S. Kontopoulos and G. Drakopoulos. A space efficient scheme for graph representation. In *ICTAI*. IEEE, 2014.

R. K. V. Kothuri, S. Ravada, and D. Abugov. Quadtree and R-tree indexes in Oracle Spatial: A comparison using GIS data. In *ICDM*, pages 546–557. ACM, 2002.

I. Markovsky and S. Van Huffel. Overview of total least squares methods. *Signal Processing*, 87(10):2283–2302, 2007.

D. W. Marquardt. An algorithm for least-squares estimation of nonlinear parameters. *JSIAM*, 11(2):431–441, 1963.

M. Newman, A.-L. Barabasi, and D. J. Watts. *The structure and dynamics of networks*. Princeton University Press, 2011.

H. Nurmi. Comparing voting systems, volume 3. *Springer Science & Business Media*, 2012.

H. Samet. *The design and analysis of spatial data structures*. Addison-Wesley, Reading, MA, 1990.

R. Schneider, B. Seeger, N. Beckmann, and H. Kriegel. The R*-tree: An efficient and robust access method for points and rectangles. In *ICDM*, pages 322–331. ACM, 1990.

IDENTIFICATION OF SERIOUS GAMES FOR MUSEUM'S EDUCATIONAL ROLES: A CRITICAL LITERATURE REVIEW

Meng Wang and Miguel Baptista Nunes[1]
School of Information Management, Sun Yat-sen University, R.P. China

ABSTRACT

This paper presents a critical literature review aiming at mapping educational roles of museums against serious educational games used in support of those roles. The study is made in the specific context of informal learning in museums. The paper first proposes a conceptual visualization of the multidisciplinary area of use of serious games in support of the educational roles in museums. This conceptualization supported the search query construction for the study. This is a reusable and transferable conceptualization that can help others engaged in similar literature reviews. The paper then proposes two# clear and detailed typologies of educational roles and serious games types for informal learning. Both typologies contain explicit links to the sources that originated each of the types and subtypes. The classification of educational roles proposes 5 main categories. For each category there is a clear indication of the corresponding learning activities that may be used to put them into practice. The typology of serious educational games for informal learning in museums contains 3 main categories and 13 subcategories. For each of these sub-categories there is also a clear indication of the corresponding features of each game type. Finally, the paper proposes a conceptual map of the intersection of informal educational roles with serious educational games. This concept map, if used in conjunction with the two classifications proposed, can support serious game designers, museum educators and educational practitioners in their decision making, concerning choice of game types, customization and content design to support specific educational activities.

KEYWORDS

Informal Learning, Museum Educational Roles, Educational Games, Educational Games for Museums, Museum Experiences

1. INTRODUCTION

Museums are regarded as informal learning places that can complement school education, enhance knowledge of history, enable understandings of cultural backgrounds and environments, enrich awareness of natural and artificial surroundings (Efird, 2014; Guzin *et al.*, 2017). Different types of museums display different exhibitions with different learning atmospheres. Researchers have proposed that the visualization, closeness and even manipulation of emulated items displayed in museums are fundamental to visitors' learning processes and have more impact on understanding and knowledge acquisition (de Rijcke and Beaulieu, 2011). Visiting experiences and close interaction with artefacts, often in a more inclusive environment, are deemed to have a deeper and longer-lasting effect on learning than classrooms and textbooks (Guzin *et al.*, 2017, Kang *et al.*, 2009, Terreni, 2015, Williams, 1996). In fact, young children's life experiences, both in and out of school, have a profound effect on their learning, intellectual growth and capacity to function in society (Resnick, 1987). Therefore, education is often a primary focus for museums, since it can influence and benefit their social environment more directly than preservation, conservation, collection, or research (Ames, 1988).

To fully understand young children's learning, it is necessary to look not only at learning that takes place in the school but also at learning that takes place out-of-school as part of children's life experiences and their interactions with their social and economic environment (Eshach, 2006). In this sense, "in-school learning is

[1] Corresponding Author

formal learning and out of school is informal *tout court*" (Eshach, 2006). Therefore, informal learning is the sum of activities that comprise the time individuals are not in a classroom in the presence of a teacher who is imposing a formal and well-defined curriculum (Gerber *et al.*, 2001). Since informal learning occurs outside formal learning settings, it is more spontaneous, and often occurs in places which form part of the child's day-to-day routine, such as homes, yard, parks or streets, leisure environments (such as digital multi-player games, social media or traditional sports clubs) and even museums, galleries and exhibitions (Eshach, 2006). Informal learning is therefore a learner-centered and learner-directed process, where the learner has agency over what is being learned, on how it is learned, how it is being understood and contextualized, and on how it is evaluated and used in the future(Russo *et al.*, 2009). Hands-on exhibits, physical demonstrations, and guided discovery tours are familiar examples of informal learning activities in museums (Ramey-Gassertet *et al.*, 1994).

This paper does not focus specifically on young people, but on informal learning in general. This study aimed at identifying the full range of educational games that can enhance informal learning and turn a museum into an enjoyable and rich informal learning environment. Thus, the paper identifies museum educational roles, then identifies types of suitable serious educational games and finally proposes an edutainment match between educational roles and types of games for informal learning. The findings proposed are integrative and collate proposals and models that have been discussed in isolation for more than 4 decades.

2. RESEARCH OBJECTIVES

The following objectives were set for the research project reported in this paper:

1. To identify and classify the educational roles of museums;
2. To identify and classify serious educational games suitable for use in museum environments;
3. To map games to each of the educational roles identified, according to each game's suitability.

3. RESEARCH METHODOLOGY AND DESIGN

The study aimed to identify and classify museums' informal educational roles through reviewing published and unpublished materials. It also sought to identify and classify types of educational games and to evaluate their respective weaknesses and strengths. Finally, the study mapped types of games against informal educational roles.

A critical literature review approach was used as defined by Saunders *et al.,* (2016), which requires an extensive search of relevant literature, combined with assessment of its quality and content (Booth *et. al*, 2009). A critical literature review should be a constructive critical analysis of existing literature aimed at developing a clear, well-informed and well-argued narrative (Wallace and Wray, 2011; Saunders *et al*, 2016). Therefore, a critical literature review should seek to identify seminal works on the topic of the research, related current debate, and theoretical propositions to solve these debates and critical analysis of these propositions.

During the critical review process undertaken by this research, conceptual and theoretical theories were analyzed, representative research cases are identified, trends explained and illustrated and resulting understandings integrated to form an argumentation of the *state-of-the-art* in the research topic and the potential contributions that can be made by future studies (Banerjee, 2011). This approach helped to make sense of a multidisciplinary, disparate and heterogeneous body of knowledge dispersed in articles published in journals covering disciplines ranging from computer science to museology. Sources could be loosely understood to be in the realm of digital humanities but sources from other areas are required as well.

3.1 Research Design

According to Saunders *et al.* (2006) a critical literature review can achieve much more than a mere description and synthesis of the existing literature in a given research topic. It should be a constructively critical analysis of theoretical propositions, models and frameworks, including rejections and acceptance of different theoretical constructs, integrated into a well-argued and logically structured narrative. Therefore, this type of review requires processes to find, select, read and evaluate the literature (Saunders *et. al*, 2016). More specifically, it should present key academic theories, opinions of key authors, the newest theories, use of concepts, weaknesses and strengths of existing work (Saunders *et al*, 2016).

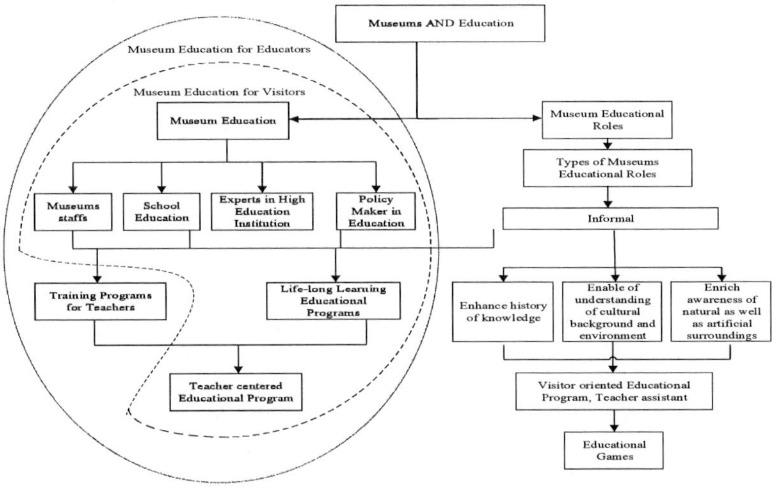

Figure 1. Conceptual Visualization of the Multidisciplinary Area of Use of Serious Games in Support of the Educational Roles in Museums

This study explores the intersection between literature about museums' educational roles and literature about educational games. Therefore, the selection of key articles and key authors was dictated by this intersection and the need to respond to the research objectives stated above. A critical process of decomposition and synthesis was used for this research that resulted in an integrated research topic map as shown in Figure 1. A top-down decomposition from the basic intersection area of (*museums AND education) AND educational games AND informal learning*. The initial search and subsequent iterations were conducted in the following academic databases: EBSCO, Web of Science, Google Scholar and Elsevier Science, following the evolving keyword and search string strategy.

4. DISCUSSION OF FINDINGS

4.1 Educational Roles of Museums

The educational roles of museums fall into three categories: formal, informal and complementary. Since the late 1880s, museums have adopted formal teaching practices to educate visitors about their collections and displays, as well as their preservation strategies and methods (Joel and Mintz, 2018). A good example of such formal learning can be found in natural history museums (Bamberger and Tal, 2006) where museums act as a place for learning about aspects of natural science ranging from environments to evolution (Hooper-Greenhill, 2007; Williams, 1996). However, the year 1992 is commonly regarded as a changing point in museum education due to the publication of a report from the American Association of Museums (Anastasia and Sylaiou, 2015; Nuzzaci, 2006; Williams, 1996). "Excellence and Equity: Education and the Public Dimension of Museums" pointed the way for museums to expand their role as educational institutions, without the traditional boundaries of walls and collections, in order to reach broader audiences. The aim was to provide guidance in supporting and increasing educational opportunities for people of every age (American Association of Museums, 1992). Before that time, museums were seen as rigid physical spaces for the display, presentation and education of collection-based knowledge. After 1992, museums started attempting to assist schools by supplementing formal syllabuses, educational programs and other initiatives (Williams, 1996; Efird, 2014). Consequently, the next wave of educational development in museums saw researchers designing educational programs in school-museum mode. However, these designs were mostly teacher-led, with teachers designing and controlling the learning processes. These initiatives have achieved very positive learning outcomes, and better engagement of the museum with society through the schools and universities, but the teacher-led pedagogical strategy was still essentially based on a passive learning mode which reduces student motivation and engagement (Efird, 2014; Kang et al., 2009; Bracun, 2015).

Conversely, informal learning is a learner-centered and learner-directed process, where the learner has agency over what is being learned, and how it is learned Eshach (2007). The process requires reflection and evaluation of the learning, and increases the likelihood that the learning will be used in future. As such, it has become a hot topic for the development of museums' educational roles (Russo et al., 2009). Attempts to generate experiential learning based on visitor experiences (Choi, 2016, Kawalilak and Groen, 2016, Zingone, 2011) through using games strategy, therefore, only pay attention to a learner-centered informal learning education in museums.

As shown in Table 1, there are different specific types of informal educational roles for museums, and each of these roles may realize a number of different educational strategies and activities.

Firstly, one of the informal educational roles is promote awareness of museum environment. Museum environment refers to the physical environment and physical sub-spaces that compose the museum as whole. These spaces can range from a reception desk to a specialized collection room. The physical environment also includes: the architectural features of the building; the museum organization; the physical interpretation of the museum's nature and mission; and the different interaction modes with visitors (such as: audio guides, sign posting, mobile phone guides and features, etc) (Shostack, 1985; Bateson, 1991; Goulding, 1999). The museum environment influences, even determines the informal learning activities that take place within it, and how the collections, artefacts and exhibitions are understood (Zukin, 1991; Delaney, 1992). Therefore, there is a strong need to provide informal educational activities to develop visitor's the awareness of museum's environment. Workshops, family fieldtrip events and special museum or collection days are often organized by the museum's educators in an attempt to provide a natural and contextualized understanding of real museum environment, aiming to enrich visitor's experiences and stimulated further visits (Ramey-Gassert, 1994; Efird, 2014; Sun, 2013; Liu, 2014). Some of the museum's informal education is targeted at primary school students and it can include materials appropriate for that age range, such as vivid cartoon, movies and hands-on games. These type of targeted materials aim to engage and encourage them to learning about and engage with the collection and exhibitions (Efird, 2014; Mortar et al., 2014; Xie, 2017; Sun, 2013).

Secondly, another informal role is promoting awareness and understanding of special collections, areas of science, specific cultures and other areas of interest to the museum that are related to its collection. Since the museums are opened to the public and aim at having a social impact beyond their walls, it has become common practice in the most advanced museums worldwide to try and transform the museum to provide learning about the social contexts, historical events, cultural backgrounds and scientific knowledge that are related to their collections (Efird, 2014, Guzin *et al.*, 2017; Ames, 1988). In this sense, educational roles of museums have changed beyond formal and complementary learning spaces to more fun and gamified informal learning that is based on both internal and external knowledge to the institution (Efird, 2014, Guzin *et al.*, 2017; Sun, 2013; Resinck, 1987; Eshach, 2006; Liu, 2018).

Thirdly, informal educational roles may aim at changing social and visitor's habits from "cultural consumption to cultural production". This was initially the base philosophy adopted by the United Kingdom to develop cultural and creative industries (O'Connor, 2006). It draws on a particular model argued by Lash and Urry (1994) that cultural production moved from its character as marginal or exceptional to a role of cutting edge or new "template" in a postmodernist society. It has since been used as the main focus of cultural industry strategies in the United Kingdom and elsewhere (O'Connor 2004b; Leadbeater & Oakley 1999; Landry 2000). In the context of museums, the shift from "cultural consumption to cultural production" is a complex and challenging concept and usually associated with the concept of participation. Within the wider discourse on arts and media, this notion of participation implies both an active and passive aspect, both taking part and creating, within transitionary processes of cultural consumption and cultural production (Bryant and Pozdeev 2010; McCarthy & Jinnett 2001; Duncombe 1997; O'Hagan 1996; Bourdieu 1993; Schuster 1991).

Table 1. Informal Educational Roles of Museums for Visitors

Informal Educational Roles	Informal Educational Activities for Visitors	Sources
Promote awareness of Museum Environment	• Workshops; • Family days and fieldtrips; • Children's learning materials (cartoons, movies; games); • Learning by doing activities, including active learning, interactive, self-learning, and experiential learning (simulations, educational games, real-life sensoria experiences; role play games; studio activities, workshop activities, practical exercise); • Challenge activities (puzzles; trivia quizzes, treasure-hunt games, team-working); • Structured sessions and lectures (individual activities, paired activities and small group activities); • Online (web) and virtual activities.	Mortar et, al., 2014; Sun, 2013; Ramey-Gassert, 1994; Hooper-Greenhill, 1999; Kang et al., 2009; Russo et al., 2009; Proctor, 2011; Mortara et al., 2014; Smart et al., 2007; Murphy, 2018; Fu, 2009; Liu, 2014;Xie, 2017; Wang 2018; Cao et al., 2018.
Promote awareness and understanding of special collections, areas of science, specific cultures and other areas of interest to the museum that are related to its collection	• Workshops, lectures, case study presentations, special exhibitions & performances; • Children's Learning Materials (Cartoons, movies; games); • Learning by doing activities including active learning, interactive, self-learning, and experiential learning (participatory design workshops, simulations, educational games, real-life sensoria experiences; role play games; studio activities, workshop activities, practical exercises); • Specialized visits and collection-related stories, narrative and accounts; • Thematic educational games (mini games, puzzles, trivia quizzes, treasure hunt games, role play games); • Skill development activities (artistic skills, IT skills, management, soft skills activities, co-creation practices and role play); • Structured sessions and lectures (individual activities, paired activities and small group activities); • Online courses, online activities, online threaded discussions and other virtual actives; • Artificial Intelligence man-machine dialogues.	Mortar et al., 2014; Sun, 2013; Hooper-Greenhill, 1999; Kang et al., 2009; Efird, 2014; Russo et al., 2009; Proctor, 2011; Eshach, 2006; Mortara et al., 2014; Smart et al., 2007; Murphy, 2018; Smart et al., 2007; Guzin *et al.*, 2017; Ramey-Gassert, 1994; Resinck, 1987; Ames, 1988; Shen 2008; Cao 2005; Fu, 2009; Xie, 2017; Liu, 2018.
Changing social and visitor habits from cultural consumption to cultural production.	• "Continuum of interactivity" activities throughout visits and beyond; • Thematic blogging and online forums for experience sharing and knowledge exchange; • Problem-based learning (knowledge creation competition, thematic activities); • Design and production workshops (pottery, Chinese traditional fan, paper cut).	Mortar et, al., 2014; Sun, 2013; Hooper-Greenhill, 1999; Kang et al., 2009; Proctor, 2011; Mortara et al., 2014; Smart et al., 2007; Murphy, 2018; Wang, 2017; Xu, 2013; Qin, 2017; Gong and Zhong, 2017.
Changing social and visitor habits from individual visit and knowledge acquisition to sharing of learning experiences and social negotiation of meanings.	• In-site group discussions, meetings and group activities; • Group activities led by the museum elsewhere (schools, universities, research institutes, culture dissemination institutes, etc); • "Continuum of interactivity" activities throughout visits and beyond; • Thematic blogging and online forums for experience sharing and knowledge exchange; • Role-play activities, performances and games; • Solicit constructive feedback, act on the feedback and pro-actively demonstrate the use of that feedback to visitors.	Mortar et al., 2014; Sun, 2013; Kiili, 2005; Rennie and McPherson, 2007; Russo et al., 2009; Proctor, 2011; Smart et al., 2007; Murphy, 2018; Cao et al., 2018; Liu, 2014; Wang, 2018; Zhu, 2017; Qin, 2013.
Contextualizing museum visits and knowledge with prior experiences, knowledge and learning	• Elective customized workshops; • Customized learning by doing activities, including active learning, interactive, self-learning, and experiential learning (simulations, educational games, real-life sensoria experiences; role play games; studio activities, workshop activities, practical exercise); • Customized specialized visits and collection related stories, narrative and accounts; • Customized thematic educational games (mini games, puzzles, trivia's, treasure hunt games, role play games); • Customized in-site and off-site multi-sense exploration, discovery and learning activities; • Popularization of science, history or culture activities (knowledge creation competitions, art competition, writing competitions, fieldtrip and artifact collection special groups).	Bracun, 2015; Sun, 2013; Murphy, 2018; Guzin et al., 2017; Li, 2018; Kang et al., 2009; Russo et, al., 2009; Proctor, 2011; Mortara et al., 2014; Smart et al., 2007; Perez-Valle et al., 2014;Murphy, 2018; Cao et al., 2018; Xie, 2017; Zheng 2017; Gong and Zhong, 2017; Tang et al, 2012; Shen, 2008, Li, 2008.

Fourthly, another important informal role responds to the challenge of changing societal and visitor's habits from individual visit to a more social and shared experience. Individual and quasi-isolated knowledge acquisition is to change to a more constructivist process though the sharing of learning experiences and social negotiation of meanings. Visitors understanding and existing knowledge of phenomena can be influence and be changed by others through sharing experiences, discussion and social communication (Hooper-Greenhill, 1999; Mortar *et al.*, 2014; Cao et al., 2018; Liu, 2014). For instance, sharing of experiences and social negotiation of meanings can be implemented by thematic blogging and online forums, which create an environment for the expression of individual opinions, as well as comments on those opinions by other visitors (Kiili, 2005; Rennie and McPherson, 2007; Russo, *et al.,* 2009; Wang, 2018; Zhu, 2017). These social networks allow learning and interest of visitors to last far beyond the specific time of their visits. In addition, museums are now actively soliciting constructive feedback, acting on this feedback and also pro-actively demonstrating the use of that feedback to visitors (Mortar *et al.*, 2014; Smart *et al.*, 2007; Sun, 2013; Qin, 2013; Wang, 2017).

Fifthly, the final important educational role is related the aim of contextualizing museum visits and knowledge with prior experiences, knowledge and learning. The aim here is to highlight the value of the museum experience as well as give visitors a sense how their learning in the museum can be transferred beyond the physical walls of the museum. Through exploring collections in museums, visitors can not only improve their knowledge about what they experienced and learned before, but also contextualize this knowledge in their professional, leisure and special interest experiences (Bracun, 2015; Kang *et al.*, 2009; Gong and Zhong, 2017). Prior knowledge acquired in formal learning and real life experiences can be related with the museum's collections and therefore shape visitors' depth and usefulness of understanding (Murphy, 2018; Guzin *et al.*, 2017; Li, 2018; Shen, 2008).

4.2 Serious Educational Games for Museums

As illustrated in Table 1, a large range of serious games has implemented for museums in order to serve very specific educational programs. There are many different criteria for classifying types of games, like scope, marketing and purpose (Breuer *et, al.,* 2010). Djaouti *et al.* (2011) propose a very useful classification of games into:

- Gameplay;
- Purpose;
- Scope.

As shown in Table 2, there are three purposes of designing serious educational games, one is for broadcasting specific messages about the museum environment, collections, special exhibitions, historical periods related to artefacts in the collection, cultural background associated with the collection, etc. This type of game usually includes a more formal educative content that is informative, persuasive and subjective but transmits explicit information and has a clearly stated aim. Another type of game aims at training learners for the acquisition of intellectual and physical skills normally associated with the "from cultural consumption to cultural production" paradigm. In line with the same paradigm, the third type of game aims at providing interactive and complex serious game environment and challenges to support experiential learning. Examples of these experiential games are: exploration games, role-play games, adventure games, strategy games, location based games, pervasive games and live action role-playing games. All these types of games provide an active learning environment to serious game player (Kiili, 2005; Breuer *et, al.,* 2010; Mortara *et, al.*, 2014).

The critical and common feature of all these serious educational games are the focus around museum related learning objectives, which through different reward mechanisms and other encouraging features enhance the learning process and engage the visitors (Borges *et al.*, 2014). These games usually include, faithful information reconstructions, explicit informative knowledge, designed pedagogical strategies, pervasive environments, as well as challenging and subjective quests. 3D and virtual reality games have the advantage of being able to offer a more stimulating real-life like scenario and sometimes even build virtual environments that allow learners to do things that cannot be done in reality (e.g. the reading of ancient books that are otherwise not accessible to the public) (Gee, 2004; Kiili, 2005; Mortara *et, al.*, 2014).

Table 2. Types of Educational Games

Purpose of game	Types of games	Features of games	Sources
Message broadcasting	Documentary games	Informative, educative, persuasive; subjective, Informative, subjective; clues. Rewards/Achievement systems.	Mortara et al., 2014; Borges et al., 2014
	Treasure hunt/Scavenger hunt motif	Informative, educative, subjective; learning objectives; Rewards/Achievement systems.	Kohen-Vacs et al., 2012; Heykoop, 1999; Wicks, 1999; Sporgis, 2001.
Training for the acquisition of intellectual and physical skills	Puzzle	Informative, educative, persuasive, subject; Logical intelligent tasks.	Granic, 2014; Bonanno, 2014; Park, 2010
	Trivia	Learning objectives; question and answers.	Smith, 2004; Hartsock, Lange, 1987; Mann, 2002.
	Mini games	Learning objectives; Rewards/Achievement systems.	Ray and Coulter, 2010; Prensky, 2008; Bakker et al., 2012.
	Individual or group task games	Rewards/Achievement systems; Motivation affordance; learning objectives; learning instructions.	Garris et al., 2002; Deesri, 2002; Stathis. and Sergot, 1996.
	Stimulation games	Reflective and experiential and interpretive; stimulate a reality.	Killi, 2005; Borges et al., 2014; Hartsock, Lange, 1987;
Experiential learning (active learning/ interactive learning/ learning by doing)	Exploration games	Reflective and experiential and interpretive; Failed, advice, suggest and help; MED framework; Game interaction; Rewards/Achievement systems; Internal motivation /cognition.	Gee, 2004; Borges et al., 2014; Bakker et al., 2012; Garris et al., 2002; Hamari and Eranti, 2011; Thompson et al., 2010; Dickey, 2006.
	Role play games	Informative, educative, persuasive; motivation affordance; cognitive behavior; game-play mechanic.	Mortara et al., 2014; Borges et al., 2014; Thompson et al., 2010; Van Ments, 1999; De Freitas and Griffiths, 2009.
	Adventure games	Informative, educative, persuasive; motivation affordance; affect and moral related; cognitive behavior; game-play mechanic.	Dickey 2006; Borges et al., 2014; Hamlen, 2011; Cornillie et al.,, 2011
	Strategy games	Skillful and careful thinking and planning	Hsieh and Sun, 2008; Borges et al., 2014; Amory et al., 1999.
	Location based games	Navigation system, informative information of items	Sporgis, 2001; Rashid 2006; Borges et al., 2014; Avouris; 2012.
	Pervasive games/ Live action role-playing game	It's Alive, game takes place blends with the physical world	Magerkurth et al., 2005; Borges et al., 2014; Jegers, 2007; Garrido et al. 2011, Avouris; 2012; Bonis et al., 2009; Lepouras and Vassilakis, 2004.

4.3 Mapping Educational Role with Educational Games

As discussed throughout the paper, museum learning activities have a cognitive (Annis, 2000) and a social dimension (Hooper-Greenhill, 2000). Therefore, informal learning museums' existing pedagogic programs aims to contribute to the improvement of visitors' knowledge acquisition and corresponding change in cognitive behaviors and thinking, by providing an interactive and immersive hand-on approach (Terreni, 2015). Specifically, these educational programs includes attempts to develop customized learning experiences to different audience groups (e.g children, teenager, adult or expert versus amateur). This customization and specialization has been particularly successful with developing primary children's skills (such as intellectual, speech and communication skills) and interests (Bracun, 2015) and expanding their horizons by observing and experimenting things that are beyond their daily experiences. Other customization efforts have focused on aspects of cultural identity, social inclusion, and understanding of aesthetic (Guzin *et al.*, 2017).

Modern interactive, 3D and VR serious games use the strong features of digital games, such as digital interaction with immersive environments, online synchronous communication with other learners and realistic landscaping to reproduce physical worlds and natural environments in museums (Xhembulla *et al.*, 2014). Educators, researchers and practitioners are now paying more attention to visitors' physical and virtual experiences, and use the knowledge transfer potential of serious games experiences to fulfill the museums educational roles (Sanchez *et, al.*, 2015). Serious games are fulfilling some of the pioneer thoughts of Dewey (1916), as cited by Hein (2013) that proposed that learners need “intellectual tools to integrate the experiences into understanding”. Serious games can be an exemplary of such intellectual tools.

However, serious games are difficult to design because as pointed out by Hein (2013) designers must consider users’ emotion as key factors. These emotional states can influence users’ usage of educational games. It is therefore important to monitor players’ emotional status during playing. For instance, when and if a player fails to tackle problem too many times, it is necessary to give advice on how to recover and if necessary to rebuild the learner’s confident in completion of the educational game and in doing so maintaining their commitment and interest (Terreni, 2015, Hein, 2013). Additionally, research into gameplay has found that players like to stay in comfort zones with which they are familiar with and guarantee them that they have the capability to handle challenges and uncertainties (Landers et al., 2015). Furthermore, Green and Hannon (2006) suggested that there are four key components to learning: finding information and knowledge, doing something with it, sharing it with an audience and reflecting on it. Thus, a suitable serious game for a museum should be designed to have processes to provide: explicit information; exchange data, information, heuristics and experiences; as well as create and generate knowledge through learning by doing and experiencing.

All of these characteristics of serious games need to meet the specific and customized need of museums, their aims, their espoused educational roles, their audiences, their socio-political surroundings and their social role as expressed in Figure 2.

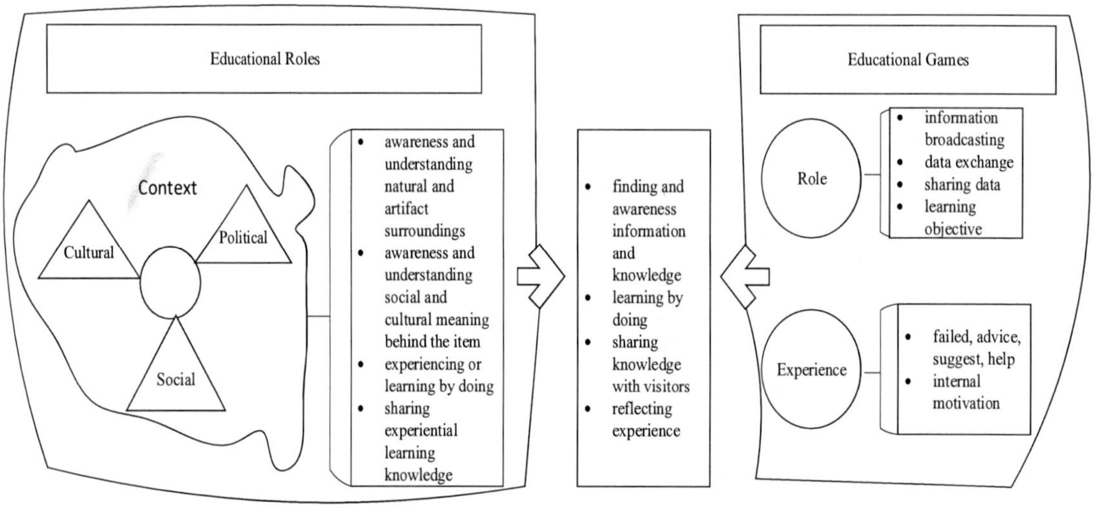

Figure 2. Conceptual Map of Intersection Informal Educational Roles with Serious Educational Games

5. CONCLUSION

This paper results from a critical literature review that explored the types of museum’s educational roles and typology of educational games, in an attempt to identify what serious games are suitable to support museums educational roles. The explicitly stated lens of the study is informal learning and therefore the classification of games proposed is based on this specific purpose. Therefore, this study identified and classified the educational roles in museums, and identified and classified the educational games for informal learning in museums. The effort of matching these two classification resulted a conceptual map that if use together with the two classifications proposed, can support serious game designers, museum educators and educational practitioners in their decision making concerning choice of game types, customization and content design to support specific educational activities.

Future work for this research group will comprise specific studies for different supporting technologies of serious games (such as mobile games or Virtual Reality games); evaluation studies of the applicability, efficiency and usefulness of these serious games in the real practice of museums and studies of efficiency and adequacy of pedagogical strategies embedded in these games.

REFERENCES

Amory, A., et al, 1999. The use of computer games as an educational tool: identification of appropriate game types and game elements. British Journal of Educational Technology, 30(4), pp.311-321.

Avouris, N.M. and Yiannoutsou, N., 2012. A review of mobile location-based games for learning across physical and virtual spaces. J. UCS, 18(15), pp.2120-2142.

Bakker, M., et al, 2012. Effects of mini-games for enhancing multiplicative abilities: A first exploration. In Serious games: The challenge (pp. 53-57). Springer, Berlin, Heidelberg.

Bichen, H. and Kocakoyun, S., 2017. Determination of University Students' Most Preferred Mobile Application for Gamification. World Journal on Educational Technology Current Issues, Vol. 9, No. 1, pp. 18-23.

Bonanno, P. and Kommers, P.A., 2005. Gender differences and styles in the use of digital games. Educational Psychology, 25(1), pp.13-41.

Bonis, B., et al, 2009. A platform for virtual museums with personalized content. Multimedia tools and applications, 42(2), pp.139-159.

Borges, S. D. S. et al, 2014. A systematic mapping on gamification applied to education. Proceeding of ACM Symposium on Applied Computing. Springer, US, pp. 216-222.

Bracun Sova, R., 2015. Art Appreciation as a Learned Competence: A Museum-based Qualitative Study of Adult Art Specialist and Art Non-Specialist Visitors. Journal of Center for Educational Policy Studies, Vol. 5, No. 4, pp. 141-157.

Breuer, J.S. and Bente, G., 2010. Why so serious? On the relation of serious games and learning. Eludamos. Journal for computer game culture, 4(1), pp.7-24.

Bryant, P. and Pozdeev, N., 2010. The impacts of governance practices and knowledge on the participation of grassroots communities in community broadcasting. In Proceeding of International Association of Media and Communications Research Annual Conference.

Cornillie, F., et al, 2011. Vocabulary treatment in adventure and role-playing games: A playground for adaptation and adaptivity. In Interdisciplinary Approaches to Adaptive Learning. A Look at the Neighbours (pp. 131-146). Springer, Berlin, Heidelberg.

De Freitas, S. and Griffiths, M., 2009. Massively multiplayer online role-play games for learning. Handbook of research on effective electronic gaming in education, 1, pp.779-793.

De Rijcke, S. and Beaulieu, A., 2011. Image as interface: consequences for users of museum knowledge. library trends, 59(4), pp.663-685.

Deesri, A., 2002. Games in the ESL and EFL class. The Internet TESL Journal, 8(9), pp.1-5.

Dickey, M.D., 2006. Game design narrative for learning: Appropriating adventure game design narrative devices and techniques for the design of interactive learning environments. Educational Technology Research and Development, 54(3), pp.245-263.

Djaouti, D., et al, 2011. Classifying serious games: the G/P/S model. Handbook of research on improving learning and motivation through educational games: Multidisciplinary approaches, 2, pp.118-136.

Efird, R., 2014. Learning places and 'little volunteers': an assessment of place- and community-based education in China. Environmental Education Research, Vol. 21, No. 8, pp. 1143-1154.

Eshach, H., 2006. Bridging In-school and Out-of-school Learning: Formal, Non-Formal, and Informal Education. Journal of Science Education and Technology, Vol. 16, No. 2, pp. 171-190.

Filippoupoliti, A. and Sylaiou, S., 2015. Museum & Society special issue. Museum education today: creative synergies and pedagogic innovations in multicultural contexts. Journal of Museum & Society, Vol. 13, No. 2, pp. 119-122.

Garrido, P., et al, 2011, February. Use of NFC-based pervasive games for encouraging learning and student motivation. In Near Field Communication (NFC), 2011 3rd International Workshop on (pp. 32-37). IEEE.

Garris, R., et al, 2002. Games, motivation, and learning: A research and practice model. Simulation & gaming, 33(4), pp.441-467.

Gee, J.P., 2004. Learning by design: Games as learning machines. Interactive educational multimedia: IEM, (8), pp.15-23.

Gerber, B. L. et al, 2001. Development of an informal learning opportunities assay. International Journal of Science Education, Vol. 23, No. 6, pp. 569-583.

Granic, I., et al, 2014. The benefits of playing video games. American psychologist, 69(1), p.66.

Green, H., & Hannon, C. (2006). Their space. Education for a digital foundation. London: DEMOS. Retreived January, 2007, from http://www.demos.co.uk/publications/theirspace

Guzin, O. A. et al, 2017. An alternative educational method in early childhood: Museum education. Educational Research and Reviews, Vol. 12, No. 14, pp. 688-694.
Hamari, J. and Eranti, V., 2011, September. Framework for Designing and Evaluating Game Achievements. In Digra Conference.
Hamari, J. et al, 2014. Does Gamification Work? -- A Literature Review of Empirical Studies on Gamification. Hawaii International Conference on System Sciences, Hawaii, USA, pp. 3025-3034.
Hamlen, K.R., 2011. Children's choices and strategies in video games. Computers in Human Behavior, 27(1), pp.532-539.
Hartsock, J.M. and Lange, R.H., 1987. Trivia games: stimulating student learning. Nurse Educator, 12(1), pp.24-27.
Hein, G. E., 2013. Progressive Museum Education: Examples from the 1960s. Journal of Progressive Education, Vol. 9, No. 2, pp. 61-76.
Heykoop, N., et al, 1999. Pirates treasure hunt game and method of playing same. U.S. Patent 5,924,695.
Hooper-Greenhill, E., 1999. Education, communication and interpretation: towards a critical pedagogy in museums. The educational role of the museum, 2, pp.3-27.
Hooper-Greenhill, E., 2007. Museum and Education. Abingdon, Routledge.
Hsieh, J. and Sun, C., 2008, June. Building a player strategy model by analyzing replays of real-time strategy games. In Neural Networks, 2008. IJCNN 2008. (IEEE World Congress on Computational Intelligence). IEEE International Joint Conference on (pp. 3106-3111). IEEE.
Huotari, K. and Hamari, J., 2012. Defining gamification: a service marketing perspective. Proceeding of the International Academic Mindtrek Conference, pp. 17-22.
Jegers, K., 2007. Pervasive game flow: understanding player enjoyment in pervasive gaming. Computers in Entertainment (CIE), 5(1), p.9.
Kang, C. et al, 2009. Chinese perceptions of the interface between school and museum education. Cultural Studies of Science Education, Vol. 5, No. 3, pp. 665-684.
Kiili, K., 2005. Digital game-based learning: Towards an experiential gaming model. The Internet and higher education, 8(1), pp.13-24.
Kohen-Vacs, D., et al, 2012. Mobile treasure hunt games for outdoor learning. Bulletin of the IEEE Technical Committee on Learning Technology, 14(4), pp.24-26.
Lam, A.S., 2004. A Comparative Study of the Educational Role of Public Art Museums in Three Asian Cities: Taipei, Singapore and Shanghai. 教育學報.
Landers, R. N. et al, 2015. Psychological Theory and the Gamification of Learning. Springer International Publishing.
Law, K.Y., 2014. The red line over European colonialism: comparison of the Macao museum and Hong Kong museum of history after their return to China. International Journal of Heritage Studies, 20(5), pp.534-555.
Lepouras, G. and Vassilakis, C., 2004. Virtual museums for all: employing game technology for edutainment. Virtual reality, 8(2), pp.96-106.
Magerkurth, C., et al, 2005. Pervasive games: bringing computer entertainment back to the real world. Computers in Entertainment (CIE), 3(3), pp.4-4.
Mann, G.S., 2002, August. Learning how to answer questions using trivia games. In Proceedings of the 19th international conference on Computational linguistics-Volume 1 (pp. 1-7). Association for Computational Linguistics.
Mortara, M. et al,, 2014. Learning cultural heritage by serious games. Journal of Cultural Heritage, 15(3), pp.318-325.
Murphy, M., 2018. "Blending" Docent Learning: Using Google Forms Quizzes to Increase Efficiency in Interpreter Education at Fort Henry. Journal of Museum Education, Vol. 43, No. 1, pp. 47-54.
Nunes, M. and McPherson, M., 2007. Why designers cannot be agnostic about pedagogy: The influence of constructivist thinking in design of e-learning for HE. In Evolution of Teaching and Learning Paradigms in Intelligent Environment (pp. 7-30). Springer, Berlin, Heidelberg.
Park, E. and Park, Y., 2010. A hierarchical interface design of a puzzle game for elementary education. International Journal of u-and e-Service, Science and Technology, 3(2), pp.43-49.
Perez-Valle, A., et al, 2014, November. Playhist: play and learn history. Learning with a historical game vs an interactive film. In Euro-Mediterranean Conference (pp. 546-554). Springer, Cham.
Prensky, M., 2008. Students as designers and creators of educational computer games: Who else?. British Journal of Educational Technology, 39(6), pp.1004-1019.
Primo, J., 2007. To think museology today.
Proctor, N. ed., 2011. Mobile apps for museums: The AAM guide to planning and strategy. AAM Press.
Ramey-Gassert, L. and Prather, J., 1994. Planning productive field trips: Enhancing student learning. Workshop presented at the annual meeting of the Association of Educators of Teachers of Science. El Paso, TX, USA.
Rashid, O., et al, 2006. Extending cyberspace: location based games using cellular phones. Computers in Entertainment (CIE), 4(1), p.4.
Ray, B. and Coulter, G., 2010. Perceptions of the value of digital mini-games: Implications for middle school classrooms. Journal of Computing in Teacher Education, 26(3), pp.92-100.

Rennie, L. and McClafferty, T., 1995. Using visits to interactive science and technology centers, museums, aquaria, and zoos to promote learning in science. Journal of Science Teacher Education, 6(4), pp.175-185.
Resnick, L., 1987. The 1987 presidential address learning in school and out. Educational researcher, Vol. 16, No. 9, pp. 13-54.
Russo, A. et al, 2009. The impact of social media on informal learning in museums. Educational Media International, Vol. 46, No. 2, pp. 153-166.
Sanchez, E. and Pierroux, P., 2015. Gamifying the Museum: Teaching for Games Based 'Informal' Learning. In: Munkvold, R. and Kolas, L. (eds.) Proceedings of the 9th European Conference on Games Based Learning.
Schiller, S.Z., 2009. Practicing learner-centered teaching: Pedagogical design and assessment of a second life project. Journal of Information Systems Education, 20(3), p.369.
Smart, K. and Csapo, N., 2007. Learning by doing: Engaging students through learner-centered activities. Business Communication Quarterly, 70(4), pp.451-457.
Smith, T., 2004. Method and system for playing trivia games. U.S. Patent 6,752,396.
Spence, I. and Feng, J., 2010. Video games and spatial cognition. Review of General Psychology, 14(2), p.92.
Sporgis, P., 2001. Treasure hunt game utilizing GPS equipped wireless communications devices. U.S. Patent 6,320,495.
Stathis, K. and Sergot, M., 1996. Games as a metaphor for interactive systems. In People and Computers XI (pp. 19-33). Springer, London.
Terreni, L., 2015. Young children's learning in art museums: a review of New Zealand and international literature. European Early Childhood Education Research Journal, Vol. 23, No. 5, pp. 720-742.
Thompson, D. et al, 2010. Serious video games for health: How behavioral science guided the development of a serious video game. Simulation & gaming, 41(4), pp.587-606.
Van Ments, M., 1999. The effective use of role-play: Practical techniques for improving learning. Kogan Page Publishers.
Wicks, J.E., 1999. Treasure hunt game using pager and paging system. U.S. Patent 5,942,969. Sony Corp and Sony Electronics Inc.
Williams, B., 1996. An Examination of Art Museum Education Practices Since 1984. Journal of National Art Education Association, Vol. 38, No. 1, pp. 34-49.
Cao, C., et al., 2018. 博物馆互动体验式教学及其反馈. 改革与开放, vol.1, pp.131-132.
Fu, X., 2009. 多媒体交互技术在展示设计中的应用研究. 天津美术学院, 天津.
Gong, M. and Zhong, Q., 2017. 地质博物馆6-12岁儿童科普教育活动的设计探讨——以湖南省地质博物馆为例. 中国博物馆, vol.3, pp.60-64.
Li, S., 2018. 浅谈博物馆教育中手工活动的教学设计——以中小学生的考古学普及教育为例. 遗产与保护研究, vol.1, pp.93-94.
Liu, F., 2018. 博物馆教育与学校教育的整合与利用问题研究. 科教导刊(下旬), vol.2, pp.178-185.
Liu, Q., 2014. 关于博物馆互动体验活动的再思考. 中国纪念馆研究, vol.01, pp.192-197.
Qin, W., 2017. 到博物馆玩去吧:浅论博物馆的幼儿教育活动策划. 教育现代化, vol.42, pp.188-189.
Shen, Z., 2008. 以"人"为核心的当代中国博物馆传播理念与展示设计研究. 江南大学, 江苏无锡,
Wang, J., 2018.浅谈科技发展对博物馆教育工作的影响. 博物馆研究, vol.1, pp.29-34.
Wang, L., 2017. 启智.互动.体验.创新——浅谈博物馆青少年教育活动的策划与开展. 科技风, vol.14, pp.263.
Wang, T., 2018.将建构主义理论应用于博物馆教育活动的研究——基于广东省博物馆教育活动实践的思考. 客家文博, vol.1, pp.31-35.
Xie, J., 2017. 山区博物馆儿童教育项目探索——以中国客家博物馆为例. 客家文博, vol.4, pp.45-50.
Xu, Z. , 2013. 高黎贡手工造纸博物馆. 世界建筑, vol.11, pp.90-93.
Zheng, Y., 2017. 博物馆教育的出发点:创设让人"快乐"的体验, 中国博物馆, vol.4, pp.1-6.
Zhu, Y., 2017. 科技博物馆应用VR/AR技术的特殊需求与策略. 科普研究

ESTABLISHING A DIGITAL MEMORY APPLICATION MODEL FROM AN INFORMATION SCIENCE PERSPECTIVE: A SYSTEMATIC LITERATURE REVIEW

Ting Yu Luo, Miguel Baptista Nunes and Jia Zheng Liang
School of Information Management, Sun Yat-Sen University, No. 135, Xingang Xi Road, Guangzhou, 510275, China

ABSTRACT

The term "digital memory" is widely used with multiple definitions in different contexts ranging from hardware design to archival studies. Considering that definition is critical for conducting academic research as it decides the research boundary and research scope, this paper aims to first answer the question of how digital memory should be defined from the perspective of information science. Followed by a proposal of a theoretical model of digital memory applications according to an information science perspective, including purpose, organization structure and information architecture scheme. This research is conducted using a series of sequential and connected critical review of literature processes, using Google Scholar as the source of data. The final result was the definition of digital memory in the context of digital humanities is general, information science in particular. Additionally, the research also explored potential roles of information scientists in digital memory studies. Finally, this paper proposes a new model of digital memory from the perspective of information science that can be further used both in practice and applied research in the future.

KEYWORDS

Digital Memory, Digital Humanities, Information Science

1. INTRODUCTION

The term "digital memory" has different meanings in different disciplines. By searching "Digital memory" in Google Scholar Database, there were over 50,000 results. The disciplines covered range from Electronic Engineering to Library and Information Science. This paper focus on the later science and in particular on an information science perspective. In general, memory study is an interdisciplinary research area which involves psychology, sociology, computer science, history, communication and culture study (Shao, 2014). Memory can be classified into cultural memory, historical memory, individual memory, collective memory, social memory, media memory and digital memory according to different disciplinary lenses. The digital nature of the memory applications means that these can be easily changed, adapted and extended therefore reflecting continuous developing social memory and trends. Hence, digital memory applications, similarly to society and human activity systems, rather than stay static, become more dynamic which is continually in progress (Bond *et al.*, 2016). Brockmeier (2015) has rethought the very notion of memory as information storage, as "an archive of the past". The author argued that even though it was widely accepted for a long term that memory is a distinctive human capacity which enable us to "encode," "store," and "recall" past experiences, a new picture is emerging with different cultural forms of remembering and forgetting, embedded in a broad spectrum of human activities and artifacts. It comes with a number of alternatives to archival memory which not only permits us to explore the storied weave of our most personal, namely, autobiographical form of remembering; it also sheds new light on the interrelations among memory collections, self, and culture (Brockmeier, 2015).

Digital Memory is discussed in this paper a term widely mentioned in digital humanities studies. Online mementos, digital photography, memorable web pages, digital shrines, text messages repositories, digital archives, online museums, online condolence message boards, virtual candles, social networking and alumni websites, digital television news broadcasts of major events, broadcaster websites of archival material, blogs, digital storytelling, computer games based on past wars, fan sites and digital scrapbooks are all examples of digital memory (Garde-Hansen *et al.*2009). Although there is no unified definition, it is generally accepted in the literature in the field that digital memory is not only a collection of electronic files or databases, or merely digital presentation of history or memory facts, it is more related to the combination of facts and ways that individuals, groups and societies understand and accept these facts. Understandings, acceptance and use of digital memory is also related and affected by the information technologies that instantiate it. The same as collected physical souvenirs and mementos, digital memory also serves to "control" access to past time periods, as well as processes of recollection, grief and trauma (Broderick & Gibson, 2005). However, the description and interpretation of the memory facts can vary by applying flexible and customizable technologies, which allow the implementation of different information organization schemes and structures.

As a result of the development of information technologies such as Web and multi-media, it is possible to present the relationship between past, present and future in a variety of ways according to different user groups' information behavior, information needs and customized narrative paths. All of these aspects are closely related to information science. It is suggested by Berry (2011) that information technology has innovated new processes of communication, and with recent developments in networked media and social network, collaborative thinking is rising. With these technologies, it is possible to create collective digital memory platforms, which can take us beyond the trends of individual blogs and comments. Generating new ideas, modes of thoughts, theories and practices (Berry, 2011). Therefore, the introspection on definition of digital memory from information science perspective gives the possibility for researchers to discuss the relationship between information (historical facts), people (users) and information technology involved in digital memory systems. In addition, similarly to any other information systems, digital memory applications face further fundamental questions prior to their development, such as: What goals should digital memory system serve? And how can human memory be activated and supported by technology?

This paper tries to make sense of a very diverse and heterogeneous body of literature review, using the specific lens of information science. The primary objective of this paper is to review concepts of digital memory and to discuss their relevance to information science research. Through a critically review on current literatures to give a definition and a model to digital memory, which can guide information scientists to do future research.

2. RESEARCH METHODOLOGY

2.1 Research Aims

The aims of the research presented in this paper were as discussed above to define a digital memory application theoretical model from an information science perspective. Disciplinary lenses on cross-disciplinary concepts are fundamental to the success of any social science project and the definition of research boundaries, scope, potential outputs. Additionally, the research had an additional aim of exploring potential roles of information scientists in digital memory studies unparticular and digital humanities in general.

2.2 Research Design

The research was conducted by doing a systematic and critical literature review according to the methodology proposed by Nunes et.al (2009) and Saunders, Lewis, & Thornhill (2009). Systematic review is adopted due to its rigorous scientific nature, that is to identify, select and retrieve relevant scholarly literature from well-established academic databases by following a precise protocol and a systematic data extraction strategy (Nunes *et al.*, 2017). In addition, a critical analysis is adopted in order to use opinions, case-studies and theoretical propositions in the literature to help the understandings extracted from the systematic literature review. This critical analysis of literature makes it also easier to establish and discuss similarities and differences between different definitions. In addition, it also provides a critical insight into models of digital memory, which can be considered into building the model that is the second objective of this research.

The research design for this research included 6 steps: Keywords Identification, Search Queries Production, Inclusion and Exclusion Criteria Definition, Database Identification, Document Selection and

Critically Analysis of Literatures. A detailed description of how each step is fulfilled will be revealed in the following sections.

2.2.1 Identifying Keywords and Query

The term "digital memory" have different meanings in different disciplines. By searching "Digital memory" in full text in English in google scholar, there are 57300 documents founded. By searching "digital memory" in research Topic in Web of Science database, there are 4024 documents founded. (Both of the data is collected by May 2018). These covered more than 200 subjects according to Web of Science research area classification, ranging from Engineering Electrical Electronic to Information Science Library Science and including such disparate things as Biology and Urban Studies. In order to focus the sample and taking in consideration that the general area of research of this study is in "digital humanities", the basic search queries have been further decided as follower:

"Digital Memory" AND "Digital humanities"

"Digital Memory" AND "Digital humanities" AND ("Purpose" OR "Objective" OR "Aim" OR "Use")

"Digital Memory" AND "Library and Information Science"

"Digital Memory" AND ("archive" OR "archives" OR "archival")

"Digital Memory" AND ("Web Applications" OR "Web Designs" OR "Web Architectures")

"Digital Memory" AND ("Information Architecture" OR "Information Organization" OR "Information Structure")

"Digital Memory" AND ("Model" OR "Structure")

2.2.2 Backward and Forward Search

Structured search was conducted for all the English published records. In addition, backward and forward search based on structured search output was conducted manually to avoid missing related literatures. This critical literature review is down based on works published in English language. Title and abstract of those studies are browsed first, followed by analysis of the main content of the paper.

2.2.3 Inclusion and Exclusion Criteria

By using search queries listed in section 2.2.1, records published in English after year 2000 are included in the review. The limit of the year 2000 was established since before this date there is very little relevant research in the area of digital memory. The following exclusion criteria were used and added to the search strings, since these terms reflect areas of research that are not the focus of this study: personal data management, human digital memory, objects memory, robotic memory, digital prosthetic memory and digital legacy. In addition, literature which discussed how "digital memory" is defined was included; whilst literature that simply mentioned "digital memory" but provided no further explanation or useful definition of digital memory was excluded. The relevance of each article was decided upon analysis of relevance in the title, key words, abstract and full text sequentially.

3. DISCUSSION OF FINDINGS

The findings of the systematic and critical literature review are presented here in two different ways. First, a descriptive statistics of the results is presented, followed by a synthesis of the reviewed literature. Finally, an integrative conceptual model for digital memory from an Information Science perspective is proposed. This model results from the literature reviewed and a critical analysis using an information science lens.

3.1 Descriptive Presentation of Findings

The systematic literature review was conducted using two main sources: Google Scholar (https://scholar.google.co.uk/) and the Web of Science database (apps.webofknowledge.com/). The results of the structured and manual searches described above were subject to a descriptive analysis that is summarized in Table 1, Table 2, Figure 1 and Figure 2.

Table 1 generally shows the results of the systematic review in stages and reflects the modus operandi used for each of the search queries. Each query returned a total number of documents that were then filtered down according to relevance to the study and removal of redundancies. From the Table 1 it is also very evident that numbers of relevant documents returned by the same query in Web of Science and Google Scholar are significantly different.

Table 1. Results Distribution for each of the Search Queries and Databases

Key	Search Query	Number of Documents	Relevant Documents	Relevant non-Redundant Documents	Web of Science	Google Scholar
Q1	"Digital Memory" AND "Digital humanities"	257	61	59	1 (1)	256 (60)
Q2	("Digital Memory") AND "Digital humanities" AND ("Purpose" OR "Objective" OR "Aim" OR "Use")	251	51	49	0 (0)	251 (51)
Q3	"Digital Memory" AND "Library and Information Science"	162	30	28	9 (3)	153 (27)
Q4	"information science" "digital memory" -"computer science" -"cognitive neuroscience" -electronic -neurobiology -forestry	126	23	22	9 (1)	125 (22)
Q5	archive OR archival "digital memory" -hardware -"storage device" -semiconductor -"flash memory" -"solid state" -microprocessor -autobiographical -pedagogical -biomedical -"image sensor" -physiology -microscopic	1002	180	158	38 (21)	964 (159)
Q6	"Digital Memory" AND ("Web Applications" OR "Web Designs" OR "Web Architectures")	164	13	12	0 (0)	164 (12)
Q7	"Digital Memory" AND ("Model" OR "Structure") -"digital memory card" -Accelerator -mechanical -teach -"computer science" -electronic -neurobiology -forestry -quantum -device -Symposium -engineering -biological -neuroscience	907	35	32	441(2)	466 (33)

Table 2. Totals of Documents Returned

	Web of Science	Google Scholar	Number of Relevant Documents	Number of Unique Relevant Documents
Total Documents Returned	22	268	290	279

As can be seen in Table 2, queries in Web of Science resulted in 22 unique documents, whilst the same set of queries in Google Scholar returned 268 relevant documents. This is due to the fact that Google Scholar indexes a wider variety of journals, books and conference proceedings in comparison to a more traditionally closed selection of journals contained in Web of Science. This can also be explained by the fact that digital memory is still an emergent concept, mostly discussed in archives, digital humanities and digital heritage conferences and journals, many of which are not indexed by the Web of Science. This is not uncommon in humanities and social science emerging topics and areas of research. Finally, the total of unique and relevant documents analyzed was 279. It is important to notice that this figure does not represent the totality of digital memory documents available in the literature or even in the two main sources of data. This figure represents unique and relevant documents that help responding to the set research questions for this study.

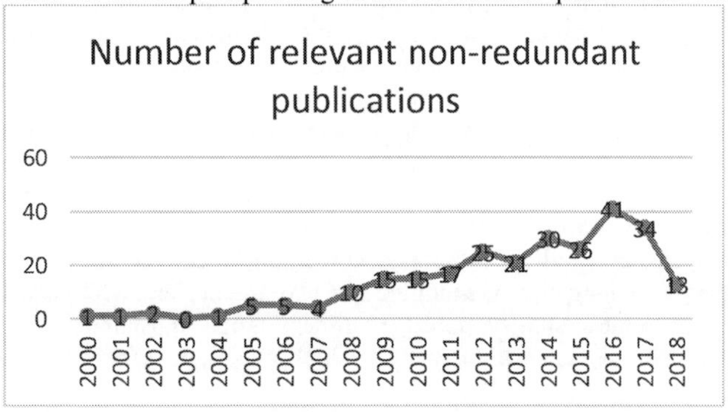

Figure 1. Evolution of Returned Documents Per Year by May 2018

In terms of further analysis, Figure 1 shows a trend in digital memory research has been increasing with some fluctuation since 2000. After 2008, the number of published research increased significantly and then kept increasing steadily and quicker than before. The peak is achieved in 2016 with 41 relevant publications. The numbers for 2018 are not significant as this study was completed before May 2018. However, this provide a good indication that interest in the area is still growing and indicate a good possibility of exceeding previous years.

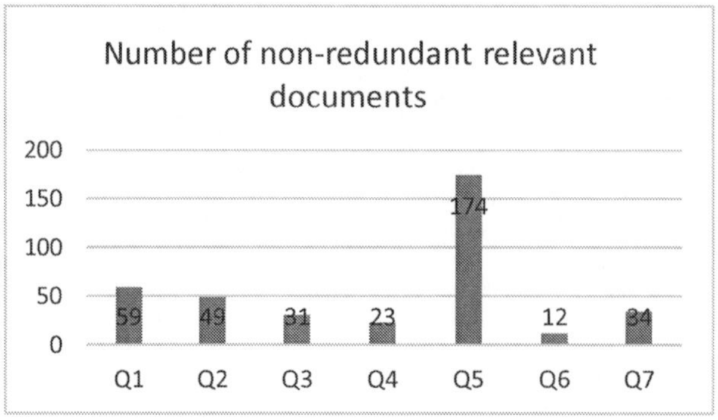

Figure 2. Distribution of Returned Documents Per Search Query

From an analysis of Figure 2, it is very clear that the queries (Q1, Q2 and Q5) associating digital memory with digital humanities and archives are the ones that returned a higher number of results. The association with digital humanities is a natural one and links digital memory with broader areas such history, heritage management as well as cultural and media studies. This was predictable form the onset of this study and proved extremely useful and fruitful for the research. On the contrary, the vast predominance of archives research dedicated to this area was both surprising and an important contribution of this study. Archivist and archives researchers have lately invested a significant effort in developing this area of research and the overwhelming amount of documents returned for Q5 (60.3%) is a clear evidence of that investment. Conversely, the association with information science was as expected rather disappointing 8.2%. This confirmed an expected gap of engagement of the information science community with digital memory research and vindicated the objective of this study.

3.2 Digital Memory Discussions in the Different Disciplines

Garde-Hansen et.al (2009) in their work of "save…as digital memory" mentioned that digital memory deals with the relationship between past and present through digital media technology and engaged in the deferral of death, the deferral of ending and the deferral of memory. Definitions of digital memory in the literature are usually based on the mediating technology, the data on which it is based and the form of representation of the memory. Researchers agree that the unifying difference in digital memory studies is that data is collected, stored and made available in digital form through digital media, instead of traditional forms of digital preservation such as cuneiform tablets, scroll and printed pages (Rumsey, 2016; Bay-Cheng, 2017; Rutten et al., 2013). A good example of these new forms is provided by Dib's (2013) research on designing digital recording devices aiming to maintain, extend, and commemorate life. The author mentioned that in his research digital memory is collected through digital recording devices including texts, photos, sounds, webpages, notes, pulse rates, temperature, locations and so on. This collected information shaped by time allows fragments of memory to be retraced. However, Bay-Cheng (2017) suggested that more than recording, digital memory also shapes mechanisms for documentation and communication. Rutten et al (2013) emphasized that internet is dramatically changing the way memories travel between different communities and generations due to its accessibility, speed of dissemination and anonymity. Memories are no longer merely consumed but also generated by audience, which makes the process of remembering more shared, negotiated and creative.

Any memory process is not mere recollection and includes a meaning-making process. Ciula (2018), a researcher in King's Digital Lab of King's College London, suggested that digital memory is strongly related to the concept of Humanities, which can be viewed as a process of meaning-making practices that connect past and present. Furthermore, Lurk et al. (2012) suggested that the modeling of any cultural product, such as a digital memory, is affected by how it is created and used. Consequently, Lurk et al. (2012) also claimed, that in culture heritage preservation studies, digital memory is more than just storing files, contents and assets, but also recording and linking contexts and structure of actions. They also propose that objects and their action-based context are necessary in terms of restoring "original knowledge".

Ernst (2013: pp 81), one of the most influential scholars who studied digital memory from an archaeology media perspective, has compared classical archives with digital memory and started his own way of understanding the integration of media, history and archives. The emergence of digital recording and digital storage has changed the concept of memory and archives in a very significant way. Ernst thought that through digital technology such as photography, what has been recorded digitally becomes technologically neutral rather than a subjective discourse. New archives when compared with classical archives are more flexible and more able to comply with current needs through the linking of networked technology and digital resources. New search methods also go beyond the traditional rigid index, which allows learning from user queries. Therefore, the original object-oriented archives can turn into more use-oriented dynamic archives (Ernst, 2013: pp 82). Kirschenbaum (2013) also referred to an earlier conceptualization of digital memory from Ernest (2002) to propose that in the digital realm, archives can only be the keepers of truthful and genuine documentation. Archives offer and enable the potential of reconstruction and reconstitution. This process of reconstruction and reconstitution can therefore be implemented by information science.

3.3 A Proposal of an Information Science Conceptual Model of Digital Memory

Information science is a discipline that focuses on studying processes related with the management and processing of information, including collecting, storing, analyzing, classifying, manipulating, retrieving and disseminating and securing information, in the context of society, organization and individuals (Stock & Stock, 2013; Saracevic, 2009). Information science pays particular attention to users' information behaviors related to their use information systems to resolve their information needs. In addition, information scientists also care about how to represent and organize information so that information retrieval can be optimized.

To some extent, memory itself can be viewed as a kind of information, which stored in our brain. In the digital age, as mentioned above, memories can also be stored digitally as binary codes, into forms such as photo, video, audio and text. With the development of information technology such as machine learning and artificial intelligence, this data can be filtered, clustered and processed automatically. Furthermore, in the context of social media and online communities, users have more flexibility to create their own shared and negotiated understandings, memories and narratives.

Figure 3 proposes a conceptual model for digital memory applications from an Information Science perspective. From this perspective, digital memory is not focused on the management and curation of digital resources but rather the generation of the memory narrative driven from those resources. The key issue is the generation of the application that allows access to the digital resources in a structured way that reconstructs and reconstitutes memory according to the perspectives, socio-technical environment, needs and even recollections of the users. It is in the generation of the content of these memory representations, narrative and reconstitutions that information can be more useful by using data-driven, hierarchical of predetermined path approaches. Hierarchical and predetermined path approaches can be further customized for different user groups according to age, expertise, interests as well as social and ethnic groups.

The content of the digital memory narrative can be retrieved, analyzed, manipulated and represented according to user's expectations, cultural and social anticipations, and personal perspectives. Furthermore, the way the system organizes the memory representation can depend on specific user group skills, capabilities and purposes. From a technological interface point of view, the digital memory system should provide an integrative way to portray and record memory. Additionally, these user interfaces must provide interactive functions to dynamically reflect the relationships between the different data resources (archives, community resources, personal resources, testimonies, and even crowdsourcing) as well as past and present lenses.

Based on the conceptual model presented in Figure 3, it is possible to synthesize a general information science model for digital memory applications development. This model presents and integrates the main agents and components in a digital memory application and is adapted from a generic information system socio-technical model proposed by Nunes (2000) and presented in Figure 4.

According to the synthetic Information Science model for digital memory applications development presented in Figure 4, the components of digital memory application include goals, digital memory application, digital memory resources (data), interaction with the user and the role of supporting information and communication technology. The goal of a digital memory application as mentioned by several researchers (Garde-Hansen et.al., 2009; Dib's, 2013) is to present the relationship between past and present interactively and engage with the deferral of death by guaranteeing the preservation of memory. The model suggests that there are at least three aspects that information scientist should engage with users' information behavior in regards to digital memory applications; content generation models to support digital memory representations and the use, design and development information technology in order satisfy users' demands.

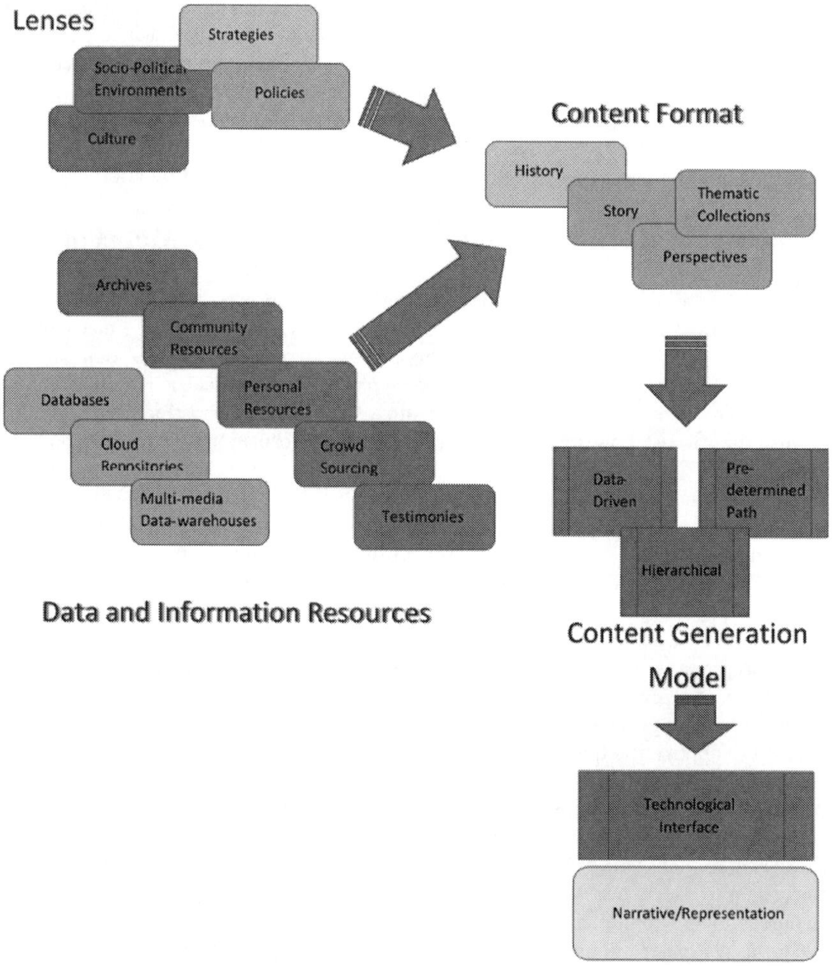

Figure 3. A Conceptual Model for a Digital Memory Application from an Information Science Perspective

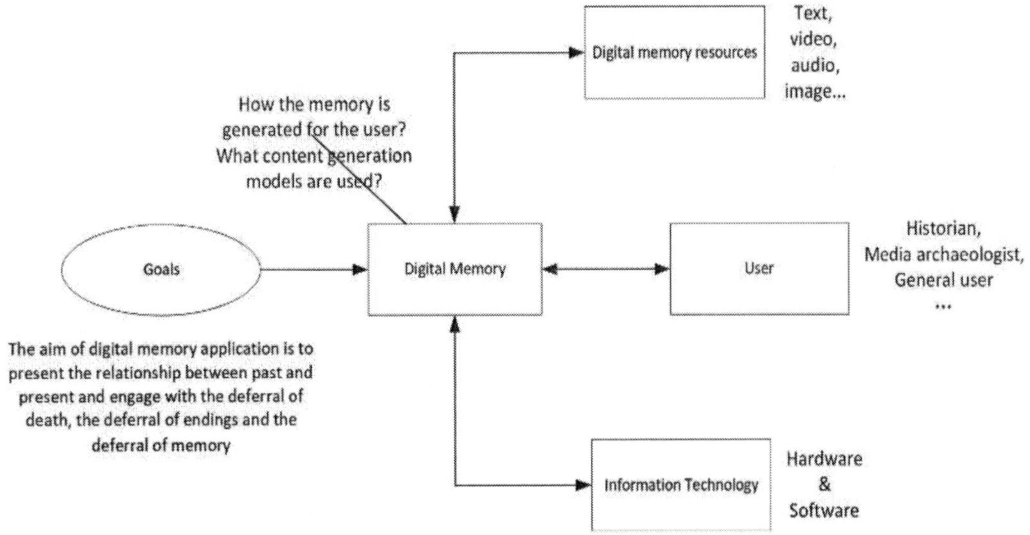

Figure 4. Synthetic Information Science Model for Digital Memory Applications Development

Information scientists can be extremely influential in a complementary and multidisciplinary effort with other researchers to achieve the dream of a digital memory that is dynamic, use-oriented, reconstructed accurately and unforgettable. The theoretical framework and model proposed aim at clearly pointing to the areas where this research effort by information scientist may be more effective.

4. CONCLUSION

The research on digital memory has become increasingly popular with the development of digital humanities and the seizing of opportunities afforded by the rapid development of networked information technologies. There are hundreds, if not thousands, of digital memory applications that have been created in recent years to commemorate, celebrate and recollect the past as well as record what is happening now. Most digital memory application development is multidisciplinary including experts from the humanities, history, archivists, database administrators, computer scientist and information scientists. Therefore, it is important to discuss the role of each of these expert researchers may have. This paper tried to identify the potential roles of information science in the development of digital memory applications in terms of retrieval and integrations of digital memory resources design of interactions with the user and design and development of information technology solutions to support the narrative and representation of the digital memory. To illustrate and explain this proposition that paper proposes a conceptual model for a digital memory application from an information science perspective. This model can help both academic researchers and digital humanities practitioners to better understand their roles and interactions and better plan this type multidisciplinary project.

REFERENCES

Bay-cheng, S. (2017) 'Pixelated Memories : Performance, Media and Digital Technology Pixelated Memories : Performance , Media , and Digital Technology', *Contemporary Theatre Review*. Routledge, 27(3), pp. 324–339.

Berry, D. M. (2011) 'The computatonal turn: Thinking about the digital humanities', *Culture Machine*, 12, pp. 1–22.

Bond, L., Craps, S. and Vermeulen, P. (2016) 'Memory on the Move', in Bond, L., Craps, S., and Vermeuien, P. (eds) Memory Unbound: *Tracing the Dynamics of Memory Studies*. Berghahn Books.

Brockmeier, J. (2015) Beyond the Archive: Memory, Narrative, and the Autobiographical Process. New York: Oxford University Press.

Broderick, M. and Gibson, M. (2005) 'Mourning, Monomyth and Memorabilia', in Heller, D. (ed.) THE SELLING OF 9/11. New York: PALGRAVE MACMILLAN, p. 207.

Ciula, A. (2017) 'Digital humanities and practical memory: modelling textuality', *Estudos em Comunicação n,* pp. 7–17.

Dib, L. (2013) Memory as Concept and Design in Digital Recording Devices. Rice University.

Ernst, W. (2002) 'Agencies of Cultural Feedback: The Infrastructure of Memory'. In Neville, B. and Villeneuve, J. (eds.), 'Waste-Site Stories: The Recycling of Memory'. Albany: Suny Press, 2002. pp. 107-120.

Ernst, W. (2013) *'Digital Memory and the Archive'*, In Parikka, J. (ed.) 'Digital Memory and the Archive'. Minneapolis: University of Minnesota Press.

Garde-Hansen, J., Hoskins, A. and Reading, A. (eds) (2009) Save as...Digital Memory. London: PALGRAVE MACMILLAN.

Kirschenbaum, M. (2013) 'The .txtual Condition : Digital Humanities, Born-Digital Archives, and the Future Literary', DHQ: Digital Humanities Quarterly, 7(1).

Lurk, T., Espenschied, D. and Enge, J. (2012) 'Emulation in the context of digital art and cultural heritage preservation', PIK - Praxis der Informationsverarbeitung und Kommunikation, 35(4), pp. 245–254.

Nunes, J.M., Kanwal, S., & Arif, M. (2017). Knowledge Management Practices in Higher Education Institutions: a Systematic Literature Review. *In Ifla Wlic 2017*, pp. 1–16.

Nunes, M. (2000) New Organisational Forms and CSCW (part 2) ,Lecture on 14/03/2000, Department of Information Studies, University of Sheffield, UK.

Nunes, M. et al. (2009) 'The use of e- learning in the workplace: a systematic literature review.', *Journal of Applied Research in Workplace E-learning*, 1(1), pp. 97–112.

Rumsey, A. S. (2016) *When We Are No More: How Digital Memory Is Shaping Our Future*. Bloomsbury Publishing USA.

Rutten, E., Fedor, J. and Zvereva, V. (2013) Memory, Conflict and New Media: Web Wars in Post-Socialist States. Routledge.

Saracevic, T. (2009) 'Information Science', Encyclopedia of Library and Information Sciences, pp. 2570–2585.

Saunders, M., Lewis, P. and Thornhill, A. (2009) for Business Students Fi Fth Edition.

Shao, P. (2014). 媒介作为人类记忆的研究——以媒介记忆理论为视角.

Stock, W. G. and Stock, M. (2013) Handbook of Information Science. Walter de Gruyter.

AN INVESTIGATION OF THE USE OF VR BOOKS TO RESOLVE DIFFICULTIES WITH ACCESS AND READING OF CHINESE ANCIENT BOOKS

Ning Zhang, Miguel Baptista Nunes[1] and Junyang Li
School of Information Management, Sun Yat-sen University
No.132, Waihuan Dong Road, Guangzhou,510006,P.R. China

ABSTRACT

The study reported in this paper aimed at investigating the viability and usefulness of Virtual Reality (VR) book applications in resolving some of the very significant barriers involved in accessing and reading Chinese Ancient Books. Ancient books in China, as in other cultures, are difficult to read due to the differences in writing system, in social and historical environment, in the embedded culture and in the capacity to access and retrieve information on these books. This research used a critical literature review and a social media consultation to identify and classify the barriers and to understand current solutions based on traditional books and e-books. Finally, the study compared these solutions with the potential offered by VR technologies. The study concludes with the proposition of a generic model for the development of VR books in the future.

KEYWORDS

Chinese Ancient Books (CAB), Virtual Reality, Immersive Reading, Experience education, Reading Promotion

1. INTRODUCTION

China is making large efforts to encourage citizens and international users to read Chinese Ancient Books (CAB). However, there are many barriers in reading these CAB that not only derive from their rarity and difficulty of access, but are mainly related to the nature of these books, namely: the writing in ancient Chinese; the use of different characters in different historical periods; the historical and social background that characterizes the narratives; the objects mentioned; the technologies described; the historical events; and human characters. This has led Chinese legislators, librarians, linguists and historians to make efforts in trying to resolve these barriers through the use of modern technologies that characterize modern digital humanities. Among efforts to implement e-books and Virtual Reality (VR) books, Chinese researchers have come up with theoretical concepts for immersive CAB reading using VR technologies to overcome the aforementioned access barriers (Liu 2014; Hu & Ma 2017; Zhang 2017). However, other than theoretical but simplistic propositions and very simple applications that allow the turning of pages and the illustration of ancient characters, there are no substantive theories or applications on how VR can overcome CAB barriers and help non-experts to understand and read these books.

Virtual Reality (VR) is now one of the most cutting-edge research and application directions of science and technology. It involves computer technology as the core of a modern simulation technology, collection of computer graphics, man-machine interface technology, multimedia technology, sensor technology and other technologies to simulate a 3D Virtual environment (Pujol & Economou 2009). VR achieves artificial immersive experiences, though the use of specific hardware equipment (such as vision goggles, VR headsets, audio technologies and touch sensors) and specialized software components for landscaping, object design, artefact design as well as interactions with the virtual environment, virtual objects, virtual characters and other users. Therefore there are three main features of virtual reality are immersion, interaction and imagination (Carrozzino & Bergamasco 2010).

[1] Corresponding author

VR is not a new technology, but the high specification demands of this technology made it not very efficient until recently. Advances in networked and audio-visual 3D technologies have allowed VR applications to become feasible in a rapidly growing range of multidisciplinary areas, including gaming, medical informatics, education and industrial simulations (Noh et al. 2009). Virtual Reading is one of the computer-based interactive applications of virtual reality that creates a visual representation of a book's contents, together with relevant artifacts, helping to make the book, along with its culture, accessible to global audiences.

This research project on which this paper reports, aims firstly to identify and classify all the barriers associated with the reading and understanding of CAB, then to identify VR-based solutions to these barriers. It is expected that the findings of the research will improve the future design of VR book applications, helping non-experts readers to access CAB.

2. LITERATURE REVIEW

2.1 Policies and Practice of Nationwide Reading Promotion

Promotion of nationwide reading in China has been a high priority since 2012 (see Figure 1), with information and communication technologies (ICT) being suggested to support and sustain reading habits. Specifically, a recent series of National polices identified the use of VR and other ICT-enabled solutions as contributing to new ways of reading (including reading Chinese ancient books). These new technology-based approaches are seen to be crucial for the development of heritage and cultural innovative products and artifacts for culture promotion (The State Council Information Office of the People's Republic of China, 2017).

Therefore, making CAB accessible is clearly part of the promotion of Chinese traditional culture and CABs are considered as a special collection in Chinese libraries. In line with these national policies and the requirements of CAB preservation and management institutions (e.g. libraries, archives, museums and cultural centers), leading researchers, database companies, publishers and government departments are investing in developing and cooperating in CAB digitalization. This is the socio-political environment of the research presented in this paper and seen as a National priority in China. This paper focuses on VR as a digitalization and open access solution that not only provides digital preservation but also aims at enhancing the reading and understanding of the content and message of CAB.

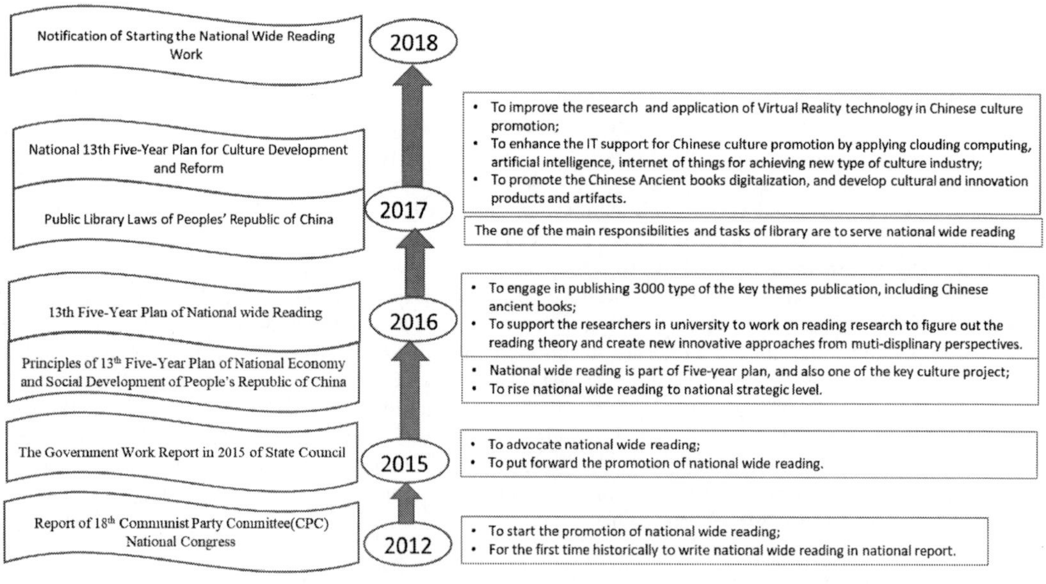

Figure 1. Chinese National Polices about Reading Promotion Released Since 2012

2.2 Access to CAB

Physical preservation and conservation of CAB is often in conflict with the reading and use of these books. Librarians and archivists restrict access to precious, rare and often deteriorating CAB to guarantee conservation and prevent damage. In doing so they prevent the public from using these important heritage and cultural artifacts. The recognition of this problem has resulted in world-wide efforts of digitization by public, non-for-profit and private organizations. In China, this has resulted in two types of approach: simple scanning and e-books (Li 2010;Lu & Cheng 2017). Simple scanning is just basic digitization of the CAB without any special enhancing features. E-books, by contrast, have added features such as annotation, direct searching, explanatory notes (Polding et al. 2008), which all help to make the digitize CAB more accessible. Representative databases, such as the Database of Chinese Fundamental Ancient Books (www.Er.07.com 2005), enable readers to access digitized Chinese ancient books online, solving access barriers to CAB to some extent. However, while it is relatively easy to make the contents of a CAB accessible, overcoming the barriers to understanding and reading them is a greater challenge.

2.3 Chinese Ancient Books and e-Books

As defined in the 2009 Catalogue of Rules for Chinese Ancient Books, published by the National Chinese Ancient Books Preservation Center, a CAB is defined as a document that was "produced before 1912 and exists in the form of original book drafts, hand-copied books, printed books, rubbings from stone and other sculpted documents and so on." (National Chinese Ancient Books Preservation Center 2009). CAB are usually kept in appropriate conditions within Libraries Ancient Book Section collections, Ancient Book Archives and Museums. Due to access restrictions, their fragility and rarity, CABs are often scanned or turned into e-books. Scanned copies add very little in features to the actual CAB. However, e-books may contain very many functionalities that may help in resolving comprehension and reading of CAB. Therefore, it is important at this stage of the research to understand and describe e-book as a commonly available and used technology to provide access to CAB and in ways that identify their strengths and shortcomings.

An e-book is defined as "a book composed in or converted to a digital format, including text, graphics, video, animation, and/or sound, for display on a computer screen or hand- held device" (Merriam-Webster.com 2014;Hsiao & D 2017). E-books are downloadable or web-based and sometimes require a special reading device (Hernon et al. 2007). According to Polding *et al* (2008), they offer the following benefits:

- Increased portability and easy storage;
- Platform and location independence as well as availability for downloads;
- Text searching;
- Multimedia facilities;
- A reduction in environmental impact (i.e. no paper, ink etc.);
- Less chance of damaging a copy than with a printed book;
- Several e-books on one device, preventing the need to carrying several volumes in print, even entire libraries can be carried on one device if storage allows;
- Customization of form and content, expanded and updated by a user;
- Annotation notes can be written in the margins;
- Interactive dictionaries to allow for easy reference;
- Backlighting can allow users to read in the dark.

The traditional technologies used for CAB use different and very specialized modes of writing (e.g., carving, brush handwriting, typography and engraving printing) on materials that changed over time such as turtle shell, bones, bronze bars, bamboo, silk, rice paper, or paper). Whilst e-book are digital representations that are portable and transferable to many digital formats, CAB have unique features associated with calligraphy styles, quality of craft, type of binding, nature of the layout and the use of language. E-books, by contrast, are based on integrative graphic user interface (GUI) designs and a variety of options available through menu systems, such as navigation, glossaries, dictionaries, content interpretation, content retrieval, referencing and annotation, occasionally supplemented by audiovisual materials and effects. Users' experiences of the two kind of book are very different, with CABs providing a more tactile and physical experience with the actual book. An experience associated with the base technologies and the materials that support it were used to create the original book. Because of their age and rarity access and manipulation of CAB is generally restricted, even in the way different parts of the book are read, i.e. even in the turning of a page or the opening of a silk roll may be subject to many limitations and regulations. No such problems apply to e-books, which provide more flexibility, transportability and mobility. Finally, e-books have the capacity to enhance readability and add interest by incorporating features that motivate readers due their capacity to illustrate, explain and even entertain and therefore adding to the interest and usefulness of the book contents (Bricken 1991).

2.4 VR Book Applications

Reading CAB with VR, using 3D modelling techniques, enables most, if not all the enhancing features of e-books. It also has the potential to add power by offering immersed experiences, illustrations, simulations and user interfaces. VR book instantiations of CAB can recreate precious ancient books as 3D representations that can replicate some of the tactile experiences of reading a physical CAB, while adding explanatory features and direct searching and illustrations. Amongst the pioneers of this kind of practice are the British Library and the Chinese National Library (Zhang 2017), which created virtual books in which pages can be turned and annotations made. The Chinese National Library developed the "Virtual Reading Station" enabled by 3D animation and a human gesture recognition system, which allows readers to turn the pages of CAB through body actions and gestures without touching the real books. The China Academic Digital Associative Library (CADAL) has developed a writing and appreciation of Chinese Calligraphy system with VR technologies to simulate the writing of ancient Chinese with virtual brushes on rice paper (Lu 2017). However, in reality the research on the application of VR in reading CAB is still limited and is in its initial theoretical stages. Current applications only reflect very simple simulations, such as the turning of pages.

VR book applications have the power to do much more. They can enable readers to engage in fully immersed reading experiences by integrating touch, sound, and vision. VR book applications have fascinating potential in terms of functionality, content presentation, technology support and user experience. This research aims to go beyond the fascination and promise and clearly identify, classify and establish, which aspects of VR books should be included in future applications of VR CABs.

3. RESEARCH AIMS, QUESTIONS AND DESIGN

As discussed above, the aim of the research reported here is to address the following research questions:

- **RQ1:** What are the differences between the features offered by CAB, E-books, and VR books?
- **RQ2:** What are the barriers for non-experts readers in reading CAB?
- **RQ3:** What can VR technology add that will help non-experts readers to read CAB?

The research design developed for this study uses a multi-method approach as illustrated in Figure 3. This multi-method approach which consists of: a critical literature review; a CAB state-of-the-art and application of technology survey; and consultation of experts through social media (Figure 3). The critical literature review was used to identify barriers to the reading of CAB, and differences between the formats of CAB, e-books and VR books. The literature search used three different and well-established academic resources: Web of Science, China National Knowledge Infrastructure (CNKI) and Google Scholar.

The CAB state-of-the-art and application of technology survey explored features and limitations of current digitalization of CAB. This study investigated and compared digital CABs, such as the two shown in Figure 2. These e-books are stored in three well-established Chinese CAB databases: the Database of Chinese Fundamental Ancient Books (www.Er.07.com 2005), Book Reservation of Han Tang (Hytung Books 2003), and Compilation of History Written on Stone in Chinese History (Unihan Digital Technology 2004). Researchers studied functionalities, types of digitization, retrieval and reading modalities.

The expert consultation targeted librarians and scholars in University Academic Libraries and Provincial Public Libraries with experience of CAB preservation, conservation, management and reading promotion. This consultation aimed to identify the existence of current solutions for the digitalization, digital preservation and promotion of CAB. Open interviews were conducted using WeChat and QQ, the two most common social media tools in China.

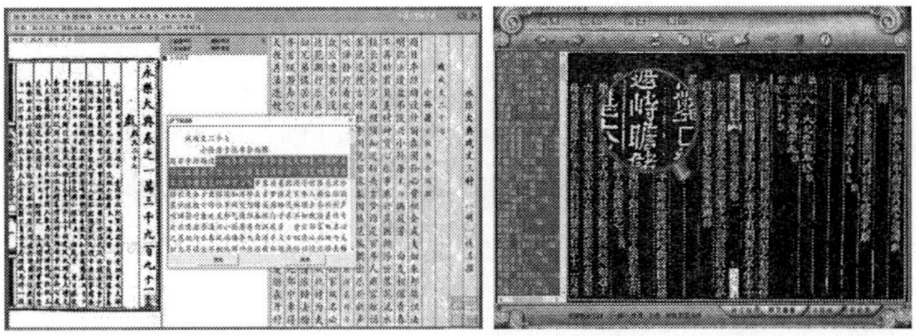

Figure 2. Two examples of CAB e-books: a 17th century Ming dynasty book and a rubbing of carving inscriptions

This project description was advertised on library special interest groups on QQ and WeChat calling the volunteers for interview. Additionally, targeted messages were sent to expert librarians in institutions with CAB collections. Because, this is a very specialized area of librarianship, numbers of participants were low: 11 professional librarians from 11 different libraries. In addition, researchers interviewed 4 professors and associate professors of library science, specializing in the study of ancient books. Data was collected through logs of the WeChat conversations, and subjected to a thematic analysis.

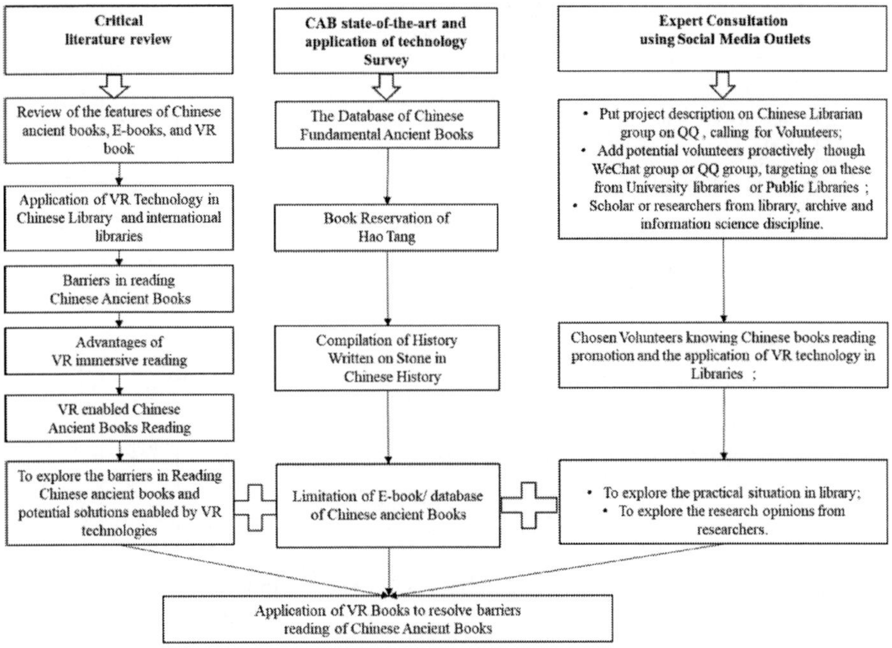

Figure 3. Research Design

4. FINDINGS

4.1 A Comparison of CAB, e-Book and VR Book Formats

As identified in the critical literature review, the comparison between CAB, e-book and VR book formats included the following basic criteria: basic format; book functions; content presentation; technology support and user experience (Bricken 1991; Hernon et al. 2007; Elias et al. 2012; Zhang 2011; Jie 2015). The comparison using these criteria is shown in Table 1.

From Table 1, it is clear that VR books have all the advantages and features of e-books, but provide additional useful and important ones. For instance, in terms of the original forms, VR books can simulate gestures of use and manipulation relevant to the different types of books, including different craft textures, original styles and even mountings and bindings. E-books cannot really support these types of features.

Table 1. Comparison among CAB, e-Book and VR Book

Criteria	**Sub-Criteria**	**Description**	**CAB**	**E-book**	**VR book**
Basic Format	Form/format	Turtle shell/bones/bronze bar bamboo/silk/rice paper/paper	Yes		
		Digital transferable format		Yes	Yes
		VR simulation of other physical formats			Yes
	Craft and style of book	Textures, printing, painting, calligraphy	Yes		Yes
	Binding and layout	Booklet, scrolls, book pages, thread book	Yes		Yes
	Language	Ancient Chinese	Yes	Yes	Yes
		Multilingual version (Chinese or other languages)		Yes	Yes
	Book mounting	Frames	Yes	Yes	Yes
	Interface design	Menu systems		Yes	Yes
		Gesture Recognition			Yes
Book Functions	Navigation/ exploration	For right book, right version, reference		Yes	Yes
	Annotation			Yes	Yes
	Glossary/dictionaries			Yes	Yes
	Content illustration			Yes	Yes
	Content retrieval			Yes	Yes
	Reference			Yes	Yes
	Context control	Control time, scale, and physics			Yes
	Interaction with objects	Move, talk, gesture;			Yes
Content Presentation	Text/picture/ table		Yes	Yes	Yes
	Audio/ video			Yes	Yes
	3D effects	3D objects			Yes
	Informative contents	Providing related knowledge in context			Yes
	Immersive experience				Yes
Technology Support	Techniques &printings	Carving/ brush handwriting/ typography/ engraving printing	Yes		
	ICTs	Internet/networked access		Yes	Yes
		Scanning and digitalization		Yes	Yes
		Specialized VR equipment			Yes
Users Experience	Concentration	Vulnerable to distractions from real world	Yes	Yes	
		Immersing in virtual world			Yes
	Learning motivation	From the interest / usefulness of contents	Yes	Yes	Yes
		Relying on fun experiences			Yes
	Customization	To readers' characteristics		Yes	Yes
	Natural interaction with content	Natural interaction with the book in its original form	Yes		Yes
	Flexibility/mobility	No space and time restrictions		Yes	Yes
	Safety	Prevents damage to the book		Yes	Yes
	Ease of access			Yes	Yes
	Enhances reading and understanding				Yes
	Enhances motivation to read	Enhances motivation through fun experiences			Yes
		Shared experiences with others			Yes

The interface design of VR book is also more functional than that of e-books. It allows better book and reading control, through features such as the ability to customize scales and modify textures, or jump to environments associated with the CAB and interact with objects and other readers. The content presentation features of VR books are the most promising advantages. VR enables immersive environments that contribute to better concentration and more customized reading. 3D visual effects contribute to a richer and more entertaining experience that can increase motivation to read, and aid understanding of complex content. However, such features require time-consuming development and expensive VR equipment. Immersion can also have negative aspects such as early fatigue and indispositions (such as nausea).

VR books can also provide additional immersive active experiences, educational games, simulations and interaction with explanatory context, audio-visual narratives and active interaction with other readers. These are not necessarily part of the traditional reading process but they can help users gain understanding though active experimentation and interaction. Readers' experiences can become learning experience and "experience is the best teacher" (Bacon & Tertium 1993). Finally, VR books can offer flexibility of access and mobility for readers without restrictions of time and space. VR books can be read anywhere and at any time, they do not need to be read in the library at specific times.

4.2 Identification of Reading Barriers of CAB

Although the e-book formats have removed some of traditional barriers of access to rare CAB, it is still difficult for non-expert readers to read and understand this type of book. In China, CAB promotion is not exclusively directed to expert readers (e.g. Chinese linguists, historians, anthropologists, archeologists, ancient book librarians, archivists, etc.) but also to non-expert enthusiasts, students, and other interested researchers. It is for this later group that the barriers to the reading of CAB are considerable and, at times, unsurmountable.

> "Only a small group of readers are actually reading Chinese Ancient Books. They are researchers of those who have professional requirements. Other people do not read them because they see no value in it for them." (Library and Information Science Professor 3).
>
> "The Library cannot help with the reading of these books. It is heavily reliant on expert skills that cannot be taught fast" (Library and Information Science Professor 4).

Figure 3 presents a synthesis of all the barriers identified from both the expert consultation and the literature review. Retrieval skills are associated with the capacity to finding the right CAB for particular special interests. Non-expert readers generally lack knowledge of ancient literature retrieval (Xu 1998). For example, non-expert readers are not familiar with the classification of ancient books, lack the knowledge of the characteristics of different editions (i.e. books which have been reproduced in different historical periods with different character sets, calligraphies, interpretations, versions of events and even biases in their narratives). Therefore, it is often difficult for non-experts to choose the best version or edition to read.

As shown in Figure 3, readers may have problems with the content of the book, the social and historical contexts, and the language skills necessary to understand deep meanings and narratives. These obstacles are particularly significant for non-expert readers as they may have the knowledge or the skills to overcome one or more of these barriers (Li & Xia 2010). In the past, surmounting these barriers meant seeking support from professional historians and linguists, and frequent discussions of interpretations and understandings (Qiu 1986). Additionally, researchers frequently need to refer to dictionaries, glossaries and other CAB (Xu 1998). All this professional help is rarely available and cannot be offered on an individual basis. VR books could incorporate considerable amounts of this knowledge and fulfil an important role not only in supporting readers, but also in knowledge transfer.

Additionally, non-experts readers also encounter linguistic barriers related to ancient grammar, lexicons, rhetoric and narrative styles, phonological representations, ancient characters and calligraphies, and attribution of meanings to words in Chinese (Liu 2016; Li & Xia 2010; Zhen 2013). These problems may sometimes have simple causes, but create significant barriers to reading. Non-experts readers are normally not familiar with ancient sentence construction (which was unpunctuated. This causes all kinds of misunderstandings of argumentation, meanings and context and is one of the major reasons for demotivation of readers (Xu 1998). Problems of motivation are in fact one of the major obstacles in CAB promotion, and the combination of all the factors above contributes to a feeling that CAB are too complex and boring and not worth the effort needed to read them.

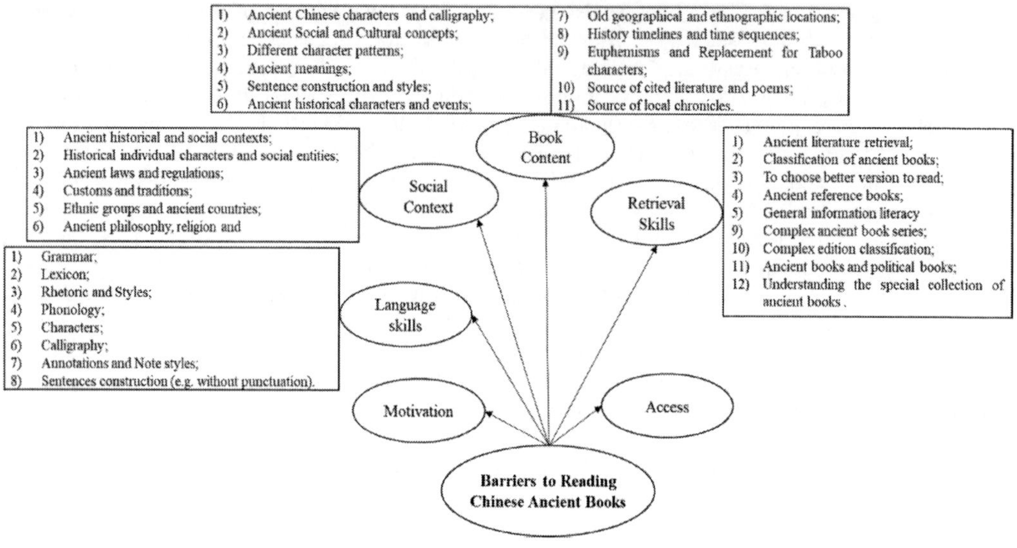

Figure 3. Barriers for non-experts in reading Chinese Ancient Books

4.3 How VR Technology can help Resolve CAB Barriers

The final part of this research investigated how the features of VR can help to reduce barriers to the reading of CAB. This section synthesizes the knowledge presented in Figure 3 and maps it against the positive features identified in Table 1.

Table 2. Potential VR Book Contributions to Solve the CAB Barriers

CAB barriers	VR book features	Potential VR technology contributions
Access restrictions	• Internet/networked access; • Flexibility/mobility; • No geographical, space and time restrictions • Digital transferable format; • Access to digitalized books with specialized VR equipment; • Safe access, prevent damage to the book.	• Breaks with geographical, space, time, and book preservation restrictions. • Provides downloadable, transferable, safe access to CAB. • Provides VR immersive reading.
Language skills	• Ancient Chinese and multilingual version with simplified Chinese or other language; • Glossary/ Dictionaries; • Reference.	Contributes to overcoming barriers relating to grammar, lexicon, rhetoric and styles, phonology presentation, characters, calligraphy, annotations and note styles, and sentence construction.
Social-cultural context of the book	• Informative content; • Knowledge in Context; • Content illustration; • Audio/video illustrations; • Reference.	Helps readers understand historical and social context, individual characters and social entities, laws and regulations, customs and traditions, ethnic groups and ancient countries, philosophy and religion knowledge.
Book content	• Immersive experience with audio, video and 3D effects; • Shared reading experiences; • Annotation; • Customization; • Enhanced Interaction with content ; • Natural interaction with emulated CAB different original forms (booklet, scrolls, book pages,	• Presents the book content in an immersive VR Environment; • Allows all good features of e-books; • Allows natural interaction with the book in a simulation of its original format; • Allows enhanced illustrations, and explanations of ancient Chinese language, social context, related ancient literature, ancient history, etc.

	thread book) and textures (turtle shell, bones, bronze bar, bamboo, silk, rice paper, paper); • Natural interaction with the emulated book features prints, paintings, calligraphy, binding and layout, book mounting. • Natural appreciation of book writing carving, brush handwriting, engraved printing, typographies.	
Retrieval skills	• Navigation & exploration aids; • Menu Systems; • Classification Schemes; • Book versions, edition catalogues and explanations; • Content retrieval; • Direct Search; • Thematic Search; • Reference.	Helps overcome the barriers to retrieval and classification of ancient literature, and difficulties arising from lack of book version knowledge, general information literacy and so on.
Motivation	• Fun reading experiences; • Enhanced attention on the reading; • Context control; • Interface design with menu systems and gesture recognition; • Personal customization; • Enhancing reading and understanding; • Natural interaction with other readers and objects in the VR environment (move, talk, gesture); • Shared experiences with others • Shared understandings and meanings with others.	Provides fun experiences in an immersive environment, such as engaging in explanatory and descriptive experiences of concepts in the book; offers emulations of social environments, events of historical facts; allows interaction with objects and controlling of the surrounding context through gestures. VR book can provide customized learning, and a simulated book with many ofl the characteristic of a real CAB.

The analysis presented in Table 1 shows that issues relating to digital preservation, access, mobility and portability have been resolved by the previous generation of e-books. Therefore, the impact of VR should be most significant in helping to resolve the other remaining five categories. Table 2 illustrates how specific features of VR may contribute to enhancing the experiences of reading CAB.

The promise of VR books is clearly apparent from Table 2. The potential features and functionalities of a VR book may have an important impact on the promotion of CAB, by removing barriers to access, and barriers arising from linguistic and contextual issues and by increasing the motivation to read CAB. Features such as VR's ability to immerse readers in an environment where they can interact naturally with simulated CAB, may change reading habits and reading interests. Internal links from the illustrations, simulations and audio-video materials may help not only to overcome reading barriers, but also to enhance motivation and interest in these books.

5. CONCLUSIONS

This paper results from an effort to review and integrate areas of research that would otherwise be isolated, and to try to bridge the gap between traditionally non-communicating disciplines such computer science, library studies and ancient book management. Furthermore, the paper contextualizes the study in the Chinese Ancient book community and its very specific needs, legislation and promotion strategies. The research reported in this paper identifies barriers to the reading of CAB by non-experts, analyses previous generations of digital books and compares these with the specific features of VR. The findings of the paper result from the mapping of CAB barriers onto VR features. The paper argues that VR CAB can provide readers with immersive and interactive reading environments that, through features such as customization, annotation and illustration, can offer significant linguistic, auditory, spatial, kinesthetic, interpersonal and intrapersonal dimensions to the reading experience. VR CABs should therefore be strongly considered in CAB promotion policies, strategies and programs. Future work in this area should provide clear design and implementation models to enable the implementation of the theoretical propositions made by this paper.

REFERENCES

Bacon, R. & Tertium, O., 1993. Applying Virtual Reality in Education: A Prototypical Virtual Physics Laboratory. In *IEEE Symposium on Virtual Reality*. pp. 67–74.

Bricken, M., 1991. Virtual reality learning environments: potentials and challenges. *ACM SIGGRAPH Computer Graphics*, 25(3), pp.178–184. Available at: http://portal.acm.org/citation.cfm?doid=126640.126657.

Carrozzino, M. & Bergamasco, M., 2010. Beyond virtual museums : Experiencing immersive virtual reality in real museums. *Journal of Cultural Heritage*, 11(4), pp.452–458.

Elias, E.C. et al., 2012. E-books in the classroom : A survey of students and faculty at a school of pharmacy. *CPTL*, 4(4), p.262–266.e1. Available at: http://dx.doi.org/10.1016/j.cptl.2012.05.009.

General Office of Central Committee of the Communist Party of China, 2017. National 13th Five-Year Plan for Culture Development and Reform(国家"十三五"时期文化发展改革规划纲要). *The State Council Information Office of the People's Republic of China(中华人民共和国国务院新闻办公室)*. Available at: www.scio.gov.cn/zxbd/wz/Document/1550922/1550922.htm.

Hernon, P. et al., 2007. E-book Use by Students: Undergraduates in Economics, Literature, and Nursing. , 33(1), pp.3–13.

Hsiao, K. & Chen, C., 2017. Telematics and Informatics Value-based adoption of e-book subscription services : The roles of environmental concerns and reading habits. *Telematics and Informatics*, 34(5), pp.434–448.

Hu, G. & Ma, L., 2017. 虚拟现实和增强现实在智慧图书馆的应用. 图书馆工作与研究, 9, pp.50–54.

Hytung Books, 2003. 翰堂典藏 *Beijing Times Hantang Technology Co., Ltd.* Available at: http://shdatatrans.com/pd.jsp?id=20 [Accessed May 20, 2018].

Jie, W., Y., 2015. 古籍图书在博物馆陈列展览中的作用. 安徽文学(下半月), 5, pp.155–160.

Li, 2010. 中华再造善本工程的意义及其验收利用对策. 图书馆学刊, 7, pp.38–41.

Li, P. & Xia, C., 2010. 中医古籍数字化多功能阅读环境模型构建 *Information on Traditional Chinese Medicine*, 27(1), pp.118–12.

Liu, G., 2014. 虚拟现实技术及其在高校数字图书馆服务方面的应用研究. 图书馆论坛, 13, pp.247–249.

Liu, X., 2016. 浅析古代汉语教学中的"深入浅出" *Course Education Research*, 36, p.122.

Lu, Y., 2017. 我国图书馆虚拟现实应用及研究述评. 图书与情报, 5, pp.120–127.

Lu, Y. & Cheng, L., 2017. 基于虚拟现实技术的图书馆信息资源建设与服务创新研究——以CADAL为例. 图书与情报, 4, pp.8–12.

Merriam-Webster.com, 2014. E-book. *Merriam-Webster.com*.

National Chinese Ancient Books Preservation Center, 2009. 中华古籍总目编目规则, Available at: https://wenku.baidu.com/view/63328bafdd3383c4bb4cd257.html.

Noh, Z., Sunar, M.S. & Pan, Z., 2009. A Review on Augmented Reality for Virtual Heritage. *Lecture Notes in Computer Science*, pp.50–61.

Polding, R., Baptista Nunes, J.M. & Kingston, B., 2008. Assessing e-book model sustainability. *Journal of Librarianship and Information Science*, 40(4), pp.255–268.

Pujol, L. & Economou, M., 2009. Worth a Thousand Words ? The Usefulness of Immersive Virtual Reality for Learning in Cultural Heritage Settings. *International Journal of Architectural Computing*, (August 2014), pp.157–176.

Qiu, X., 1986. 阅读古籍要重视考古资料. 文史知识, 8, pp.3–9. Available at: http://kns.cnki.net/kns/ViewPage/viewsave.aspx.

Unihan Digital Technology, 2004. Compilation of History Written on Stone in Chinese History(历代石刻史料汇编). Available at: http://library.sysu.edu.cn/red_theme/e-resource-detail.html?id=212 [Accessed May 18, 2018].

www.Er.07.com, 2005. Database of Chinese Fundamental Ancient Books(中国基本古籍库). Available at: http://library.sysu.edu.cn/red_theme/e-resource-detail.html?id=213 [Accessed May 18, 2018].

Xu, J., 1998. 高校新生如何阅读古籍文献. 津图学刊, 2, pp.132–135.

Zhang, H.Q., 2011. 浅谈古籍图书的装裱与修复. 晋中日报, 004, pp.1–2.

Zhang, Y., 2017. 图书馆虚拟现实技术应用研究演进的可视化分析. 国家图书馆学刊, 26(04), pp.64–74.

Zhen, J., 2013. 用联想法教学文言文词汇. 经验交流, p.2013.

CONTINGENCY AND ACTIVITY FRAMEWORK FOR BUSINESS INTELLIGENCE IN FINANCIAL BANKING ENVIRONMENT

Tshegofatso Lesole[1] and Ray M Kekwaletswe[2]
[1]*Graduate School of Business Leadership, University of South Africa, Midrand, South Africa*
[2]*School of Economic and Business Sciences, University of the Witwatersrand, Johannesburg, South Africa*

ABSTRACT

This paper is about contingencies and activities necessary for Business intelligence in the context of financial bank environment. Business intelligence is still a key issue and a critical phenomenon for both information systems practitioners and researchers. Nonetheless, business intelligence is still without challenges and issues as organizations struggle to achieve their projected business targets and desired goals. In this paper, the argument is that a profound analysis of activities and contingencies is key to improving business intelligence. Following an interpretive philosophy and a qualitative approach, this paper explores and describes the activities and contingencies enabling a business intelligence environment. Business intelligence is seen as a complex social activity system. To this point, Activity and Contingency theories are used as lenses to analyze the activity system. The paper conceptualizes a business intelligence activity framework for improving and enhancing performance, in a financial bank.

KEYWORDS

Activity Theory, Business Intelligence, Business Analytics, Contingency Theory, Master Data Management, Data Analytics

1. INTRODUCTION

For decades, Business intelligence has remained one of the key issues for information systems researchers and practitioners. This is because it is seemingly still seen as a key asset to many organizational efficiency in today's dynamic and competitive environment - especially with the high demand for real time delivery of quality operational information for reporting and analytics. Organizations, especially financial banks, work with high volumes of data (also referred to as "Big Data") regularly. Big data need proper management and care in order to produce best data quality for on-demand analytics and reporting. However, the way organizations have understood Business intelligence has proven to be more complex than meets the eye, with financial banks continually experiencing challenges. This paper uses Activity and Contingency theories to understand (1) what and how Business intelligence manifests in a financial bank, and (2) ways to transform business intelligence in financial banks.

2. THEORETICAL FOUNDATIONS

Organizations, especially banks, have a high demand to perform timely analysis and decision making with relevant, accurate, adequate, correct and complete data. However, they still face challenges because there is little understanding of the contingencies necessary for enhancing Business Intelligence. These contingencies include maintaining and creating quality data and data management for adequate analysis and decision making. This section discusses terminologies associated with BI – that is, Data Mining, Master data Management, Data Analytics and Business Analytics.

Business Intelligence (BI) is a discipline that has evolved as one of the critical applications in organizations to provide useful insight, support decision-making and drives organizational performance

(Ramakrishnan et al., 2011). Lonnqvist and Pirttimaki (2006) add on to explain BI to refer to systems and systemic processes, by which organizations obtain, analyse and distribute the information for making decisions and business operations.

On the other hand, Negash (2004) explains Business Intelligence (BI) to be the ability to combine data gathering, data storage and knowledge management with analytical tools to present complex internal and competitive information to decision makers. This is through possible due to the communities that bring clarity and reasoning to unreadable data to empower good decision making. IT It is also through data that has been rationalized, centralized and mastered for reporting on the business (Rudin and Cressy, 2003). Langseth and Vivatrat (2003) list some of the essential contingencies that fulfill BI to be: (1) Real- time data warehousing, (2) Data mining, (3) Automatic learning and refinement, and (4) Data visualization. Another contingency that could be added to this list is Big Data Analytics.

Burnay et al.,(2014) labelled various other entities as part of 'BI requirements' , that pertinent to be looked at to satisfy reporting needs, that is: (1) – What data is needed? – Fields, (2) How will data be obtained? – Sources, (3) How will data be organized? - Schemas, (4) What computations will be performed? – Indicators, (5) What outputs will be shown? – Analytics. These entities highlight that the ultimate delivery of business value is centred on the data that is acquired. Decision making is aided by data, its management and ability to analyze the latter therefore it is valuable for organizations to manage their data , irrespective of its size, to gain tremendous results when performing business analytics and reporting. Data is the lowest level of abstraction, information is the next level, and finally, knowledge is the highest level among all three (Mitra, 2011). He further explains that though quality data is important, alone it does not carry any meaning; it needs to be interpreted to be valuable information. In the early 2000s a new term that also hit IS research wave emerged - Big Data.

Davis et al., (2011) explains Big Data as enormous amounts of data, measured in peta bytes and higher, produced in high performance applications and/or systems. Big data can be characterized by the four V's, says Gani et al., (2014): (1)Velocity - is the data processed, stored and managed real time, near real time and/or batch; (2) Volume - huge amount of transactional and/or records of data measured over terabytes; (3) Variety - structured, semi structured and/or unstructured types of data; (4) Value - discovering the worth of this huge data sets. Organizations are challenged with managing and exploring these datasets that grow daily, which also is difficult to work with normal database management systems. This can be referred to as Master Data Management and Data Analytics.

2.1 Data Mining

Many organizations may not be able to to perform effective market analysis, compare customer feedback on similar products, discover strengths and weaknesses of their competitors retain valuable customers and make smart decisions, if they do not make data mining the core of BI (Pei et al., 2011). Diamantoulakis (2015) adds on to describe data mining as the standard process to harvest information from a stream of data and transform it to an understandable structure for further use.

2.2 Master Data Management

Haselden and Wolter (2006) say master data management is the technology, tools, and processes required to create and maintain consistent and accurate lists of master data. Some of the processes in MDM include source identification, data collection, data transformation, normalization, rule administration, error detection and correction, data consolidation, data storage, data distribution, and data governance. The tools include usage of data networks, file systems, a data warehouse, data marts, an operational data store, data mining, data analysis, data virtualization, data federation and data visualization. Loshin (2009) says Master Data Management is a collection of best data management practices. It should be a system of business processes and technology components that ensures information about business objects, such as materials, products, employees, customers, suppliers, and assets (Swanton et al., 2007). To have a more solid MDM that would assist improve BI, both the below MDM approaches should be practiced simultaneously. The Managerial Approach would look at a plan of action on MDM activities and processes with the analogy of the organization and what it needs. It will focus on the Business Objectives, Business Process Policies, Business Rules, Change Management and Resource (people) Roles and

Responsibilities in ensuring and controlling a set MDM strategy. The Technical Approach would look at the technologies, tools and technical rules and guidelines that will ensure a single truth of the data 'mastered' and an efficient MDM. The advantage of using the two approaches simultaneously is the commonality that they both promote data governance and data quality, which is a mandatory outcome for a good MDM strategy. Combining the two approaches will tackle the people, processes and technology aspect to MDM.

2.3 Data Analytics

Elragal & Elgendy (2014) say having piles of data on hand is no longer enough to make efficient decisions at the right time. The data needs to be analysed. With the multitude flow of data in and out of organizations, it became pertinent to find faster and efficient ways to analyse this data in order to unlock its importance and/or value. "Data analytics is a technology-enabled strategy for gaining richer, deeper, and more accurate insights into customers, partners and the business and ultimately gaining competitive advantage", says Kumar et al., (2013). It can also be defined as making time sensitive decisions and monitoring trends through the use of automated models. This can be made possible in the following three major processes, explains Power (2013):

1. Retrospective Data Analysis – this is using historic data and quantitative tools to understand patterns and results to make decisions about the future.
2. Predictive Data Analysis – this is using simulation models to generate scenarios based on historic data to understand the future.
3. Prescriptive Data Analysis – this is using planned, quantitative analyses of real time data to trigger events. These models recommend actions.

2.4 Business Analytics

It is often that the words "Business analytics" and "Data Analytics" are used interchangeability, but the question is, are the terms the same? Many researchers use Data Analytics and Business Analytics interchangeably in their studies, like Vahn (2014) on Business Analytics in the age of Big Data. Question is what is Business Analytics? Usually Business analytics (BA) is defined by Seddona et. al. (2017) the use of data to make sounder, more evidence-based business decisions. It includes gathering, processing, and analyzing data using a variety of methods and communicating the results of the analysis, matching them to key strategic goals (Cegielski and Jones-Farmer, 2016). Vahn (2014) says Business Analytics can be classified into the following:

1. Descriptive Analytics – this is taking data to describe what is happening, with the important aspect being the presentation of information. Required human intervention to interpret results.
2. Predictive Analytics – this is using past data to forecast the future in all aspects of a business. Required human intervention to interpret results.
3. Prescriptive Data Analysis – this is using past data and optimization (decision) model to reach actionable recommendations. Allows for automated decision making.

Although data analytics and business analytics may seem to have a common goal of using data to for informative decision making, business analytics goes a step further than data analytics to also learn competitor strategies, innovation and technology in order to move an organization in the competitive marketplace.

3. BACKGROUND TO THE RESEARCH PROBLEM

3.1 Context and Study Location

Two major financial institutions whose main component is to allocate funds from investors/savers to borrowers efficiently are studied. These banks provide personal, commercial and corporate banking services to more than 6 million customers across South Africa (SA). Both banks make every effort to remain on the cutting edge of banking by offering their customers with innovative and yet competitive services and products. Financial banking sector in South Africa have seen an exponential shift and growth of data through innovative services provided to customers in this environment, there is a high demand in the for real time delivery of quality operational information for reporting, analytics and to enhance customer service.

3.2 Challenges and Issues Experienced by These Banks

A preliminary observation has shown that organizations are faced with challenges of how to manage and analyze data - which is increasing exponentially. However the demand to use this data to deliver accurate information in real time for Business intelligence is as crucial. This poses a challenge to organizations to gain insight on how to make business agile, through several contingencies that may also include Data and Business Analytics. South African banks have the fastest growing use and demand for quality data; with challenges often hindering business intelligence (BI). The commonly experienced issues and/or contingencies are data inaccessibility, data duplication, poor data quality, data inaccuracy, data incompleteness and data irrelevance are being experienced, particularly in business decision making and reporting.

Some of the commonly experienced challenges experienced by these banks are: (1) inadequate data governance therefore causing redundant data storage, inconsistent data due to multiple data storages, that may also be duplicated, in the organization and incorrect data sourcing for analysis and reporting; (2) the business needs and the chosen MDM methodologies and technologies used are not in sync. Inappropriate methodologies and technologies to assist with data analytics are often used. The study was about how contingencies and BI activities happening in the banking locations could be addressed and alleviated to improve BI. The next section discusses the theoretical frameworks.

4. THEORETICAL FRAMEWORKS

This section discusses the theories used as lenses in studying the business intelligence phenomenon in a financial bank.

4.1 Activity Theory and Contingency Theory

For the study, the Activity and Contingency theories were seen as appropriate theoretical lenses. Activity theory is a conceptual framework based on the idea that activity is primary, that doing precedes thinking and that goals, images, cognitive models, intentions and abstract notions like 'definition' and 'determinant' grow out of people doing things' (Morf and Weber, 2000). The Activity theory has four (4) elements which help guide a research: tools in use, subject of study, objective of study and the outcome. Informed by Engestrom (2001) Expanded Activity Theory Model below, Lesole Bank Ltd used the expansion elements that are optional to in Activity Theory to deepen the understanding of the activities at the bank, these elements were: definition of rules, community and division of labour.

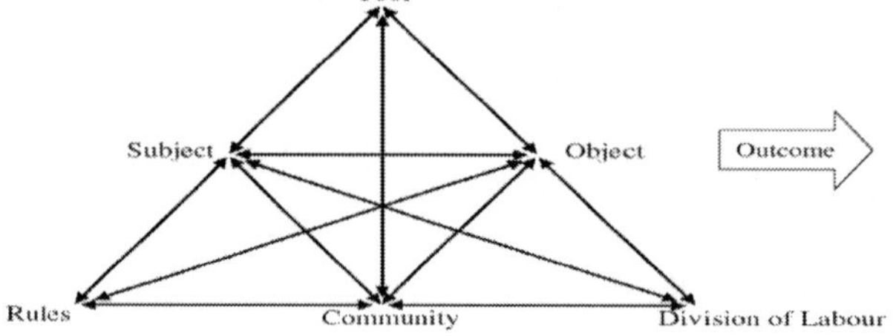

Figure 1. The Expanded Activity Theory Model (Engestrom, 2001)

Iterating Kaptelinin et al (2005) three levels of activity, for this study the conceptual framework included the elements below of the activity system:

a) Activity towards an objective (goal) – a conceptual framework of BI done by the BI communities. The framework would ease the BI community in knowing that they are reporting on the best sourced data. The motive of this being that it answers the "Why" question.
b) Action towards a specific goal (conscious), done by an individual or a group possible goals and sub-goals, critical goals - answers the "What" question.
c) Operation structure of activity normally automated and not conscious concrete way of implementing an action in according with the specific conditions surrounding the goal – answers the "How" question.

In addition to Activity theory, the study also used Contingency theory as a lens to explore and describe contingencies necessary to transform BI in a banking environment. Daft (2011) says contingency theory is a theory meaning one thing depends on other things. The word contingency refers to conditions that something will strive or fail. This theory will help understand what needs to be in place to improve business intelligence in a banking environment. Organizations can create strategies that may assist with improving their economic position in the market and profitability, Blanton et al., 1992 suggested the below contingency model (figure 2) could assist with this. The components of the contingency system were: structure, environment, strategies.

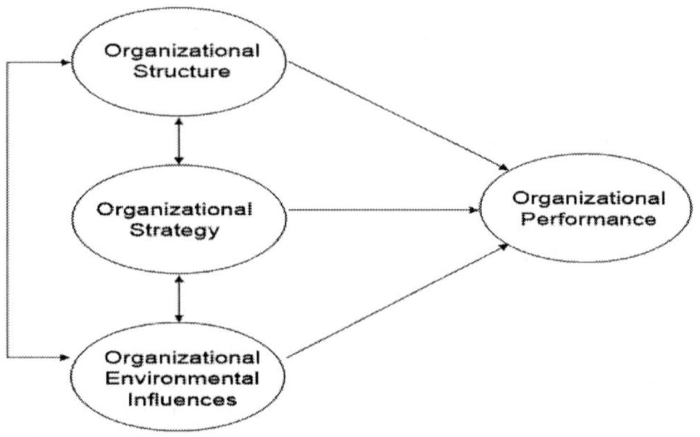

Figure 2. The Contingency Theory Model (Blanton et al., 1992)

Components of Contingency theory are briefly explained as follows:

a) Organizational Structure: this is the organizational hierarchical flow, that is, employee arrangement, how tasks are shared within and across different levels and/or business departments and each business' workflow.
b) Organizational Strategy: This is the basis with which a business exists and/or operates. Organizational strategies direct the way in which they work and ultimately perform – in other words, directs the goal of organizations.
c) Environmental Influences: these have been well documented by organizational literatures from Thompsons (1967), Lawrence & Lorsch (1967) and Pfeffer & Salancik (1978). Boyd et al., (1993) suggest that environmental characteristics have implications for most aspects of management of organizations. Furthermore, characteristics of the environments within which organizations function affect the level of managerial discretion (Hambrick & Finkelstein, 1987). The environment with which an organization operates influences they want level of success of a business, in particularly working conditions of employees.

5. RESEARCH METHODS

This section discusses the research methodology and the approach undertaken for the study. It discusses the research paradigm, strategy, design, as well as the data collection techniques of research. A qualitative approach was used for this study, which entailed a lot of researcher involvement in data collection; in order to have a deeper understanding of the meaning and significance of BI. Obrien (1998) explains interpretive paradigm as an emphasis on the relationship between socially-engendered concept formation and language. Interpretive paradigm is underpinned by observation and interpretation, thus to observe is to collect information about events, while to interpret is to make meaning of that information by drawing inferences or by judging the match between the information and some abstract pattern (Aikenhead, 1997). Interpretive paradigm was found appropriate to follow in order to understand the subjective meanings experienced by participants of the case study.

5.1 Research Strategy

This was a case study of two of the leading banks in South Africa. Yin (2003) defines the Case Study Research Method as an "empirical inquiry that investigates a contemporary phenomenon within its real-life context. The case study units possesses large volumes of data and data warehouse and has a complex BI environment; as a case study, helped draw views of a banking industry in South Africa and provided a more realistic approach to understanding issues and problems with respect to BI activities.

5.2 Research Design

5.2.1 Participants

Research participants were from the following case study units: (1) The Data Warehouse Team - which provide data to the bank's reporting team and should ideally hold the master data, (2) The Business Intelligence Team – these are the report generators/creators, (3) The Analytics Team in the different business units (core banking services) – this is a group of individuals whose mandate is to analyse how the bank is financially and operationally performing and give suggestive activities to improve business, (4) The Production Team - manages end of day data feeds into the warehouse or warehouses, (5) The Branch Consultants – they capture data on to the case study's systems at branch level.

5.2.2 Data Collection Techniques

In collecting empirical evidence for the study, the following were the techniques used:

Interviews: For this study, semi-structured face to face interviews were individually conducted with 14 participants whose average number of year that the respondents held at their current role and within the bank was seven (7) years that work and were affected by Data Management and Business Intelligence in the bank.

Observation: A naturalistic observation was done in terms of thick description of BI activities, e.g., how data is captured, stored, managed and used for reporting and analytics, including who is acting in which role, when and where, etc. There was engagement with limited interaction and intervention with research participants and events; interventions only occurred when clarity of activities & actions was needed.

6. CONTINGENCY AND ACTIVITY FRAMEWORK FOR BUSINESS INTELLIGENCE

Having discussed the theories underpinning the study and the methods used to study the phenomenon, the framework is discussed in this section. The framework shows a high level view of the elements that make up the contingency and activity system for business intelligence.

Figure 3 is the framework meant to inform the "know what" and the "know how" to business intelligence transformation.

The framework describes a dialectic process between subjectivity and objectivity, learning and doing, individual and collective, technical and social, and also tacit and explicit knowledge (Crawford & Hansan, 2006). Grifford and Enyedy (1999) say this theory is a framework that involves models of knowledge building, perspectives and artefacts that guide the design of collaborative activities and how people can participate. The framework also shows actions established through operations which are concerned with conditions, Kaptelinin et al., (2005). Actions involve tools, which differentiates them from being operations: while actions are connected to conscious goals, operations are related to routine behaviors performed automatically, without including the same level of `consciousness, (McMichael, 1999).

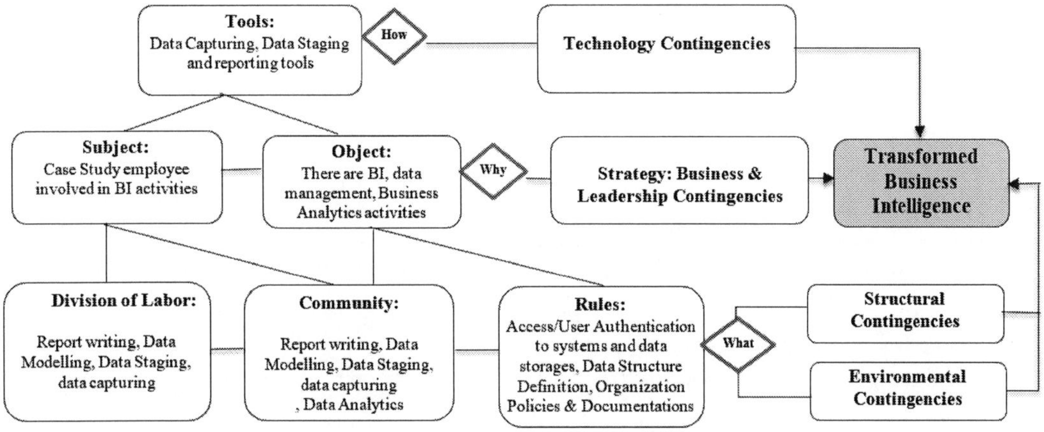

Figure 3. Contingency and Activity Framework for Transforming Business Intelligence

The framework, following Moosavi et al (2013), also shows how to best understand the "Know How" status quo of business intelligence in a banking environment:

i) Organizational Strategy - Knowing the organization's goals and targets may help in acquiring and storing correct data. The proposed conceptual framework will assist with exploring if key business fundamentals are aligned to the rules that are built to capture and store data for reporting. In understanding the results hoped for by the organization, the business objectives could be the guidelines that become the foundation of the BI activities and practices. The motive of this would be answers the "Why" question;

ii) Organizational Structure – This answers the "What" and "Why" questions. It focuses on the organizational hierarchy and the task they are responsible for. This helps identify information to be captured and/or required within the business intelligence activities.

iii) Technology - This answers the "How" question. Utilized technologies have a potential impact on the information or data types provided for business intelligence.

iv) Organizational Environment - Environmental factors, internal and external, may also play an important role in identifying any socio-cultural forces and institutions that may determine the success of business intelligence.

7. CONCLUSION

The paper showed how business intelligence is a complex social activity system. This activity system compromises of varied processes and actions that need to be understood profoundly in order to bridge the continuing challenges faced by the vibrant financial bank environments. The paper also concludes that using Activity and Contingency theories as lenses may be the best and appropriate way to profoundly analyze business intelligence as an activity system. These lenses offer insights into the what and the how of the phenomenon.

REFERENCES

Aikenhead, G.S., 1997, *Toward a First Nations cross-cultural science and technology curriculum*, Science Education, 81, pg 217-238.

Aikenhead, G.S., & Huntley, B., 1997, *Science and culture nexus: A research report. Regina*, Saskatchewan, Canada: Saskatchewan Education.

Alboali S., Esmail Hamid E., and Moosavi S. A., 2013, The Study of Contingency Components Roles in the Design of Municipals' Accounting Systems: A Case Study, *Journal of Business and Management Sciences*, pg 96-104. DOI: 10.12691/jbms-1-5-3

Blanton, J.E., Watson, H.J., Moody, J., 2007, Towards a better Understanding of Information Technology Organization: A Comparative Case Study, *MIS Quarterly*, 16(4), pg 531-555

Burnay, C., Jureta I. J., Linden I., and Faulkner S., 2014, A framework for the operationalization of monitoring in business intelligence requirements engineering, *Software and Systems modeling*, volume 15, 2nd Issue, pg 531–552.

Daft, R., Kendrick, M., and Vershinina, I., 2010, Innovative Management for Tubulent times, Management, 9th Ed, pg 8.

Diamantoulakis, P.D., Kapinas, V.M, and Karagiannidis, G.K., 2015, Big Data Analytics for Dynamic Energy Management in Smart Grids, Big Data Research, *Elsevier*

Engestr¨om, Y. (1990) Learning, Working and Imagining: Twelve Studies in Activity Theory, Helsinki: Orienta-Konsultit.

Engestr¨om, Y. (1996) 'Interobjectivity, Ideality, and Dialectics', *Mind, Culture, and Activity* 3: 259–65.

Engeström, Y., 2001, Expansive learning at work: towards an activity theoretical reconceptualization, *Journal of Education and Work*, 14(1), 133-156

Haselden, K., and Wolter, R., 2006, What is Data Management, The What, Why, and How of Master Data Management, *Microsoft Research.*

Hashem, I.A.T., Yaqoob, I., Anuar, N.B., Mokhtar, S., Gani, A., and Khan, S.U., 2014, The Rise of Big Data on Cloud Computing: Review and Open Research issues, *Elsevier: Information Systems*, 47 edition, pg 95-115.

Juran, J.M., 1964, Data Importance, *Managerial Breakthrough*, New York: McGraw-Hill

Langseth, J. and Vivatrat N., 2003, Why Proactive Business Intelligence is a Hallmark of the Real-Time Enterprise: Outward Bound, *Intelligent Enterprise*, Pg 34-41.

Loshin, D. 2009, What is master data management, *Master Data Management*, 1st Edition.

Lönnqvist, A., & Pirttimäki, V., 2006, Measurement of Business Intelligence, *Information Systems Man-agement*, Winter issue.

Mosley, M., 2007, Introduction & Project Status, *Data Management Body of Knowledge Guide Presentation*, Pg 6

Pei, J., Kamber, M., and Han, J., 2011, *Data Mining: Concepts and Techniques*, University of Illinois at Urbana- Champaign, 3rd Edition, pg 1-27.

Kaptelinin, V., Kuutti, K., Bannon, L., 2005, Activity Theory: Basic Concepts and Applications, *In Blumenthal et al. (Eds.) Human-Computer Interaction*, Lecture Notes in Computer Science, Springer.

Karel, R., Moore, C. and Coyne, S., 2011, Master Data Management: Customer Maturity Takes A Great Leap Forward Inquiry Spotlight On Master Data Management, *For Business Process Professionals.*

Kuutti, K., 1991, Activity Theory, Activity theory as a challenge to systems design, *H-E.*

Morf, M. E., & Weber, W. G.,2000, I/O Psychology and the Bridging Potential of A. N, *Leont'ev's Activity Theory*, Canadian Psychology , pg 81-93.

Mitra, A., 2011, Classifying data for successful modelling, Available on http:www.dwbiconcepts.com/data-warehousing/12-data-modelling/101-classifying-data-for-sucessful-modeling.html, accessed April 01, 2018.

Mosley, M., 2007, Introduction & Project Status, *Data Management Body of Knowledge Guide Presentation*, Pg 6

Moss, L.T., 2003, Nontechnical Infrastructure of BI Applications, *DM Review*, Pg 42-45.

Negash, S., 2004, Business Intelligence, Communications of the Association for Information Systems, Vol. 13, Article 15, Available at: http://aisel.aisnet.org/cais/vol13/iss1/15

Obrien, R., 1998, An Overview of the Methodological Approach of Action Research.

Ramakrishnan, T., Jones, M. C., and Sidorova, A., 2012, Factors influencing business intelligence (BI) data collection strategies: An investigation, *Decision Support Systems*, Elsevier.

Rudin, K. and Cressy D., 2003, Will the Real Analytic Application Please Stand Up?, *DM Review*, Pg 30-34.

Song, Y., Cuzzocrea, A., and Davis, K.C., 2011, Analytics over large-scale multidimensional data: The big data revolution, *DOLAP.*

Swanton, B. , Hagerty, J. and Cecere, L., 2007, MDM Strategies for Enterprise Applications, *AMR Research.*

Thomsen, E., 2003, BI's Promised Land, *Intelligent Enterprise*, Pg 21-25.

Willen, C., 2002, Airborne Opportunities, *Intelligent Enterprise.*

Yin, R. K., 2003, Case study research: Design and methods, *CA: Sage,* Newbury Park.

A LEARNING-BASED KEYWORD EXTRACTION TECHNIQUE FOR MULTILINGUAL SOCIAL MEDIA COMMENTS

Kishan Kumar Ganguly, Amit Seal Ami and Md. Mahbubul Alam Joarder
Institute of Information Technology, University of Dhaka,
Dhaka, Bangladesh

ABSTRACT

Due to the large user base in social media, social media posts and comments hold discussions on trending issues. Different Government and Non-Government organizations actively maintain social media accounts where many people engage and discuss their problems. This provides a great opportunity to systematically arrange and present these problems topic-wise which helps to direct these into appropriate recipients. Previous works provide techniques on topic extraction, sentiment classification, opinion mining from natural language text. However, these techniques are neither applied to social media nor do consider the multilingual aspect of social media. In this paper, a learning-based technique is proposed that extracts important keywords from multilingual social media interactions and provides a ranked list of these under predefined topics. The proposed technique is evaluated on Bangla, English and French comments from well-known and/or verified Facebook pages. It was observed that a high accuracy both in case of keyword extraction and ranking can be achieved by utilizing the proposed approach.

KEYWORDS

Facebook, Ranking, Decision Tree, Social Media

1. INTRODUCTION

The widespread use of social media has made it one of the most used sources of public opinion on social, political and economic issues. A social networking statistics by Statista shows that there are about 22 important social media sites that hold about 2.789 billion social media users (Statista, 2018). Moreover, since January, 2016, active social media users are increasing by 21% annually (Statista, 2018). These users spend about 135 minutes on average (Statista, 2018). Additionally, social media has become an active platform for discussion on trending topics around the world. Issues raised in social media spread very fast and may have impactful social consequences. An example of this include the famous 'Arab Spring' incident. The sensitive and significant statuses of Tunisian users' greatly influenced this event (Rocque, 2015). According to the statistics by Statista, 71 countries around the world use social media for providing government services. Recently, the government of People's Republic of Bangladesh has declared that each ministry and department must actively maintain a Facebook page. As a result, interactions on these pages are valuable source of public opinion that can help to drive government activities.

The public opinions and issues from the interactions are generally extracted manually. However, this is laborious and time-consuming. This is because not only the number of comments is large but also a user may engage about a topic at an irrelevant social media page. Therefore, an automated approach is required that extracts topics from comments and sends these to the intended recipient, ministries for example. There are namely three challenges that needs to be addressed in this case. Firstly, the comments from social media are often short, informal and lacks proper grammatical structure. Secondly, important but unpopular comments must be also extracted and sent to appropriate departments. Thirdly, the comments can be multilingual. For example, we extracted the comments from the ICT Division, Bangladesh, and found out that people comment in both Bengali and English. Two comments in two different languages may have similar meanings. These should be categorized under the same topic.

In this paper, a learning-based problem topic extraction technique from social media comments is proposed. We have used a learning-based approach to address the aforementioned problems. If appropriate features are used, learning-based approaches discover the rules under the training data to provide accurate results. The input of the technique is comments made by users in natural language format in social media and the output is a ranked list of problem topics under predefined problem keywords. At first, unigram (single word), bigram (pair of words) and trigrams (three words) are extracted from the comments as candidate keywords. These are manually labelled as actual keyword or non-keyword (1 or 0). Several Features, such as term frequency (TF), length, and closeness from the end of the sentence are calculated from the comments. The candidate keywords, their label, and extracted features constitute the training data. We use the Decision Tree machine learning technique to derive a training model from the training data. This model helps to extract keywords from a list of candidate keywords. These keywords are matched with a predefined list of problem topics to categorize these. Levenshtein distance is used for this matching and a ranked list is returned (Lavenshtein, 1966).

Among the social media sites, Facebook is the most popular one which holds about 1.87 billion active users where about 76% of social media users log in daily (Statista, 2018). . Therefore, we applied the proposed technique on Facebook to evaluate our approach. We have extracted the comments from verified public service based Facebook pages and applied the technique. The accuracy, precision and recall of the technique are 88.72%, 88.5% and 88.62%. The produced ranked list has been compared to a manually derived ranked list to calculate its Mean Reciprocal Rank (MRR), which is significantly high (0.71). Based on our evaluation, our results, indicate that the proposed technique is effective.

2. RELATED WORK

In the literature, works on topic extraction, keyword extraction, opinion mining etc. have been proposed. Tsur et al. presented an algorithm for clustering Twitter tweets. The clustering task is broken into two stages namely batch clustering of annotated data by the users, and online clustering of a tweet stream (Tsur, Littman, & Rappoport, 2013). In the first stage the algorithm bootstrap on the tags for clustering of a large pool of hash tagged tweets. The stable clusters achieved in the first stage are used for online clustering of a stream of tag less messages which are more common (Tsur, Littman, & Rappoport, 2013). Maynard et al. proposed a technique for extracting opinions from social media and rating interestingness of these opinions has been discussed (Maynard, Gossen, Funk, & Fisichella, 2014). After extracting named entities, terms and events opinions about these are extracted. These are then used for detecting opinion events. The total opinions over sections of text summarize sentiment about topics and documents, and also interestingness based on opinion diversity (Maynard, Gossen, Funk, & Fisichella, 2014).

Meng et al. introduced an opinion formation framework based on content analysis of social media and sociophysical system modeling (Meng, et al., 2012). Three parts of online opinion tracking and simulation are described which are automated topic, emotion and opinion detection in real-time. Akaichi et al., in their paper, discussed the use of text mining for sentiment classification from Facebook statuses (Akaichi, Dhouioui, & Pérez, 2013). It aimed to extract information, about sentiments and behaviors during of the sensitive and significant statuses on Facebook of Tunisian during the "Arabic Spring" era. The proposed technique in this paper is based on Support Vector Machine and Naïve Bayes (Akaichi, Dhouioui, & Pérez, 2013). It further builds a sentiment lexicon based on the emoticons, interjections and acronyms from status updates. Shrivastava et al. presented a technique to extract opinions of users or their reaction on concern to analyze (Shrivastava & Pant, 2012). This paper proposes a tool status puller that randomly extracts statuses from Facebook. A classifier performs classifications on the corpus prepared from these statuses. Three type of opinions which are GOOD, BAD and AVERAGE are extracted by the classifier from that statuses respectively (Shrivastava & Pant, 2012). Maynard et al. proposed another approach for extracting political leanings from a set of tweets prior to election (Maynard & Funk, 2011). It made use of different NLP techniques for extracting meaningful and high quality opinions from these tweets.

Most of the existing works either propose opinion extraction or sentiment extraction technique for social media. However, none of the techniques work on clustering the opinions into specific groups so that these can be easily represented. Specifically, extracting social problems from multilingual Facebook posts and clustering these problems into groups is a new concept that none of the papers introduce. Another problem the existing approaches do not discuss is the issue of ranking these social problems.

3. MULTILINGUAL TRENDING TOPIC KEYWORD EXTRACTION

In this paper, a technique is proposed to extract topic keywords from social media interactions that represents public opinions and issues in a language-independent way. At first, several features and candidate keywords from social media comments are extracted. Decision Tree is used to classify the candidate keywords into two classes which are accepted and rejected. A list of keywords is maintained under each predefined topic. Accepted keywords are matched with these keywords to determine their topic. Finally, top n-ranked keywords for every topic are presented as important issues. Social media features such as number of likes, and number of occurrences over time are used for ranking.

Table 1. Feature Set for Keyword Extraction

Name	**Definition**	**Data Type**
TF-IDF	The multiplication of term and inverse document frequency	Numeric
Relational Position	Distance of the first letter of the keyword from the beginning of the comment	Numeric
Length	Length of the key word	Numeric
IDF	Inverse Document Frequency	Numeric
Has Uppercase	Indication of whether the first letter of the keyword is uppercase	Boolean
Closeness from End	Distance of the first letter of the keyword from the end of the comment	Numeric
Closeness from Middle	Distance of the first letter of the keyword from the middle of the comment	Numeric
Contains Stopword	Indication of whether the keyword contains any stopword (e.g., of, an etc.)	Boolean
Word Length	Number of words in the keyword	Numeric

Input: Comment $C = t_1\ t_2\ t_3\ ...\ t_n$
Output: Candidate Keyword List, $K = t_1, t_2, t_3, t_1\ t_2,\ ...$
$K \leftarrow \varnothing$
$CL \leftarrow \varnothing$
1: $CL \leftarrow CL \cup split(C, "\ ")$
2: **for** $n = 1$ to 3 **do**
3: **for** $i = 0$ to $|C| - n + 1$ **do**
4: $K \leftarrow K \cup merge(C, i, i + n)$
5: **end for**
6: **end for**

Algorithm 1. Algorithm for Extracting Candidate Keywords

3.1 Keyword Extraction

The candidate keyword extraction process by selecting n-grams is shown in Algorithm 1. At first, the list is initialized as empty. Words are extracted from the comment with the split function (line 1). For each n-gram where n=1, 2, or 3, words of the comment are traversed and merged using the merge function (line 2-4). For example, for n=2, the ith word is merged with the $(i+1)^{th}$ one to form a bigram. Finally, all these n-grams are added to the candidate keyword list (line 4).

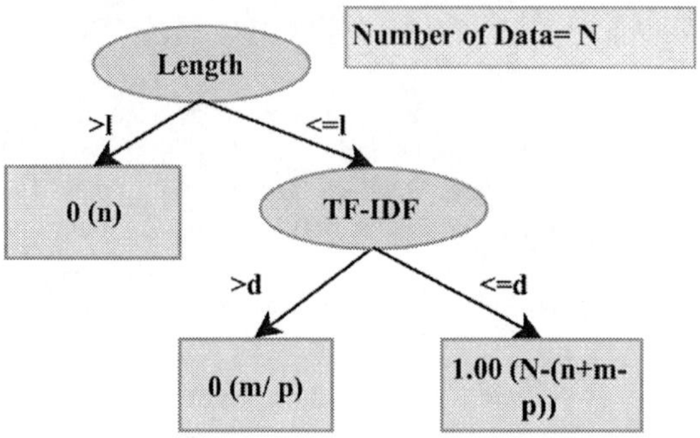

Figure 1. A Generic Decision Tree Model

Features are extracted for each candidate keyword. These are summarized in Table 1. From these features, Term Frequency-Inverse Document Frequency (TF-IDF) has been used widely in the literature to extract keywords (Gamon, Aue, Corston-Oliver, & Ringger, 2005). TF-IDF is calculated utilizing Equation i, ii and iii.

$$TF = \begin{cases} 1 + \log f_n, & f_n > 0 \\ 0, & x \geq 0 \end{cases} \tag{i}$$

$$IDF = \frac{N}{1+f_d} \tag{ii}$$

$$TF - IDF = TF \times IDF \tag{iii}$$

f_n indicates the frequency of a keyword in a comment. f_d means number of documents where the keyword appears and N is the total number of documents.

Relational position, length, uppercase are well-known features in the literature (Table 1). The last four features are used by this work additionally. The additional features, Closeness from End and Closeness from Middle are included to incorporate positional importance to detect a keyword. The 'Contains Stopword' feature, are included because keywords may consist of phrases with stopwords such as "earthquake of Bangladesh", and "Learning and Earning Project". Word length or number of words in a keyword is also considered as it is relevant to keyword detection according to Barker and Cornacchia (Barker & Cornacchia, 2000).

Training file for the keyword extraction is constructed beforehand. This file consists of sample comments, n-gram candidate keywords, features, and a Boolean value for each of the keyword that indicates whether this keyword is accepted or rejected. The training data can be represented as a tuple *(comment, candidate keyword, features, accepted/rejected)*. The proposed technique uses *Decision Tree* to learn a model from this training data. An example of a Decision Tree is shown in Figure 1.

From this figure, it is seen that decision trees have a root node that represents the feature. Depending on the values of the feature, the data is split into two classes in a way so that maximum class division is achieved. For example, as shown in Figure 1, taking length greater than l ensures that n number of instances all belong to 0 class (rejected keyword). Similarly, $TF - IDF \leq d$ and $TF - IDF \geq d$ ensure that the maximum number of instances go to class 0 and 1.

After creating the decision tree model, the candidate keywords and features are used to classify the keywords. Then, the accepted keywords are stored along with the corresponding comment.

3.2 Topic Extraction

In this step, the extracted keywords from each comment are matched with prespecified topics. These topics are clusters of keywords which are prespecified. To put the extracted keyword into one of these clusters, Levenshtein distance is used. The distance between the extracted keywords and the stored keywords are calculated and the closest one is considered. For example, the Levenshtein distance between the keyword 'Learning and Earn' and 'Learning and Earning' is 3. Similarly, the Levenshtein distance between 'Learning and Earn and 'Learning and Writing' is 6. Therefore, 'Learning and Earn' will be correctly categorized under 'Learning and Earning'.

4. EVALUATION

To evaluate this work, three languages are considered which are Bangla, English and French. Bangla and English comments are collected from the Bangladesh and U.S government ministry Facebook pages (Ministry, n.d.), (ICT Division, n.d.), (Français, n.d.). French comments are collected from the CGTN Français (State, n.d.). 2000 comments are collected in total which are all manually labeled for training. These raw comments are analyzed to collect candidate keywords. Features are extracted from these candidate keywords. The processed data is manually tagged with 1 or 0 to indicate whether the candidate keyword is keyword or not. After the training data preparation, training is conducted using the features as predictors. Topic tags and their keyword are manually defined and the final result which is the topic tag assignment is compared with a reference one. This reference topic tag assignment is constructed by manual inspection. To evaluate the trained model, a *k-fold cross validation* is performed. In this case, the whole data is randomly divided into k parts where $k-1$ parts are used for training and the other one is used for validation. The whole process is repeated k times. After this, metrics such as accuracy, precision, recall and F-measure (Friedman, Hastie, & Tibshirani, 2001) are calculated to evaluate the performance of the technique. The definitions of these metrics are given below. For all of these, higher values indicate better fit to the training model.

- Accuracy: Accuracy is calculated by percentage of correctly classified instances.

$$Accuracy = \frac{Correctly\ Classified\ Instances}{Total\ Instances} \quad \text{(iv)}$$

- Precision: Precision indicates the percentage of positive instances that were correctly classified.

$$Precision = \frac{Correctly\ Classified\ Positive\ Instances}{Total\ Positive\ Predicted\ Instances} \quad \text{(v)}$$

- Recall: Recall is the percentage of correctly classified instances from the positive instances. For example, if there are 100 correct keywords and 50 keywords are predicted correctly from these, the recall is 50%.

$$Recall = \frac{Correctly\ Classified\ Positive\ Instances}{Total\ Positive\ Instances\ in\ the\ Data\ Set} \quad \text{(vi)}$$

- F-measure: F-measure or F1-score denotes the balance between precision and recall. F-measure is calculated from the following equation.

$$\mathrm{F-measure} = 2 \times \frac{Precision * Recall}{Precision + Recall} \quad \text{(vii)}$$

To evaluate the ranking results, 200 comments with a mixture of Bangla, English and French are chosen. After keyword extraction, these keywords are manually ranked and clustered into the pre-specified problem topics manually. Thus, a ranking is defined and it is compared to the ranking provided by the proposed approach. Two metrics are used in this case which are Top N Rank and Mean Reciprocal Rank (MRR) (Voorhees, 1999).

- Top N Rank: This is the percentage of keywords that are correctly present in the top N rank according to the manually derived ranking list.
- MRR: A reciprocal rank is the multiplicative inverse of the rank of the first correct result of a query. For example, if a keyword is in rank position *n*, the reciprocal rank is $\frac{1}{n}$. MRR is the mean rank of all the reciprocal ranks. If n and ri is the number of queries and rank of a query i respectively. MRR is calculated using Equation viii.

$$MRR = \frac{1}{n}\sum_{i=1}^{n}\frac{1}{r_i} \quad \text{(viii)}$$

In the next sections, the results from this approach are presented in details.

5. RESULTS AND DISCUSSION

Figure 2 shows the boxplot of the features (except TF-IDF and IDF). Here, the boxplot with very short span is the features with Boolean values. Apart from this, the closeness from the end and closeness from the middle, features have stable values where the relational position feature has a lot of outliers. This is due to the variety of structure of the languages used.

Table 2 shows the number of instances, keywords, non-keywords and accuracy of the keyword extraction technique. In total 4763 training data was used where 2651 were keywords and 3040 were non keywords. It is mentionable that as unigram, bigram and trigrams were used, the number of non-keywords was much larger than the total number of keywords at first. This may lead to class imbalance problem where the classifier gets biased by the large frequency of a single class and attempts to classify any instance to that class only. So, some data from the higher frequency non keyword class are removed randomly to solve the class imbalance problem.

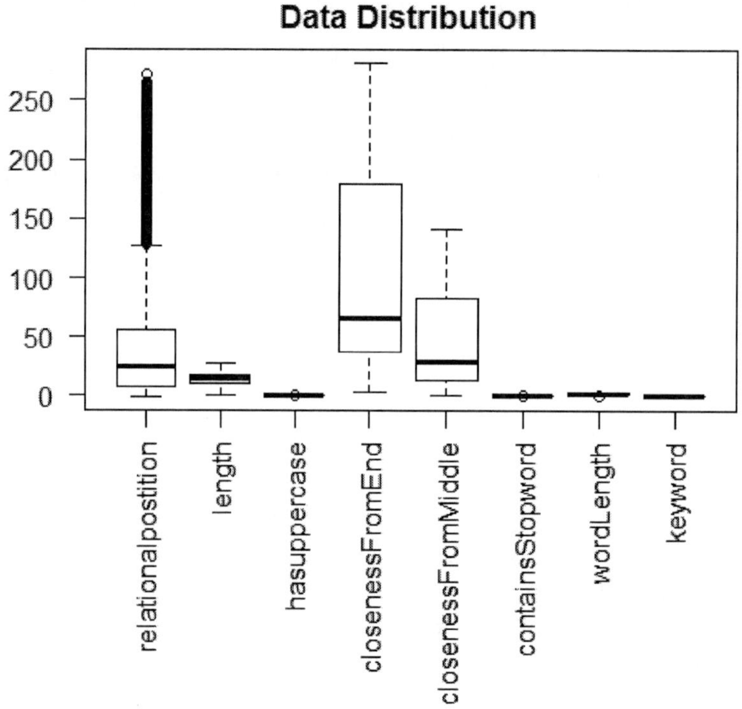

Figure 2. Boxplot of the Features and the Response

Table 2. Keyword Distribution in the Data

Total Number of Instances	Keywords	Non Keywords
4763	2651	3040

Table 3. K-Fold Cross Validation Result

Measurement Entity	Accuracy	Precision	Recall	F-Measure
Keyword Class	88.63%	89%	88%	0.9
Non Keyword Class	88.8%	88%	89.24%	0.895
Average	88.72%	88.5%	88.62%	0.8975

Table 3 shows the cross validation results. It is evident from the table that the accuracy, precision and recall measures are mostly over 88%, which is high. The F-measure is also over 0.89 which indicates better balance between precision and recall. The accuracy of the keyword class is higher than the non-keyword class to a small degree. This is due to the small percentage of error induced due to random removal of the non-keyword class values for solving the class imbalance problem. Overall, the accuracy of the keyword extraction algorithm is high in the presence of multilingual data. Table 4 shows some example of the output from the keyword extraction approach. For all the Bangla, English and French comments extracted from Facebook, it is evident that the extracted keywords by the technique match our intuition.

Table 4. Example Comments and Extracted Keywords by the Technique

Measurement Entity	Translation	Extracted Keywords
দেশের আই,সি,টি খাতে বিনিয়োগ বাড়ানো প্রয়োজন	The investment in the ICT sector of the country needs to be increased	আই,সি,টি খাত (ICT sector) বিনিয়োগ বাড়ানো (Increasing investment)
ইন্টারনেট প্যাকেজের দাম অতিরিক্ত, কমানো দরকার	The price of internet packages is too high, it needs to be reduced	ইন্টারনেট প্যাকেজ (Internet Package) দাম অতিরিক্ত (High price)
Why do we need to bring our own laptop during the Learning and Earning Project?	Why do we need to bring our own laptop during the Learning and Earning Project?	our own laptop Learning and Earning
Heureuse journée... Le cinéma Chinois est toujours célèbre!	Happy day ... Chinese cinema is always famous!	cinéma Chinois (Chinese cinema) toujours célèbre (Always famous)

Table 5. Evaluation of Ranking

Top 1	Top 5	MRR
75%	83.368%	0.81

Table 5 shows the evaluation of the ranking. It is seen that 75% of the results that are marked as top 1 is correct where it is 83.368% when top 5 is considered. The high percentage of this is due to the high accuracy of the keyword extraction algorithm and a well-defined problem topic set. It is mentionable that this ranking value highly depends on the predefined list of problem topics. So, these need to be defined in a very well-designed way. The MRR of the ranking is also high measuring 0.81. These metric values show the effectiveness of the problem topic extraction approach. So, overall the proposed approach accurately provides an informative ranked list clustered into some problem topics with even in the presence of multilingual comments.

6. CONCLUSION AND FUTURE WORK

In this paper, a multilingual problem topic clustering technique from social media has been proposed. The proposed approach extracts nine features, which are TF-IDF, length, relational position, IDF, contains stopword, has uppercase, word length, closeness from middle, and closeness from end. These are used to find keywords from the learned Decision Tree training model. These keywords are matched with a predefined list of keywords using Levenshtein distance and a ranked list is generated for each of the topics. These ranked lists present the most relevant keywords. The technique is evaluated using k-fold cross validation for training and manual ranking and inspection for the ranked list. The average accuracy, precision and recall of the training model are 88.72%, 88.5% and 88.62% respectively. For ranking, the top 1, top 10 and MRR scores are 75%, 83.368% and 0.71 respectively. These metric values indicate the effectiveness of the proposed technique.

In future, we aim to analyze the feasibility of considering grammatical features such as parts of speech of the comment to extract social media keywords. Additionally, we aim to explore sentiment analysis to find the positive and negative sentiment of a comment and denote the comment with negative sentiment a probable problem. The automation of the keyword building mechanism will be considered to improve the effectiveness of the proposed technique.

REFERENCES

Akaichi, J., Dhouioui, Z., & Pérez, M. J. (2013). Text mining facebook status updates for sentiment classification. *17th International Conference on System Theory, Control and Computing*, (pp. 640-645). Sinaia.

Barker, K., & Cornacchia, N. (2000). Using noun phrase heads to extract document keyphrases. *Conference of the Canadian Society for Computational Studies of Intelligence* (pp. 40-52). Berlin, Heidelberg: Springer.

Français, C. (n.d.). *Facebook*. Retrieved from CGTN Français: https://www.facebook.com/CGTNFrancais

Friedman, J., Hastie, T., & Tibshirani, R. (2001). *The Elements of Statistical Learning.* New York: Springer series in statistics.

Gamon, M., Aue, A., Corston-Oliver, S., & Ringger, E. (2005). Pulse: Mining customer opinions from free text. *International Symposium on Intelligent Data Analysis* (pp. 121-132). Berlin, Heidelberg: Springer.

ICT Division, B. (n.d.). *ICT Division, Bangladesh*. Retrieved from Facebook: https://www.facebook.com/ictdivisionbd

Lavenshtein, V. I. (1966). Binary codes capable of correcting deletions, insertions, and reversals. *Soviet Physics Doklady*, 707-710.

Maynard, D., & Funk, A. (2011). Automatic detection of political opinions in tweets. *Extended Semantic Web Conference*, (pp. 88-99). Berlin, Heidelberg.

Maynard, D., Gossen, G., Funk, A., & Fisichella, M. (2014). Should I care about your opinion? Detection of opinion interestingness and dynamics in social media. *Future Internet*, 457-481.

Meng, X., Wei, F., Liu, X., Zhou, M., Li, S., & Wang, H. (2012). Entity-centric topic-oriented opinion summarization in twitter. *18th ACM SIGKDD International Conference on Knowledge Discovery and Data Mining*, (pp. 379-387). Las Vegas.

Ministry, B. F. (n.d.). *Bangladesh Foreign Ministry*. Retrieved from Facebook: https://www.facebook.com/mofabdpage

Rocque, S. (2015, November 22). *The Role of Facebook in the Arab Spring*. Retrieved from GLOBAL CONNECTIONS: https://globalconnections.champlain.edu/2015/11/22/the-role-of-facebook-in-the-arab-spring/

Shrivastava, M. A., & Pant, B. (2012). Opinion extraction and classification of real time Facebook Status. *Global Journal of Computer Science and Technology*.

State, U. D. (n.d.). *U.S. Department of State*. Retrieved from Facebook: https://www.facebook.com/usdos

Statista. (2018). *Social media usage worldwide.* Statista.

Tsur, O., Littman, A., & Rappoport, A. (2013). Efficient Clustering of Short Messages into General Domains. *International AAAI Conference on Weblogs and Social Media*, (pp. 621-630). Ann Arbor.

Voorhees, E. M. (1999). The TREC-8 Question Answering Track Report. *Text Retrieval Conference*, (pp. 77-82).

DIGITAL NETWORKS OF PRACTICE: EFFECTIVE KNOWLEDGE CONSTRUCTION THROUGH NARRATIVES AND INTERACTIONS

Salam Abdallah, Anam Amin and Ashraf Khalil
Abu Dhabi University, United Arab Emirates

ABSTRACT

Successful knowledge management is considered an important factor in the success of contemporary organizations. Many organizations have adopted the use of some sort of knowledge management system, such as a network or repository to promote knowledge sharing and learning, however a great deal of them are unsuccessful. This paper investigates narratives shared on digital networks of practice and how they encourage interaction and contribute to learning. Our findings show promising signs that narratives are effective tools to facilitate learning in networks of practice.

KEYWORDS

Social Learning, Network of Practice, Narratives, Knowledge Construction, Social Web Interactions

1. INTRODUCTION

Knowledge is viewed as a fundamental resource in organisations and the importance of organizational knowledge sharing is increasingly recognized (Geiger and Schreyogg, 2012). The management of knowledge is considered a critical component in 'increasing their effectiveness, efficiency, and competitiveness' (Schultze & Leidner, 2002: 214). Tacit knowledge in particular is difficult to communicate or share and the capturing of tacit knowledge has been one of the main challenges in knowledge management (Alavi & Leidner, 2001). Narratives can convey rich tacit knowledge through their problem solving capabilities (Swap et al., 2001) and current literature shows how narratives provide 'embedded actionable problem-solving knowledge or practices' (Geiger and Schreyogg, 2012). This makes a strong case for narratives and stories as a method to impart and acquire tacit knowledge. Digital networks of practice support group interaction and creation, sharing and capturing of content (Krogh, 2012) and this social Web 2.0 technology is largely where narratives are shared.

Knowledge management involves the process of 'identifying and leveraging the collective knowledge in an organization' (von Krogh 1998 cited in Alavi & Leidner, 2001:113). Knowledge is embedded in a community and socially constructed and knowledge management systems should be developed to 'enable discussion, mutual engagement, and exchange' (Wasko and Faraj, 2000). Alavi and Leidner (2001:110) suggest that information which is 'actively processed in the mind of an individual through reflection, enlightenment or learning is useful' and by engaging in participation in discussion enables 'useful learning' and valuable tacit knowledge construction.

Tiwana and Bush (2005) concur mentioning how technical and engineering expertise is held tacitly in the minds of individuals therefore there needs to be a forum for exchange. Networks and communities of practice within virtual and online communities are just some of the computer- mediated knowledge systems that enable knowledge collaboration. They essentially facilitate contact between the person seeking knowledge and those who may have access to the knowledge. They can be either formal or informal depending on the social networking platforms in which communities congregate.

Digital networks and communities of practice and the research around these have gained traction within the last few years with the advances social media technologies. Communities of practice was a concept first introduced by Wenger and Snyder (2000) and involve tightly interlinked groups of people who know one another and seek to provide personalized tacit knowledge exchange. Networks of practice (NOP) on the other hand, as Beck et al. (2014: 1247) describe, are 'loosely interlinked and geographically distributed groups of people engaged in shared frames and practices who do not necessarily know one another or meet in person'. Networks of practice can reside both within and external to organizational boundaries. Networks of practice specifically enable useful, practice relevant content to evolve through a wide variety of collaborative processes (Krogh, 2012). Web 2.0 social technologies present themselves in the form of blogs, wikis, social networking sites, multimedia sharing tools and other related platforms (Panahi et al., 2013).

Digital networks of practice facilitate innovation and access to resources in a timely manner; as new information, expertise, novel ideas all become within easy reach (Wasko and Faraj, 2005). Furthermore participation in external networks means that individuals are not apprehensive about sharing, and afflicted with the constraints posed by organizational rules and boundaries.

1.1 Narratives as Enablers for Learning

Computer networks, electronic bulletin boards and discussion groups create a forum that facilitates contact between the person seeking knowledge and those who may have access to the knowledge. According to Weeks and Veltry (2013:25) participants in virtual communities engage with a discussion in two ways 'by sharing their experiences and knowledge through posts and by viewing and reflecting on the posts of others'; this interaction happens largely through the sharing of narratives. Wasko and Faraj (2005) concur with this view maintaining how in networks there are shared interpretations and meanings within the collective due to narratives and a shared language or vocabulary. Accordingly narratives have practical problem solving capabilities, and transfer of situation and problem-specific knowledge is mentioned in academic text (Geiger and Schreyogg, 2012; Swap et al., 2001). Some studies differentiate between stories and narratives and some use the terms interchangeably (Connell et al., 2004).

By engaging with stories and narratives that make sense of processes, events and relationships one can note the valuable tacit knowledge being conveyed (Connell et al., 2004). Stories give a detailed narrative of past management actions, employee interactions, or other intra- or extra-organizational events and these can impart rich tacit knowledge on norms, values and managerial systems and processes (Swap et al. 2001). The informality of narratives can help individuals make sense of what is going on around them, how to conduct themselves (regarding norms and values), how to source expertise and many other essential pieces of background information (Wende, 2014).

As computer networks create a forum that facilitates knowledge sharing and creation activities, and narratives reinforce the learning by allowing better assimilation of the information or knowledge, it consequently shows how valuable narratives can be on electronic networks and virtual communities. Studies also have previously shown how narratives can initially engage others in participation on online forums (Escobar et al., 2014; Wasko and Faraj, 2000). However there are very limited studies that address the process that takes place in regards to learning on networks or virtual communities, particularly studies that address how narratives can contribute to learning on DNOP's.

1.2 Research Objectives

Practice-based learning is one domain not well understood and discussed in current literature, particularly regarding individuals that come together through electronically mediated means to interact and gain new insight through learning, specifically on issues related to their practice. Therefore, we want to address whether we can analyze the learning taking place on DNOP'S and how narratives expressed through these platforms help in this learning process.

2. METHODOLOGY

The goal of our research is to examine how narratives, can facilitate learning between geographically dispersed and loosely interlinked individuals within the same practice. From previous research we can gather that participants in a virtual network are interacting to produce new knowledge or to arrive at new understandings of meaning; furthermore learning and knowledge construction appear to be inextricably linked. This firmly places us within a constructivist ontology and interpretivist epistemology as we also believe people's behaviour is influenced by the interpretations and meanings they give to social situations and they express these through the use of their narratives (Grbich, 2007: 8).

The network we will use in our study is stackexchange.com, a network of various forums where individuals largely from the software industry congregate; to share experiences, post queries, share information, collaborate to solve problems and find solutions relating to work practices and processes. It is on this platform where narratives and counter narratives are largely shared and will be examined in our study.

For this stage of the study, we examined those threads that had a long life span but were closed for discussion. This indicated that there was an answer or solution that was accepted by the original poster of the narrative, demonstrating a discussion and learning process had taken place. Furthermore, we selected three threads with a high number of votes and answers, these figures are indicated next to each thread also the title of the thread is also presented. The high votes indicated that users felt the threads and the narrative provided good research value. The content of the threads was dense and rich enough for us to analyse the process of learning taking place and the high amount of answers showed there was significant discussion, which was of interest to our study.

We used qualitative data analysis of threads to examine emerging patterns, themes, and phases related to the social construction of knowledge. The analysis process involved traditional coding that searched for themes and key ideas generated in the threads. In addition, in order to see if the narrative helped in the learning process we looked for themes in the discussion text, which linked to features in the narrative.

2.1 Case Selection

The Stack Exchange network and particularly Stack Overflow has grown into a top destination, for professional and novice programmers. We chose to focus on the 'workplace stack exchange' (selecting the software- industry tag) and 'software engineering stack exchange', which are pages or parts of the stack exchange network. From initial observation these forums or pages specifically focused on problems relating to 'know- how' of work practices, therefore tacit knowledge construction could be observed. They also provided rich narratives of workplace relevant content within the software industry context so we know the narratives were all about the same practice.

We present three thread cases in this section which will provide us with the result at the end of our research.

3. ANALYSIS

Thread case one: The narrative problem presented by the narrative poster talks about being a team lead and the difficulties he experiences in managing an intern within the organization. It also talks about the basic skill competencies that he would expect from employees within the organization, such as math skills, and finds the intern he is managing to be deficient in the skills he feels to be a necessary component of the job. (http://workplace.stackexchange.com/questions/82346/how-to-deal-with-an-interns-lack-of-basic-skills).

> *'His math knowledge was really restricted. It made me absolutely frustrated. I lost my patience and dropped several harsh remarks.' (User- Tsar)*

He asks the forum for a solution in how to manage this particular intern and he appears to be seeking affirmation for his frustrations as well. For example, a forum participant in the discussion validated the frustrations and so the narrative poster responded in accord.

'That's why I got so frustrated with him. I'll never expect someone who is doing college, and an IT course, nevertheless being unable to do simple fractions.' (User- Tsar)

Another interesting point that arises, when analyzing the respondents in the discussion, is the vast number of participants agreeing with the narrative poster about the math skills requirements in the software field. They all seem to be asserting how crucial this knowledge is, expressing how there is expectancy for developers to independently 'get up to speed' and therefore appear to be normalizing themselves to the software engineers' discourse.

'I had no need for fractions, trigonometry etc., so I had to relearn them alongside programming' (User- Pharap)...'comparison of your knowledge to your peers - these are the people which you will compete against later - if something significant is missing on your side, it's bad' (User- Sascha)

In this sense not only are they constructing tacit knowledge in regards to responding to the narrative posters problem but they are also exchanging and sharing tacit knowledge regarding norms of technical knowledge and how to acquire them that are expected in the industry. Furthermore in regards to 'dealing with the intern' a consensus is reached by the participants that aim to summarize their views and a proposal or solution is put forward by a community participant. There is a discussion after the proposal is synthesised to re- negotiate certain aspects. This proposal embraces all of the tacit knowledge dimensions that are discussed such as 'maintaining professionalism'.

'hearing that your company yelled at an intern is not good for your company (even if it makes this one leave)' (User- enderland)

Also more technical solutions such as 'checking the hiring process of interns' and 'finding work more suitable for the intern' were also presented. By analysing the text in the thread it is very evident that a learning process is taking place. However the tacit knowledge construction process is much more iterative with the negotiation and re- negotiation of meaning terms. In this case there appears to be a statement by the original poster which shows that he applied the tacit knowledge he acquired and put it into practice.

'I'm already taking a few of your suggested steps (your #4 is mostly how I roll), and I think you're correct about apologizing to him. This was the first time I was "gifted" with an intern with this skillset, so I'm still adapting to mentoring him' (User- Tsar)

The above statement shows not just his agreement with the proposed solution but also his application of the suggested solution. Also it is evident here that the compelling nature of the narrative and the shared meaning allowed the initial involvement in learning of the discussion to take place which was indicated by a 'shocked response' and sarcasm.

'Is this story even true?' (User-Joker)... 'Unacceptable. You talk about HIGH SCHOOL? Seriously? That is stuff people learn in basic school, not high school.' (User – TomTom)...Or there was an aspect of relatability which enabled some to share a counter- narrative of their own experience....'I'm not surprised at this at all to be honest. I did a HND in college and everyone in my class left with at least a pass in programming.........Until I started doing programming I had no need for fractions, trigonometry etc., so I had to relearn them alongside programming.' (User – Pharap)

Thread case Two: This narrative is more succinct than the first narrative depicted above and deals with 'how supportive an organisation is with the employee relocating' and 'unpaid overtime' as an unfair consequence of relocating. He questions whether this is an 'industry norm' and asks for a solution to this situation regarding the 'unpaid overtime' being an 'unfair' consequence in his view. Again one can note the 'seeking affirmation' component as within the narrative the question is posed 'whether or not you can consider that acceptable'. (http://workplace.stackexchange.com/questions/67563/required-to-work-unpaid-overtime-to-make-up-for-a-shorter-commute-after-moving)

Most participants respond with shock, amusement or sarcasm as they become engaged in the discussion showing the compelling aspect of the narrative, with agreement and questioning from participants. However the 'compelling' aspect of the narrative continues to pervade the discussion text of participants as sarcasm and amusement is expressed both before and after a proposal of the solution is presented.

> *'If this was posted from a new account I'd have assumed it was a troll question... I guess that says enough about how completely and utterly insane that management "decision" is. I still don't believe that your friend has seen this at 4 different companies.' (User-Lilienthal)... 'Treat it as a ridiculous joke in the first instance, which it probably is, rather than get offended and thinking about lawyers over an extra 10 hours a week. When in all probability the manager was just pulling your leg.' (User- Kilisi)*

In terms of how to approach conduct regarding this issue a summary proposal put forward by a participant of the discussion is to do the 'normal hours and go home' or to treat the unfair proposition from management with humour to show displeasure. In this sense the proposal depicts subtle forms of resistance which are considered to be a norm regarding disapproval with a managements approach. This knowledge exchange on norms regarding resistance shows the diverse nature of tacit knowledge that can be constructed and exchanged and the anonymity of the forum participants enables open and free discussion to take place on issues that perhaps would never occur within the organisational boundaries.

In terms of the learning taking place in this scenario it is only indicated by the original narrative poster accepting the highest rated summary response with a tick, which shows the tacit knowledge constructed was applicable to the narrative posters situation. The original poster did not necessarily participate actively in discussion but by reading and accepting the response that he felt is suitable in solving the problem, shows he was still active in the learning process.

Thread case three: This narrative is about managing a relationship with a 'hostile' or 'difficult' boss when asking for help. Again this narrative is somewhat terse and describes certain confrontation scenarios that took place in her employment. She appeals for advice on how to conduct herself when managing this relationship. The use of words like 'bullying' and the gender issue mentioned in the narrative invokes very compelled arguments in the discussion of the posed problem narrative.
(http://workplace.stackexchange.com/questions/59095/senior-architect-lashing-out-when-junior-developer-asks-questions-what-to-do/59105)

> *'When a man is rude to a woman of course she will think it has to do with gender especially if it's an irrational reason to be rude' (User-1261710)*

Furthermore there is felt meaning and connectedness all depicted in the initial stage. Below is a counter narrative of a participant showing comradery with the narrative poster's situation.

> *'I know how you feel, I'm in a similar position. I've been reprimanded for asking too many questions (I actually posted a question on this site about it). The people I ask aren't being openly rude to me (usually), but they do just expect me to magically know everything I need to know even though I (like you) have little experience.' (User- Gardenhead)*

The proposed summary posits specific questions or conversational techniques that could be used for mastering the developer/ boss relationship and how to conduct one 's self in this situation. For example, not reporting the situation to a senior manager and instead asking if the necessary information is in the documentation, and if questions should be addressed for a specific time. These tacit knowledge elements are largely concurred with participants, however the weighting of some elements of the proposition are re- negotiated again in the discussion following the summary.

The original narrative poster initially engages with the discussion by clarifying the situation and meaning terms but like the above mentioned thread she ticks the response she feels best fits her scenario which in turn is created with her engagement in the initial discussion with other participants. This shows learning has taken place although there is no 'Metacognitive statement' from the original poster.

4. DISCUSSIONS

After analyzing the narratives and discussion text of threads, we can note there is a process of learning taking place. It is far more iterative a process then, for example, Gunawardena et al. (1997) and their linear model of learning would depict. Also we can note that within discussion text certain interactions facilitate the learning process far more than others; perhaps the frequency of these useful learning interactions could enable us to know which threads engage in a more productive learning process and what factors within interactions allow for better overall learning and knowledge construction through discussion. An example of these useful interactions would be questioning, where participants question each other's ideas, as well as clarify and question the original narrative. The table below (table 1) gives examples of some of these valuable interactions, in thread one, and the categories with various codes assigned to describe the type of interaction taking place.

Table 1. Examples of Interactions

Code	Text
disagreement	"Since I'm the technical and team lead, it's up to me to bring the new guy up to speed with how we work and what we need to do make our software run smoothly." It's up to you make sure he is brought up to speed, not up to you to do it. Maybe someone in your team would be a good coach/mentor. (User-StuperUser)
corroborating example with solution	I'm not surprised at this at all to be honest. I did a HND in college and everyone in my class left with at least a pass in programming...(User – Pharap)
questioning each other	The question is unanswerable without the following basic context: why is your company obligated to keep employing them? and are you obligated to keep them on your team?... (User– smci)
negotiation of meaning	@smci Interns on my country are a delicate matter. You can't simply juggle them around and do whatever you please with them.(User– Tsar) But are you legally obligated to keep the same mentor for those 24 mths?(User– smci) @smci I'm the team and tech lead. While I do have some skilled coworkers, they are juniors while I'm the only senior. (User-Tsar)
statement from original poster	This answer is very good, thank you. I'm already taking a few of your suggested steps (your #4 is mostly how I roll), and I think you're correct about apologizing to him. This was the first time I was "gifted" with a intern with this skillset, so I'm still adapting to mentoring him. (User– Tsar)

4.1 Themes in Threads Reflecting on Narrative

Finally, in order to see if the narrative helped in the learning process we looked for themes in the discussion text which linked to features in the narrative. We found each post that related to the narrative did so initially in the discussion. Further on in the thread, discussion occurred between participants without reference to the narrative or the original narrative poster. Certain posts signified a certain emotion or felt meaning and below is an example of posts which were assigned descriptive codes expressing the connection to the narrative (see table 2).

Table 2. Examples of Codes Expressing Connection to the Narratives Qualities

Code	Text
sarcasm	Just to add to the many good responses here - you may want to find out where this person graduated high school, and make sure HR doesn't send you any others from there, assuming that's an option. The high school has clearly failed. (User– Don Branson)
humour	* 0.95? How about trouble multiplying two single digit whole numbers? Had that with one of my former colleagues. Good question. Basic Maths knowledge is becoming a problem these days. (User – Neolisk)
connecting with example	It could be that his interests simply lie somewhere else, I used to work as a programmer and hated it, I can't do math (it's like dyslexia with numbers) so I know I will never be a good programmer. Once I figured that out I started doing front end design instead, way less logic to program and I really excelled in this field. Maybe talk to him and see what he really likes to do. (User– dennismuys)
shock	Unacceptable. You talk about HIGH SCHOOL? Seriously? That is stuff people learn in basic school, not high school. And that is not "math" as in "demanding math" - as much as cooking spaghetti is not making you a cook. This is nonfunctional in a simple basic knowledge for everyday life. (User – TomTom)
compelling	@Sascha Not the function of an internship? It's the function of professional development and a part of every job I've ever had. Job has requirements A, B, C, D. You get a review and ratings of some sort. You are expected to fix issues. Everyone has strengths and weaknesses. Why wouldn't a formal review process - with goals, resources and a formal process of correcting your issues - be a part of internship as well? Aren't you training them to be professionals? Isn't part of the process to help them help themselves? (User – WernerCD)

These codes assigned to each post can be presented in three categories or types which are depicted below. These statements and their frequency could show how engaged participants were in the learning process but also what aesthetic features of the narrative compelled individuals to participate in the discussion.

Table 3

Factors	Examples
Hedonic statements	Enjoyment, shock, humour- 'that's interesting', 'I'm surprised'
Empathetic statements	Sharing feelings, emotion, sarcasm – 'I Feel…', 'it seems to me..'
Self- Reflective statements	Connects through own experiences- 'this happened to me', 'when I was…'

Based on our findings, we propose that narratives can be used as an appropriate instrument for learning and construction of tacit knowledge, particularly where individuals share the same practice. Narratives shared through computer mediated interaction enables communication with others from the same practice without relying solely on the proximity of their immediate social circles (Vaast and Walsham, 2009). With the help of our study of a particular network of practice, we can potentially provide deeper insights into the processes of information sharing, communication and socialization which contributes to problem solving and learning. So far, the findings of the study show that it is a practical and beneficial solution to use narratives for learning as they enable collaborative problem solving by enticing and compelling users to involve themselves in discussion through their aesthetic qualities. Furthermore, certain interactions can be seen to facilitate better learning, and certain narrative aesthetics can enable these interactions to occur.

In regards to advancing this research in the future; we can see that there is a learning process that occurs in these forums and this iterative learning process should be further explored. Furthermore, we can see how these narrative aesthetics and interaction features can be honed in on during the learning process. This will provide for a more comprehensive study which will enable us to look at the full scope of learning that occurs in these networks of practice and how narratives could prove to be beneficial. Organizations could therefore better design their own electronic network features to enable better interaction as a great many knowledge management initiatives fail (Alavi & Leidner, 2001).

REFERENCES

Alavi, M. and Leidner, D.E., 2001. Knowledge management and knowledge management systems: Conceptual foundations and research issues. *MIS quarterly*, pp.107-136.

Beck, R., Pahlke, I. and Seebach, C., 2014. Knowledge exchange and symbolic action in social media-enabled electronic networks of practice: A multilevel perspective on knowledge seekers and contributors. MIS quarterly, 38(4), pp.1245-1269.

Case, P. and Piñeiro, E., 2006. Aesthetics, performativity and resistance in the narratives of a computer programming community. Human Relations, 59(6), pp.753-782.

Connell, N.A.D., Klein, J.H. and Meyer, E., 2004. Narrative approaches to the transfer of organisational knowledge. Knowledge Management Research & Practice, 2(3), pp.184-193.

Escobar, M.L., Kommers, P.A. and Beldad, A., 2014. Using narratives as tools for channeling participation in online communities. Computers in human behavior, 37, pp.64-72.

Fadel, K.J., Meservy, T.O. and Jensen, M.L., 2015. Exploring knowledge filtering processes in electronic networks of practice. Journal of Management Information Systems, 31(4), pp.158-181.

Geiger, D. and Schreyögg, G., 2012. Narratives in knowledge sharing: challenging validity. Journal of Knowledge Management, 16(1), pp.97-113.

Grbich, C., 2012. Qualitative data analysis: An introduction. Sage.

Gunawardena, C.N., Lowe, C.A. and Anderson, T., 1997. Analysis of a global online debate and the development of an interaction analysis model for examining social construction of knowledge in computer conferencing. Journal of educational computing research, 17(4), pp.397-431.

Panahi, S., Watson, J. and Partridge, H., 2013. Towards tacit knowledge sharing over social web tools. Journal of Knowledge Management, 17(3), pp.379-397.

Schultze, U. and Leidner, D.E., 2002. Studying knowledge management in information systems research: discourses and theoretical assumptions. MIS quarterly, pp.213-242.

Swap, W., Leonard, D., Shields, M. and Abrams, L., 2001. Using mentoring and storytelling to transfer knowledge in the workplace. Journal of management information systems, 18(1), pp.95-114.

Taylor, S.S., Fisher, D. and Dufresne, R.L., 2002. The aesthetics of management storytelling: A key to organizational learning. Management Learning, 33(3), pp.313-330.

Tiwana, A. and Bush, A.A., 2005. Continuance in expertise-sharing networks: A social perspective. IEEE Transactions on Engineering Management, 52(1), pp.85-101.

Vaast, E. and Walsham, G., 2009. Trans-situated learning: supporting a network of practice with an information infrastructure. Information Systems Research, 20(4), pp.547-564.

Wasko, M.M. and Faraj, S., 2000. "It is what one does": why people participate and help others in electronic communities of practice. The Journal of Strategic Information Systems, 9(2-3), pp.155-173.

Wasko, M.M. and Faraj, S., 2005. Why should I share? Examining social capital and knowledge contribution in electronic networks of practice. MIS quarterly, pp.35-57.

Wasko, M.M., Faraj, S. and Teigland, R., 2004. Collective action and knowledge contribution in electronic networks of practice. Journal of the Association for Information Systems, 5(11), p.2.

Weeks, M.R. and Veltri, N.F., 2013. Virtual communities as narrative networks: developing a model of knowledge creation for crowdsourced environments. International Journal of Knowledge Management (IJKM), 9(1), pp.21-41.

Wende, E., King, G. and Schwabe, G., 2014. Exploring Storytelling as a Knowledge Transfer Technique in Offshore Outsourcing.

Wenger, E.C. and Snyder, W.M., 2000. Communities of practice: The organizational frontier. Harvard business review, 78(1), pp.139-146.

Websites

http://workplace.stackexchange.com/

http://softwareengineering.stackexchange.com/

http://stackexchange.com/about

EXPLORING USER'S SHARING PATTERNS ON MULTIPLE SOCIAL MEDIA PLATFORMS

Ashraf Khalil[1], Aisha Khan[1], Tasbeeh Fatima Raza[1] and Salam Abdallah[2]
[1]*Department of Computer Science and Information Technology, Abu Dhabi University, Abu Dhabi, United Arab Emirates*
[2]*College of Business, Abu Dhabi University, Abu Dhabi, United Arab Emirates*

ABSTRACT

Emergence of new features on social media platforms have become a daily occurrence. Stories, live videos, memories and various other emerging features are a crucial part of most of the popular social media platforms. Through this study, we conduct a qualitative assessment of user adaptation to these revolutionizing social media features. The aim of this study is to understand user preferences based on platforms, their adaptability and usage of these features. This study derives conclusions based on a semi-structured interview conducted on 18 users in their early twenties in the UAE. The study finds that users treat social media platforms distinctly while tending to share different contents on each platform depending on the nature of the content and the audience of each platform.

KEYWORDS

Social Media, Content Management, Privacy, Cross Sharing, Millennials

1. INTRODUCTION

Social media (SM) platforms are now an essential part of our daily communication and interaction with acquaintances. In fact, SM has evolved such that it is no more just a domain to connect with friends; users commonly interact with other social groups as well, including but not limited to their family relatives, co-workers, and old friends (Stutzman & Hartzog, 2012). With the revolutionizing use of social media platforms, a variety of platforms are competing, sometimes by creating their own new features and sometimes by replicating other's successful features like "stories". Users choose between various platforms depending on the features that platform offers (Katrin, 2016). Features can vary from status updates to stories and pictures to posts, whatever attracts the user better.

Sharing news on social media has become a trend among today's generations. The process of sharing news over social media platforms reflects on the social leadership attributes and enables users to perceive themselves as opinionated leaders, especially when they acquire a lot of friends and followers (Kümpel, et al., 2015). Expressing political views over social media platforms has become very common in the present-day scenario. Political parties have created their social media handles to address the public directly and to enable an inside view of the opinions of their followers. People take to their social media accounts to rant or praise their country's political stance (Stieglitz & Dang-Xuan, 2013). However, sometimes users regret sharing certain content, especially over Facebook as the content might depict sensitivity or reach an unintended audience (Wang, et al., 2011)

Facebook was once the primary SM platform for users. Ever since the rise of other platforms such as Snapchat and Instagram, users divide their usage time on more than one platform. This brings into question Facebook's status as the 'primary' social networking site among users. With new platforms come new features. Snapchat's popularity is due in part of the time-limited 'stories' that allows users interact more privately and intimately allowing a greater sense of binding and connection (Piwek & Joinson, 2016). On the other hand, Instagram is designed to allow users to share filtered photos and videos, making it a much more 'visual' mode of communication than Facebook. Instagram brings the 'coolness' and 'creativity' factor that is a driving source of motivation for users (Sheldon & Bryant, 2016).

With new emerging platforms, people have an opportunity to be constantly connected more than ever, yet users may exhibit significant cautiousness when selecting the people with whom they connect and when deciding what to post where. Privacy plays a crucial role in the usage of different social media platforms. A variety of distinct privacy options have been offered by various platforms like blocking, hiding, unfriending etc. This allows the user to restrict information to a certain group of individuals as per their personal choice. Studies suggest that gender plays a significant role in how users opt for privacy in their online social presence. It appears that 67% of the female users prefer keeping their profiles restricted to only friends, compared to 48% of the male users. In 2012, around 63% of users have been actively pruning their profiles, by removing unwanted friends, tags, posts and pictures (Madden, 2012). Given that the study was conducted in 2012, there is a high possibility that the percentage has increased relatively over the years. Users may have different privacy principles based on their ethnic origin and individualism (Wang, et al., 2011). Research suggests that anticipating privacy concerns related to trust based on socio-ecological factors can help social media platforms gain a better understanding of user needs in terms of online privacy (Thomson, et al., 2015). Cultural factors play a huge role in user decision of the content and type of information to be shared across (Zakaria, et al., 2003). Read receipts for messages that are provided by many social media platforms also affects the choice of use of that platform. Users experience various emotions like anxiety and negativity on discovering that their message was read by the recipient but not responded toward (Hoyle, et al., 2017). Emotions may also arise over privacy management of co-owned pictures, as (M. Such, et al., 2017) suggests that most users have Multiparty Privacy Concerns over co-owned pictures. Different users tend to perceive co-owned pictures in different ways, especially based on the audiences of co-owned pictures. Whereas, most of the conflicts get resolved amongst the users, some of them lead to unfriending and some are left unresolved.

One strategy adopted by users to maintain privacy and a sense-of-comfort involves keeping their audiences separate on different services (Lampinen, et al., 2011). (Zhao, et al., 2016) point out that different platforms provide opportunities to facilitate the diverse needs of users. Multiple SM platforms allow users to build gaps between their audiences, which allows them to post content targeting a specific group. (Piwek & Joinson, 2016) point out that 80% of the participants in their study interacted regularly with no more than 12 users on Snapchat. (Vitak & Kim, 2014) derive from their study that social approval is one of the factors that plays a role in user decisions regarding making posts. Users often share posts either when they think their friends might find interest or humor in them.

Within the context of privacy controls and cautious connection, linking between multiple SM platforms allows for more robust and more convenient social interaction. When signing up for new SM tools, an option of signing in with a Facebook username and password is often provided. This allows users to easily sign up on other social media and to link other platforms with Facebook. 'Cross-linking' refers to connecting one SM tool with another by using one's social network credentials to log in to other platforms. By linking social media platforms, users can make benefit of multiple options, including connecting with the same friends on other platforms without having to search for them. 'Cross-sharing,' a term described by (Zhao, et al., 2016), refers to sharing the same content on different platforms to target audiences on both.

Studies suggest that users choose a specific SM platform based on the task or the nature of content they are planning to share, especially with regards to how it may affect others' view of them. Self-presentation in social media is a major aspect considered by users before deciding the platform and usage. The tools offered by various SM platforms to enable the user in achieving certain perceptions by friends are crucial. Shared information may be of a sensitive nature resulting in exposure to cyber risks (Sleeper, et al., 2016) (Gross & Acquisti, 2005). Research explores that gender differences also play a role in terms of comments, likes and feedback when it comes to sharing content over Facebook. Women get more attention for sharing about their personal lives compared to males, whose sports and political content receives less attention on Facebook (Wang, et al., 2013). In another study, regarding content sharing and privacy, findings suggest that female users prefer keeping their profile setting on Facebook as private compared to male users who tend to keep their Facebook profiles public (Fiesler, et al., 2017) (A. DeVito, et al., 2017).

2. SOCIAL MEDIA PATTERN AND USAGE

The current study aims to analyze the decision-making process of users and their intentions when selecting the features, they use, and the posts they make as well as their method of managing content on multiple SM platforms. We ask participants to determine how their adaptability to multiple platforms has affected their usage time on Facebook and the role Facebook now plays in their daily SM usage. We also explore how interviewees make use of cross-linking and cross-sharing features. Overall, we cover four major aspects of social media in this study: separating audience and content on SM, cross-linking and cross-sharing on User adoption of the Story feature on Facebook, Instagram, or Snapchat:

- *Audience separation based on distinct platforms*
- *Nature of content variation based on distinct platforms*
- *Using social media as a hub for updates on latest news, gossip, public figures and products*
- *Effect on the time spent on Facebook due to popularity and preference of other platforms (e.g. Instagram, Snapchat)*
- *Cross-linking and cross-sharing social media: Convenient or unnecessary?*
- *Facebook usage norm amongst Generation Y*
- *User perception of Facebook's added features: Story, Live video & Memories*

2.1 Methodology and Findings

Data were derived by conducting semi-structured interviews of 18 participants living in the United Arab Emirates, mostly in their early twenties, falling into the category of 'young adults.' Participants were selected on the condition of their usage of more than one SM platform. Face-to-face interviews were conducted, each lasting approximately fifteen minutes to twenty minutes depending on user's responses. The interviews were recorded using a transcriber and analyzed in an iterative manner to detect recurring answers to deduce the number of interviewees selected were sufficient (Church & Oliveira, 2013).

In this research paper we use the inductive approach in thematic analysis to identify meaningful patterns within user responses in aims of deriving conclusions based on their preference of platform usage and nature of content sharing variations. (Maguire & Delahunt, 2017)

Table 1. Demographic Details of Interviewees

Gender	**Male**	8	Employment Status	**Student**	14
	Female	**10**		Employed	**4**
Age Group	17 – 20	**3**	**Education Level**	High School	**1**
	21 – 25	**12**		Undergraduate	**15**
	26 – 30	**3**		Post- graduate	**2**

2.1.1 User Adoption of the Story feature on Facebook, Instagram, or Snapchat

We asked users to identify the social platforms on which they use Story feature and to describe their decision-making process in adopting a new feature such as this. While a lot of users do incorporate the 'Story' feature into their daily usage, it is important to point out that there are some that do it to follow the norm, and not necessarily because they like it. One user said, *"I definitely use this feature, but my usage is mostly limited to Snapchat. I don't use it at all on Facebook. I really don't see the point of having it on Instagram too, but I just use it sometimes because of Boomerang."* Boomerang is a feature associated with Instagram that allows users to record short videos that continuously loop when viewed. One user mentioned following stories on Instagram, but when it comes to using it, he states, *"I don't usually put up any stories myself because I find Instagram stories inconvenient to use."* Users were asked the driving factors that lead them to put up a Story. A user said, *"Usually when I put something up [on Instagram] as a Story, it is probably because it is something new to me. For example, I go skydiving. It's a cool thing to happen, so I want people to know 'Oh look, I went skydiving and I survived.' It could also be a dish I tried, and I want people to know, so I put it up."* Another user stated, *"It is because of the filters that Snapchat provides that I use it, [otherwise] I don't use it like a normal user."* Another user said, *"I don't like the story feature on Facebook, but I love it on Instagram and I use it very often, whatever I feel is interesting I immediately story it*". Many users denied using the feature on Facebook either due to inactiveness on Facebook in general, or because it is already available on two other platforms as well. Users showed a tendency to stick to one platform when posting stories instead of adapting the feature on all three.

2.1.2 Audience Separation Based on Distinct Platforms

Having separate SM platforms gives users an opportunity to build a divide between the audiences they share their posts with. One user mentioned, *"Facebook is mostly for family and old friends from high school. Instagram is for recent friends from my university. I hardly have any family members there. For me, Snapchat is a mixture of high school and university friends, even some family members"* (M. Such, et al., 2017). This user has intentionally customized her followers and friends on her social platforms such that it allows her to target her audience based on general content and personal content (Sleeper, et al., 2016) (Gross & Acquisti, 2005). Some users even use these platforms to create a gender-gap between their audiences. *"On Snapchat I have very close friends, mostly girls. On Instagram I have cousins and even guy friends. So, I wouldn't post my pictures on Instagram because I don't want them to see me, but on Snapchat I would since it is only girls and that allows me to be private".* One user stated, *"I have Facebook since 2012 but I recently created my Instagram account and I intentionally don't add a lot of people there because I would like to keep it restricted to close people compared to the 600-random people I have added on Facebook" (*Wang, et al., 2011). We notice that Snapchat, being a platform that emerged when Facebook and Instagram were already a success, has made people more cautious about the audience they share with, and thus it tends to be more personalized. *"On Facebook and Instagram, I also have people who I don't know, so I won't be as comfortable posting my personal content on these as much as Snapchat, because the people I know on Snapchat are close ones."* When asked about the decision that leads to separating viewers, one user stated, *"Well, we have to make use of the convenience each social media tool offers. Not everyone's a close friend of yours. Some are friends of friends. Some are workplace colleagues whom you wouldn't want to get into your personal space."* There were a small number of users though, who did not feel the need to differ their viewers. *"I only accept friends [followers] who I know, so I don't worry about keeping the audience separate."*

2.1.3 Nature of Content Variation Based on Distinct Platforms

Users also take into consideration the nature of the content of posts they make on each SM platform. When asked about how they decide what to post where, a user said, *"On Facebook, typically I have the whole school and half of university friends. That's around 200 to 400 friends, so I must think a lot before I post anything. I try to be very careful about what I post on Facebook. I have more freedom when posting things on Instagram."* Another user stated, *"I keep my posts separate by keeping two accounts on Facebook and Instagram: one for family and one for friends. I would choose more wisely before I post anything on my family account."* A female user said, "*I would never share on Facebook the content I share on Instagram and Snapchat because on Facebook I have a lot of random friends and followers and I wouldn't be comfortable with sharing about my personal life on that platform."* A male user responded similarly, "*I have around 2000 friends on Facebook which includes my family members so if out late-night partying I would be sharing those posts on more personal accounts like Snapchat and Instagram but never on Facebook"* (A. DeVito, et al., 2017). Users also showed a pattern of separating their content based on the sensitivity of it. *"I will mostly post quotes on Facebook. If it's something personal, then I'd go for Instagram."* Another user stated, *"Even the smallest of things, I would post them on Snapchat. Snapchat is more personal, so no one will judge me. I would think twice before I post something on Instagram."* There are users, though, who link their social platforms for posting the same content on both, even if the audience they keep on both are separate. *"If I post something on Instagram, I select the option of putting it on Facebook. Whatever goes on Instagram goes to Facebook too."*

2.1.4 Using Social Media as a Hub for Updates on Latest News, Gossip, Public Figures and Products

Social media is not only used for communication with friends and family. Users keep up with the latest updates and happenings around the world. When it comes to following public figures, most users stated following them on Instagram. "*I use Instagram to follow my favorite celebrities, just to see what they have been up to, the kind of lifestyle that they have. I feel Instagram is more personal than Facebook, so I feel a connection to them.*" Users also use SM to stay informed about the latest products of their favorite brands available in the market. "*The brands advertise their products, so I get to know more about them. If I like a brand, I don't have much time to keep track of them, so this is a way to stay in the know. I won't go to their website to find out*" (Katrin, 2016).

Users also mentioned the helpfulness of the reviews of products they get to read through social tools. "*If I see most people disagreeing with a product, I would know not to waste my money and time on it.*" Marketing of products and user feedback on social media has become a large contributor to their success. For many users, Facebook has become a common hub to keep track of information of their interest. "*I do follow fitness and grooming pages. I follow the updates and try to get an idea of the products that are in the market and how they could benefit me.*" For example, a user in the field of business mentioned following business-related magazines and newspapers. Another user said, "*Facebook is a very public space, so I keep it just for general updates. I find Twitter very useful for real-time, by-the-second updates. I'm a sports fanatic. If I'm on the move, and there's a cricket match going on, it matters that I keep getting updates, and this is where Twitter helps.*" Another user points out the helpfulness of Twitter in terms of keeping track of the latest news. "*If I hear about something happening at NASA, all I must do is open Twitter quick, [type in] hashtag NASA, and I'd know what's happening.*" Another user stated, "*When it comes to keeping up to date with celebrity news or games I always depend on Facebook because there are so many pages and updates and on Instagram and Snapchat I would have to mostly depend on my friends whom I don't trust in this matter.*" Another user added, "*Facebook is the place for gossip and latest news on celebrities because everything goes viral there, like Sridevi died, everyone knew because of Facebook that she died before they could even get to the news channels.*"

2.1.5 Effect on the Time Spent on Facebook due to Popularity and Preference of other Platforms (e.g. Instagram, Snapchat)

Over the years, Facebook has transformed from a primary means of communication for users to one of multiple modes of connection. We analyze the effect of this on the amount of time spent on Facebook, especially since the emergence of Instagram and Snapchat. We asked users in this study how their relationship with Facebook has evolved. One user said, "*Up until six to seven years ago, people would use Facebook just to catch up and connect with their friends. Now, the importance of Facebook as a tool to communicate has reduced. The elder generation do not prefer using so many tools and still find Facebook very useful, since they've now become used to it. For them, learning new tools like Snapchat and Instagram could be troublesome. Whereas the younger generation now directly communicates through WhatsApp.*" Young adults in their twenties are more likely to adapt to emerging social platforms, which is why most users in this study acknowledged that Facebook is no longer their primary social network. "*I always prefer using Snapchat and Instagram to Facebook because I don't see the point of Facebook anymore,*" *one participant stated.* Another user pointed out, "*Based on my recent social media activity, I have hardly used Facebook at all, because it is a very public space where you've got friends, family, workplace colleagues. So, I keep it just to see what's happening out there in the world.*" One reason why users have driven away from Facebook is because other social platforms now better serve their purpose. "*Basically, what I do on Instagram now is what I used to do on Facebook. I would post on Facebook if I wanted to post about prom, or any sort of event, and I used to make an album, now, it is all on Instagram.*" A user stated that the same reason applies to sending messages to friends. "*Before, Facebook was my way to have contact with my friends. Now, if I need to talk to someone, I will either Snapchat them or send them a direct message on Instagram. I don't use Facebook for keeping in contact anymore.*" Facebook has indeed lost its appeal over the past years ever since other platforms entered the game. Users have now turned to other platforms as their primary means of sharing and communicating with friends. A user identifies, "*The idea of Facebook was to connect with friends and friends of friends. I used to get a lot of friend's requests eight years ago. Today, it has reduced to zero. Nobody sends friend requests unless they know me well.*" A user in her late teens said, "*I don't have a*

Facebook account, I never felt the need to create it as all of my friends are active users of Instagram and Snapchat and we prefer those platforms." Another user's opinion was, *"Facebook is only for official use. All my fun stuff, personal life and things I like to share with friends is done via Instagram and Snapchat."*

2.1.6 Cross-linking and Cross-sharing Social Media: Convenient or Unnecessary?

When a user first decides to join Instagram or Snapchat, there's an option available to 'Sign-in with Facebook.' This not only makes the process of joining an application easier, but also allows users to easily link their Facebook friends to other platforms as well. By linking social platforms, users can easily post on one platform, and share it on the other at the same time. This method of cross-sharing provides the convenience of bridging the gap between two or more SM platforms. In some cases, even when users sign up separately, they later link their Facebook account for the same reason. We asked participants their views on cross-linking social media to share the same posts on multiple platforms at once. According to one participant, *"I logged in to Instagram with my Facebook account because of the convenience, but I do not want what I post on Instagram to be there on Facebook, so I never connect my posts. On Facebook I have people who I cannot share personal posts with because I don't even know them."* We notice here that users' willingness to keep their content or audience separate on social media is a driving factor leading to the decision to not link them. *"I would never link my Instagram posts to Facebook. My audience on Facebook is huge. I like to keep them separate. If I'm posting it on Instagram and then sharing it on Facebook too, I'll feel like I'm just diverging [sic] both, so what would be the point?"* Cross-linking platforms does leave a door open in cases when users decide to share information on other tools; however, most prefer the convenience of having a divide between them. *"At times, it is okay to break the rules, but I keep one application for its use alone."* On the other hand, some participants also pointed out the need to share posts with friends on Facebook who are either not part of their more personalized space on other platforms, or do not use any alternate forms of social tools at all. *"I have this very cool aunt who loves knowing what I'm up to, but she is not on Instagram. So, to let them know what's up, I sometimes choose the option to also share on Facebook."*

2.1.7 Facebook Usage Norm amongst Generation Y

This study has explored the evolution of Facebook since the emergence of other SM platforms, especially Instagram and Snapchat. Most participants admitted that the time spent on Facebook has significantly decreased and acknowledged the preference of other tools to stay connected with friends. According to one participant, *"The older generation would still use it for both [information and communication], because over the years they have become used to Facebook. The younger generation is more adaptable. It is all about which tool captures the imagination of the audience: through convenience or through features?"*

Yet, we notice that most users still use Facebook despite decreased activeness on the platform. One user stated, *"Based on my recent social media activity, I have hardly used Facebook at all, because it is a very public space where you've got friends, family, workplace colleagues. So, I keep it just to see what's happening out there in the world."* Facebook has become a common center for users to get their information of interest all from a news feed. While Instagram and Snapchat have becoming the primary choice for posting personal information, users have evolved their use of Facebook such that it is now their go-to tool for information of their interest. Facebook is still used as a hub to keep track of old friends, but its communication factor has transformed such that it is only used to connect with long-distance acquaintances or people with whom users have minimal interaction. *"I sometimes do still use it [Facebook] as a communication tool. I get to stay in touch with a lot of cousins. They've always been far away. Maybe I've met them just twice. So, this is a way to stay connected."* Another user said, *"For me Facebook is mostly a one-way street, I use it to get information on friends and family and the world, when I want to communicate with someone I prefer more personal platforms like WhatsApp or even Instagram."*

2.1.8 User Perception of Facebook's Added Features: Story, Live video & Memories

Due to competition with newer platforms, Facebook has implemented several updated features. In recent years, Facebook introduced 'On This Day' Memories, a unique feature that allows users to view their posts made on the same date in a past year, with the option to share it again. In addition, Facebook followed in the footsteps of Instagram and Snapchat and introduced its Story feature as well. Users can also make 'live videos,' giving way to real-time sharing.

We asked users whether these features have helped increase their activity on Facebook. According to one participant, *"I see friends sharing old memories from their high school years. Since we are all separated from each other now, it kind of makes others want to share them too and interact like the old times."* Another user stated, *"I like the memory features a lot, it gives me nostalgia and I love re-posting what I posted 3 years ago, makes me re-live those moments, its sweet and cute."*

While the old memories feature is unique to Facebook, features offered on multiple platforms were shown to have less or even no success in drawing users to Facebook. When asked about the 'live video' feature on Facebook, one user said, *"Live videos definitely allow for more connected community."* This is a feature that allows on-the-spot sharing, further connecting users. We tried to determine its popularity on Facebook in comparison to other platforms. One user stated, *"Even though live videos were there on Facebook before they came into Instagram, I think they became more popular because of Instagram."* As for the Story feature, our participants denied using it at all on Facebook. Another user said, *"Facebook is good, not just for me, I enjoy their live video and memories feature, but the story feature doesn't work for me as it has a larger audience."*

2.2 Discussion

This study identified the choices and decision-making processes of users when posting content on SM. It also explored the evolution of Facebook with the constant emergence of other social platforms. We notice from our findings that most participants have matured in their decision-making process with the emerging platforms and features. Users are now more cautious when making a choice regarding the posts they share. These participants demonstrated a more careful approach in dealing with new social platforms (Instagram and Snapchat), especially when it comes to maintaining the privacy of their content. For most of these participants, Facebook was a public space that connected them with a large audience including friends-of-friends and long-distance acquaintances. They, therefore, opted to personalize their other SM by being more selective of their audience. This makes it easier to them to post their content on these other platforms with a level of comfort. The time-limitation of posts offered by Snapchat and 'stories' on Instagram gives users more freedom to share content. While participants admitted to active usage of the Story feature on Instagram and Snapchat, none of them use it on Facebook. This may be due to the large audience users have on Facebook, most of them being distant acquaintances. Users prefer such features when it comes to sharing it with a more personalized group of connections on other SM platforms.

SM platforms are now heavily used as a marketing tool by brands. Our analysis shows that users have a positive response toward this strategy, admitting to the usefulness of the updates and reviews they receive by following branded pages. Users also admitted to following their favorite public figures on SM, mainly Instagram, as it allows them to stay up-to-date on their latest activities. This study also found that the increased activeness of users on Instagram and Snapchat has affected the activity of users on Facebook. The participants of this study conveyed their preference for other SM tools over Facebook. The reasons leading to inactiveness on Facebook highlight the privacy factor, wherein users are choosier in sharing personal content with their connections. A larger audience drives them away from sharing on Facebook, making more customized SM such as Instagram and Snapchat their preference.

Then, why do users continue to have a Facebook account if it is not their favored form of communication anymore? We deduce from participant responses that users want to stay loosely informed of their old and distant connections. Facebook has also evolved into a hub that delivers all information of interest via a user's news feed, thereby still serving as a mode of convenience.

We explored user decisions when it comes to cross-linking and cross-sharing on social platforms. Our findings show that users are open to cross-linking their SM platforms as it allows them the convenience of logging in instantly without having to go through the sign-up process. However, users have mixed views regarding cross-sharing of posts including those using the Story and Live video features. While some find the feature unnecessary, as they would not want to share the same content on multiple platforms, others find it beneficial for instances when they would want to share their posts with a larger audience. Since, in the corporate world, Instagram comes under the ownership of Facebook, we infer that Facebook has incorporated the Story feature, and Instagram has incorporated the Live video feature, to standardize their platforms and enhance convenience for their users. However, these features have gained more popularity through Instagram than Facebook. Users are very much inclined to adopt these features, but the presence of these features on Facebook is not a contributor in increasing its popularity or usage.

Although the new players of the game like Instagram and Snapchat have taken over Facebook's audience, it still stays ahead of the game when it comes to getting brand updates, news and celebrity gossip, as posts on Facebook have a higher potential of going viral due to the prevalence of public pages and multiple share options that enable users to get updates from pages to which they have not even subscribed. Facebook has also maintained its feature of enabling high quality longer videos, which some of the new platforms have yet to adopt. Facebook has been around for nearly two decades and hence is more popular amongst the older generation, whereas millennials have shifted to the newer platforms like Snapchat and Instagram. Even though these platforms come with comparatively limited features, this is exactly what they are looking for. Millennials prefer using a more visual based platform for sharing content (Sheldon & Bryant, 2016). They tend to keep their social media updated with live stories and videos of their daily activities, hence usage of Instagram and Snapchat becomes more popular amongst them. Instagram has also been the platform where a lot of young social media influencers are gaining popularity which they wouldn't have through traditional platforms like Facebook and Twitter.

3. CONCLUSION

We conducted a qualitative analysis of semi-structured interview responses of 18 young adult users of social to identify usage patterns and their decision-making process. We conclude that users have matured with emerging popularity and presence of multiple SM platforms, giving special consideration to their privacy by managing the audience they share their posts with. Users have taken advantage of the 'new' social tools by customizing them to control their audience. In addition, users tend to have a greater sense of freedom when posting content to be viewed by their closer group of acquaintances. Instagram and Snapchat are becoming the go-to platform for users, which has led to reduced activity on Facebook, largely due to the 'publicness' factor on the latter platform. The findings of this study also deduce the positive response of users toward social media marketing strategies. In fact, users willingly follow their favorite brand pages and celebrity figures to stay more informed and connected. We also establish the convenience of cross-sharing posts on multiple SM platforms, which allows for bridging a gap in rare instances where users are more open to sharing posts with a larger audience. The study identifies that emerging SM tools and the features they come with present opportunities to users that lure them away, leading to inactiveness on the relatively older platform of Facebook. Majority of the SM users are millennials, who tend to be more fascinated by newer, image-based platforms like Instagram and Snapchat resulting in reduced popularity of older platforms like Facebook.

REFERENCES

A. DeVito, M., Birnholtz, J. & T. Hancock , J., 2017. *Platforms, People, and Perception: Using Affordances to Understand Self-Presentation on Social Media.* New York, CSCW '17 Proceedings of the 2017 ACM Conference on Computer Supported Cooperative Work and Social Computing.

Church, K. & Oliveira, R. d., 2013. *What's up with WhatsApp? Comparing Mobile Instant Messaging Behaviors with Traditional SMS.* Munich, Proceedings of the 15th international conference on Human-computer interaction with mobile devices and services.

Fiesler, C. et al., 2017. *What (or Who) Is Public?: Privacy Settings and Social Media Content Sharing.* New York, CSCW '17 Proceedings of the 2017 ACM Conference on Computer Supported Cooperative Work and Social Computing.

Gross, R. & Acquisti, A., 2005. *Information Revelation and Privacy in Online Social Networks (The Facebook case).* Virginia, ACM Workshop on Privacy in the Electronic Society (WPES).

Hoyle, R. et al., 2017. *Was my message read?: Privacy and Signaling on Facebook Messenger.* Denver, CHI '17 Proceedings of the 2017 CHI Conference on Human Factors in Computing Systems.

Katrin, W., 2016. Trying to understand social media users and usage: The forgotten features of social media platforms. *Online Information Review,* 40(2), pp. 256-264.

Kümpel, A., Karnowski, V. & Keyling, T., 2015. News Sharing in Social Media: A Review of Current Research on News Sharing Users, Content, and Networks. *Social Media + Society,* 1(2).

Lampinen, A., Lehtinen, V., Lehmuskallio, A. & Tamminen, S., 2011. *We're in It Together: Interpersonal Management of Disclosure in Social Network Services.* Vancouver, CHI '11 Proceedings of the SIGCHI Conference on Human Factors in Computing Systems.

M. Such, J., Porter, J., Preibusch, S. & Joinson, A., 2017. *Photo Privacy Conflicts in Social Media: A Large-scale Empirical Study.* Denver, CHI '17 Proceedings of the 2017 CHI Conference on Human Factors in Computing Systems.

Madden, M., 2012. *Privacy management on social media sites,* Washington D.C.: Pew Research Center's Internet & American Life Project.

Piwek, L. & Joinson, A., 2016. "What do they snapchat about?" Patterns of use in time-limited instant messaging service. *Computers in Human Behavior,* Volume 54, pp. 358-367.

Sheldon, P. & Bryant, K., 2016. Instagram: Motives for its use and relationship to narcissism and contextual age. *Computers in Human Behavior,* Volume 58, pp. 89-97.

Sleeper, M. et al., 2016. *Sharing Personal Content Online: Exploring Channel Choice and Multi-Channel Behaviors.* San Jose, CHI '16 Proceedings of the 2016 CHI Conference on Human Factors in Computing Systems.

Stieglitz, S. & Dang-Xuan, L., 2013. Social media and political communication: a social media analytics framework. *Social Network Analysis and Mining,* 3(4), p. 1277–1291. Maguire, M. & Delahunt, B., 2017. Doing a thematic analysis: A practical, step-by-step guide for learning and teaching scholars.. *AISHE-J: The All Ireland Journal of Teaching and Learning in Higher Education,* 9(3).

Stutzman, F. & Hartzog, W., 2012. *Boundary regulation in social media..* New York, Proceedings of the ACM 2012 conference on Computer Supported Cooperative Work (CSCW '12).

Thomson, R., Yuki, M. & Ito, N., 2015. A socio-ecological approach to national differences in online privacy concern: The role of relational mobility and trust. *Computers in Human Behavior,* 51(Part A), pp. 285-292.

Vitak, J. & Kim, J., 2014. *"You can't block people offline": examining how facebook's affordances shape the disclosure process.* New York, CSCW '14 Proceedings of the 17th ACM conference on Computer supported cooperative work & social computing.

Wang, Y.-C., Burke, M. & E. Kraut, R., 2013. *Gender,topic, and audience response: An analysis of user-generated content on Facebook.* New York, CHI '13 Proceedings of the SIGCHI Conference on Human Factors in Computing Systems.

Wang, Y. et al., 2011 . *"I regretted the minute I pressed share": A Qualitative Study of Regrets on Facebook.* New York, SOUPS '11 Proceedings of the Seventh Symposium on Usable Privacy and Security.

Wang, Y., Norice, G. & Cranor, L. F., 2011. *Who Is Concerned about What? A Study of American, Chinese and Indian Users' Privacy Concerns on Social Network Sites.* Pittsburgh, International Conference on Trust and Trustworthy Computing.

Zakaria, N., M. Stanton, J. & Sarkar-Barney, S. T., 2003. Designing and implementing culturally- sensitive IT applications: The interaction of culture values and privacy issues in the Middle East. *Information Technology and People,* 16(1), pp. 49-75.

Zhao, X., Lampe, C. & B. Ellison, N., 2016. *The Social Media Ecology: User Perceptions, Strategies and Challenges.* New York, CHI '16 Proceedings of the 2016 CHI Conference on Human Factors in Computing Systems.

Short Papers

ASSESSMENT OF RISK FOR ALZHEIMER PATIENTS BY MONITORING HEARTBEAT AND IDENTIFIYNG FALLS

Fernando Luís-Ferreira, Daniel Rodrigues, David Pereira, João Sarraipa and Ricardo Goncalves
CTS, UNINOVA, Dep.º de Eng.ª Electrotécnica Faculdade de Ciências e Tecnologia, Universidade Nova de Lisboa 2825-529 Caparica, Portugal

ABSTRACT

Alzheimer patients tend to forget and to lose perception of many aspects of life. A major concern is the increased probability of losing the notion of the current location and how to return to a known place. In order to address such concern, technology can provide the tools to monitor and act according to identified risky situations. The assistive system's architecture will evaluate the risk and will not disclose any data unless there is some identifiable risk. The system will monitor the persons behavior with sensors, evaluating if there is a fall as position is retrieved in real-time. The system will also monitor heartbeat and will determine if it is outside a given range. If a risky situation is identified the patient is asked to respond correcting in case of a false positive, if there is no response, the system will send a warning to a selected person sending also the current position of that patient. The system aims to provide a service to Alzheimer patients that will avoid recurrent notices that some older patient is missing, sometimes with an unpleasant result for the person with Alzheimer or other type of Dementia.

KEYWORDS

Alzheimer, Dementia, Risk Assessment, Patient Care, Wandering

1. INTRODUCTION

The advances in the health domain with new diagnosis tools and new treatments lead to an improvement in quality of life and extended life span. However, has in many other aspects of life, there is a downside and that is the emergence and prevalence of age associated pathologies. Such is the cases of people with Dementia where most of the cases refer to Alzheimer patients which have reduced ability to perform mental tasks such as memorizing, reasoning and decision making. After the sixth decade of life, the risk of dementia increases immensely taking the form, in most of the cases, of the Alzheimer pathology. The progression of Alzheimer is different from person to person and it usually starts to manifest in episodes of loss of memory, confusion and loss of references. A major consequence of such episodes is that, as they can happen anywhere at any time without previous symptoms, it may involve some type of danger for the person. Among those risks, some of major concern may be related to falling on the ground or to some dangerous place (e.g. river, cliff), being caught by a car or by another vehicle or equipment. In fact around 40–60% of falls lead to injuries (Masud & Morris 2001). Associated to those risks, variations of the heartbeat are sometimes identified, before or as a consequence of such events. Those measurements are outside normal boundaries, for a typical heartbeat at that age, or outside clinical values for that person. In order to address such problems several types of devices and services have been developed for people with mild or less severe stages of Alzheimer. The existing solutions are sometimes expensive as they require buying specific equipment and paying a monthly fee to a provider of a service that will monitor and take actions while a risky situation is identified. The solution hereby proposed proposes the usage of a smartphone with a specifically designed software that can be used, without any subscription or any other equipment, to become a supporter for a person with Alzheimer. In Section 2 is presented the adopted methodology and in section 3 the Platform and the discussion on its usage and finally in section 4 the Conclusions and future work.

2. METHODOLOGY

The identified problem relies on the risks for Alzheimer patients that can be addressed by current technology using a systematic approach. The adopted methodology takes as reference the fact that smartphones become pervasive objects that each individual carry for communication and other services. The design is based on using the sensor capacities to enable the detection of position and risk for a given person. In this case the internal accelerometers are used to determine if there was an event of the type that generates an increased acceleration, typically a falling person. There are some other works where equipment was used to detect falls at home but those need a specific architecture and a deployment of specifically designed sensors (Biros et al. 2014). Some other works try to assess the most precise definition of the kind of detected event (Anjum & Ilyas 2013). In the present work, the objective is to keep a simple solution that makes the detection of events without seeking for a precise characterization of the type of event but rather trying to get the confirmation of risky situation from the user, or by default on user's inaction. On the other side, being the risk of fall by itself a motive to take measures, it is aimed to establish a range of acceptable variations on the heartrate of the user. If the heartrate falls outside what was defined as normal, the same verification process is issued; the user is prompted and, case he does not respond a warning is issued to the registered person, communicating his location coordinates.

3. PLATFORM

The proposed platform to monitor Alzheimer patients is based on an application (App) that runs in an Android smartphone but can migrate to another operating system. The developed app uses a special algorithm designed to detect falls. This algorithm makes use of the smartphone's accelerometers detecting variations on the various acceleration axes. If a variation of acceleration is detected on, at least 2 axes above a certain threshold, it means a probable fall. Similar works have been done in detecting a person's fall (Tran et al. 2016). In order to detect false positives a confirmation is asked onscreen. In case of lack of response, in a given time, the App assumes as a risky and issues the most adequate measures. In order to accomplish that goal, a SensorEventListener instance is needed and then, to manage the sensor, it is used a SensorManager related to the SENSOR_SERVICE. The method onSensorChanged, enables the decision on what procedures to execute when the sensor (in this case, the accelerometer) changes its data. In the present case, it is aimed to read the current axis acceleration values (X, Y and Z).

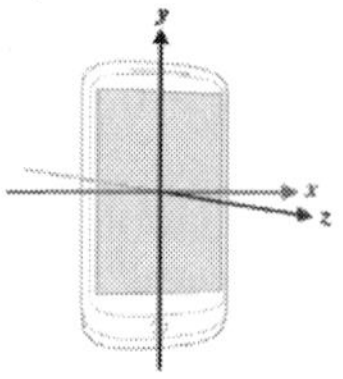

Figure 1. Coordinate System, Used by the Sensor API

The next step will be the computation of acceleration variation, based on the obtained values.

Δx = xPreviousAcceleration – xAcceleration

ΔY = yPreviousAcceleration – yAcceleration

ΔZ = zPreviousAcceleration - zAcceleration

After those resulting accelerations are calculated, the system will decide whether probably a fall occurred to the person carrying the smartphone or if variations are within normality. To accomplish that goal, it is necessary to define what is a fall. There are two major components that support such decision:

1. The intensity of the fall – represented by the threshold of acceleration difference, from which we consider as a fall;
2. The number of axis used in acceleration assessment to detect the fall – in this case two axis are used

The tests executed with two axis provided enough information about the patterns of movement and in particular to detect falls. In some studies it is demonstrated that even one sensor provides enough information to identify patterns of movement (Zdravevski et al. 2017).

3.1 HR Monitor

The HR monitoring is, in the present research, executed with the measuring device existing on the back of some smartphones or using built-in camera of the smartphone. This approach enables to demonstrate how the system performs and how to react according to abnormal reading values. In a real application scenario with patient to be permanently monitored, a bracelet, a chest strap or other wearable will be used.

The service hereby proposed is open source and depends only on a smartphone with its basic built-in sensors and connection with devices to retrieve heartbeat from an external source (e.g. Bluetooth, LTE).

Figure 2. Examples of HR Monitoring Devices, a) Bracelet, b) Chest Strap, c) Smartphone Embedded Device

3.2 Privacy

Privacy has become a central topic, these days, especially in the tech and software domains (EC, DG Internal Market, Industry 2018). People don't like to be excessively monitored and observed. For Alzheimer patients, the same holds true in particular for people with mild conditions, in early stages, when a free and normal life is desired. It becomes imperative that sensitive or private data is not disclosed without the consensus of the person or unless the system detects an almost certain risk to the patient. That is a central aspect regarded in the present work taking privacy and ethical concern as central in the development of the Application. This platform has an operational difference when regarding to many solutions on the market that depend on a subscription service and, additionally, a proprietary device to be bought and that is not open source and can only be adopted for use with the seller company or subscription provider (ITraq 2018) (AngelSense 2018).

We can consider, for instance, common applications to be used on a daily basis, like Google Navigation. If the user wants to go somewhere using this app to a calendar appointed place, the app has to access the current GPS location and calendar, potentially tracking movement and monitoring position all the time. Such data, when disclosed or shared to other users becomes a privacy issue for a person running a normal life.

The developed Application, here presented in its alpha version, uses encrypted authentication, both on the patient and the "person who monitors" side and will only disclose location data once a fall or abnormal HR is detected and only if there is no reaction from the user side. It is ensured that no other data is sent to the database, or to other individuals. Processing and decision is taken locally without the need of a cloud service or other remote services for data analysis and decision taking.

The strategy hereby adopted, mitigates possible privacy concerns as it only sends message alerts to the other person (e.g. a family member, emergency services) when, there is no response after a pop-up message is displayed on screen, asking for the patient to press a button. In this case a countdown starts giving time for the patient to confirm if he is in normal condition and thus to assess is cognitive status, in that case a false positive is signaled. When such non-risky episode happens, the system returns to normal stage and no warning is issued. However, the systems will be in surveillance, retrieving data and analyzing the occurrence of subsequent risky episodes with a flag signaling that normality could be apparent. In such conditions, the sensitivity of the system is increased resulting in more frequent sampling rates while ensuring that energy consumption remains low.

3.3 Interface

The proposed App has a major request of being intuitive and simple to use. It starts with a login screen where a registry is performed in order to ensure privacy. The interface aims to be extremely simple, as to the nature of the target public, whom due to age and other possible conditions, may be less familiar with smartphone based technologies and services. A minimum setup of email and password is required to login into the service.

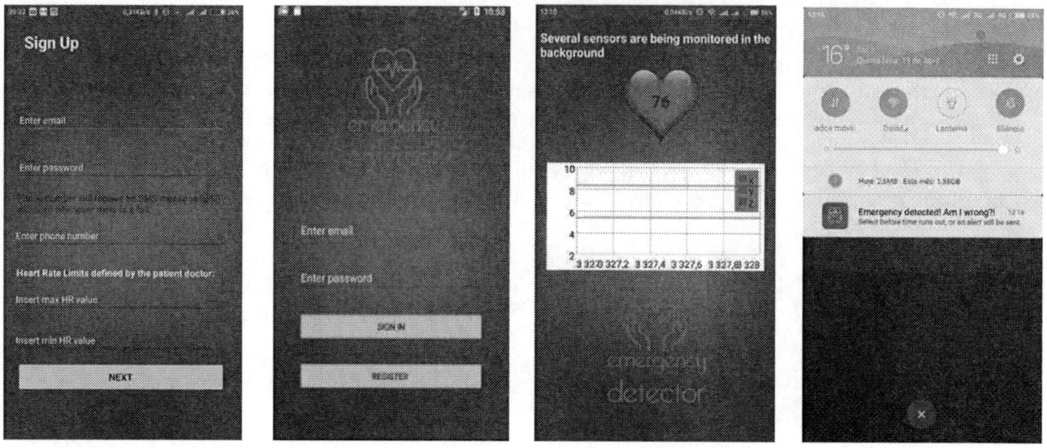

Figure 3. The Application that Monitors an Old Person with the Security Procedures of Signing Up and Login

As in almost any type of application, before the first usage of the system a user (or someone that cares) proceeds to register the user into the service. In line with the philosophy of the system, the signup interface is also minimal as can be seen in Figure *3* (left).

A cellphone number should be added, which is the selected contact person, a carer or family of that person, that when an identified risky situation occurs, will receive a customized SMS message.

The reliability of the application is ensured by the notification that alerts the user to a detected fall. The user has 10 seconds to click "YES" in case of a false positive as can be seen in Figure *3* (right). If there is no response within the 10s period an SMS message will be delivered to the selected contact person.

3.4 Database

The falls are registered in a database ordered by date of occurrence and contain three parameters as represented in Figure 4.

1. Intensity of the fall;
2. Latitude where the fall occurred;
3. Longitude where the fall occurred.

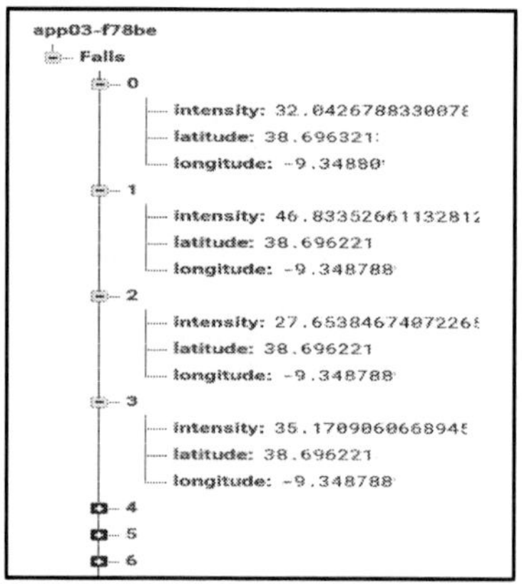

Figure 4. Falls Database

The database presented in Figure 4 stores all the detected events and where they happened. These records will provide opportunities for later reconnaissance of the places and the associated risk. It also enables the system to learn what are false positives and how detection modes can be improved so that the system behaves with better performance while maintaining readiness and safety.

4. CONCLUSION

The increasingly worrying cases of Alzheimer and other types of Dementia has becoming a health problem of major importance. The burden of the costs and life disturbance are becoming a serious concern for health authorities and for families. It urges to seek for affordable and reliable solutions that may become available for the majority of persons no matter their social stage or their computational skills.

The ongoing research was developed addressing the above mentioned concerns by providing a simple solution based on the deployment of a smartphone based solution for patients with Dementia. The system uses the smartphone's features, including the services and sensors existing in the vast majority of today's smartphones. The system ensures an active monitoring of the person's behavior and the detection of factors of risk as is the case of a fall or heartrate variations beyond what is defined as normal range. In this work, false positives were considered as a real threat to the system's efficiency as any episode that would trigger a false alarm would tag the system as faulty and would mine confidence with consequent impact on response type to emergency signals. Thus, the system addresses false positives as something to eliminate or reduce at a minimum by requesting user's collaboration and learning with time. This solution has several strong points as the privacy concerns, the inexistence of costs beyond the smartphone and the reliability of the system by empowering the user to actively participate in privacy matters and while responding to false positives. Beyond that, it is important to mention that in case of a missing person without any risk alarm emitted it is possible to use google maps to discover a person's location if that was the option of the user and family.

ACKNOWLEDGEMENT

The research leading to these results has been promoted under the project CARELINK, AAL-CALL-2016-049 funded by AAL JP, and co-funded by the European Commission and National Funding Authorities of Ireland, Belgium, Portugal and Switzerland.

REFERENCES

AngelSense, 2018. AngelSense Phone & GPS for Dementia & Alzheimer's. Available at: https://www.angelsense.com/protect/dementia/ [Accessed April 20, 2018].

Anjum, A. & Ilyas, M.U., 2013. Activity recognition using smartphone sensors. In *2013 IEEE 10th Consumer Communications and Networking Conference (CCNC)*. pp. 914–919.

Biros, O. et al., 2014. Implementation of wearable sensors for fall detection into smart household. In *2014 IEEE 12th International Symposium on Applied Machine Intelligence and Informatics (SAMI)*. IEEE, pp. 19–22. Available at: http://ieeexplore.ieee.org/document/6822406/ [Accessed April 11, 2018].

EC, DG Internal Market, Industry, E. and Sme., 2018. Data protection and privacy - Your Europe. Available at: https://europa.eu/youreurope/citizens/consumers/internet-telecoms/data-protection-privacy/index_en.htm [Accessed April 20, 2018].

ITraq, 2018. iTraq Global Location Tracking Device to find your loved ones – iTraq, Inc. Available at: https://www.itraq.com/?rfsn=608860.34035&gclid=EAIaIQobChMIwLyC8saS2gIVQiQrCh2siAz1EAAYASAAEgJimPD_BwE [Accessed April 20, 2018].

Masud, T. & Morris, R.O., 2001. Epidemiology of falls. *Age and ageing*, 30 Suppl 4, pp.3–7. Available at: http://www.ncbi.nlm.nih.gov/pubmed/11769786 [Accessed June 8, 2018].

Tran, T.D., Truong, H. & Tran, K.D., 2016. Automatic Fall Detection using Smartphone Acceleration Sensor.

Zdravevski, E. et al., 2017. Automatic machine-learning based identification of jogging periods from accelerometer measurements of adolescents under field conditions S. Taheri, ed. *PLOS ONE*, 12(9), p.e0184216. Available at: http://dx.plos.org/10.1371/journal.pone.0184216 [Accessed April 20, 2018].

STRENGTHEN CANCER SURVEILLANCE IN SRI LANKA BY IMPLEMENTING CANCER REGISTRY INFORMATICS TO ENHANCE CANCER REGISTRY DATA ACCURACY, COMPLETENESS, AND TIMELINESS

Kamal Seneviratne[1], Sudath Samaraweera[2] and Suraj N Perera[2]
[1]*Registrar in MD Health Informatics, Postgraduate Institute of Medicine, University of Colombo, Sri Lanka*
[2]*National Cancer Control Programme, Ministry of Health, Sri Lanka*

ABSTRACT

Surveillance of cancers is a process of systematic, continuous collection, storage, analysis, interpretation and dissemination of epidemiological information on cancer cases occurring in a particular geographic area. The National Cancer Control Programme (NCCP) of Sri Lanka is the national focal point for prevention and control of cancers in the country. Sri Lanka Cancer Registry (SLCR) is maintained by NCCP and plays a pivotal role in cancer control. Cancer surveillance informatics is the systematic application of information and computer science and technology to cancer surveillance practices, research, and learning. Thus, using emerging technology to incorporate automated process and electronic data exchange in cancer surveillance business is an efficient, fast and cost effective way to obtain quality, accurate and complete cancer registry data. The intended results to be expected by implementing cancer registry informatics is to convert existing cancer registry in to a modern era cancer registry. This helps Sri Lanka cancer registry stands with the global cancer registries by improving the dimensions defined by the Cancer Incidence in Five Continents (C15).

KEYWORDS

Registries, Data Collection, Health Information Systems

1. INTRODUCTION

National Cancer Control Programme (NCCP) of Sri Lanka was established in 1980 following a detailed survey of morbidity and mortality and a recommendation by the WHO in 1979. The NCCP is responsible for policy, advocacy, monitoring and evaluation of prevention and control of cancers, conducting surveillance of cancers and facilitates research related to cancer.

In 2007 there were 13,635 new cancer cases with crude cancer incidence rate (CR) of 68.0 (National Cancer Control Programme, 2013). In 2010 it rose to 16,888 new cases with CR of 82.1(National Cancer Control Programme, 2015). This increase may be due to several reasons (Bray, et al., 2012). Since a considerable number of government and private health care institutions have started diagnosis and treatment of cancer patients, collection of data from these institutions and statistics related to cancer deaths from death registrars have become absolutely essential to include all these data to fulfill the four quality indicators recommended by the International Agency for Research on Cancer (IARC) (Leal, et al., 2016).

1.1 Importance of Cancer Surveillance

Cancer surveillance provides a quantitative portrait of cancer and its determinants in a defined population. It helps epidemiologists in descriptive studies and analytical studies on cancer (Jensen, et al., 1991). Knowledge of trends in the incidence of cancer over geographical areas may be used for the projection of future incidence rates, caseloads, and the need for treatment facilities.

1.2 Cancer Registries and Cancer Registration

Cancer registry is a fundamental tool important in control of cancers in a given country. The cancer registry is the office or institution which attempts to collect, store, analyze and interpret data on persons (Rajaraman, 2012), (Jason & Whelan, 1991) with cancer. Cancer registries are an essential part of a complete cancer control program and help in finding etiological factors of cancers and also enable primary, secondary and tertiary prevention and monitoring (Leal, et al., 2016),(Bray, et al., 2014).

1.3 Cancer Surveillance Informatics

The definition of Public health informatics is given as the systematic application of information and computer science and technology to public health practice, research, and learning. It is a subdomain of the biomedical or health informatics (CDC, 2012)

The tools used in the domain include computers, information and communication systems, clinical guidelines, and formal medical terminologies (National Cancer Registrars Association, 2016). Thus, using emerging technology to incorporate automated process and electronic data exchange in cancer surveillance business is an efficient, fast and cost-effective way to obtain quality, accurate and complete cancer registry data as it increases data accuracy, completeness, timeliness, and comparability. It also minimizes manual handling of documents and manual processing of information.

1.4 Current Process of Cancer Registry

The National Cancer Control Programme has been monitoring trends in cancer and publish them since 1985. Cancer surveillance is concerned with the capture of longitudinal data from multiple and varying data sources. Therefore it collects data from various data sources (Figure 1).

The current process of cancer registry data gathering from provincial cancer treatment centres as stated above is through mix of paper-based and stand-alone electronic application. Data thus received is then fed into another open source application called CanReg5 ((IARC), 2015) for consistency check.

However, with the existing data collection method, the gap between diagnosis of cancer and the availability of data for analysis is significant. As a result the publication of cancer registry data may reflect information about patients diagnosed more than a couple of years ago. The issues that are encountered in data gathering process have to be rectified through communication and information technology. Using information echnology (IT) in order to shorten all phases of cancer registration process is well documented (Bianconi, et al., 2012).

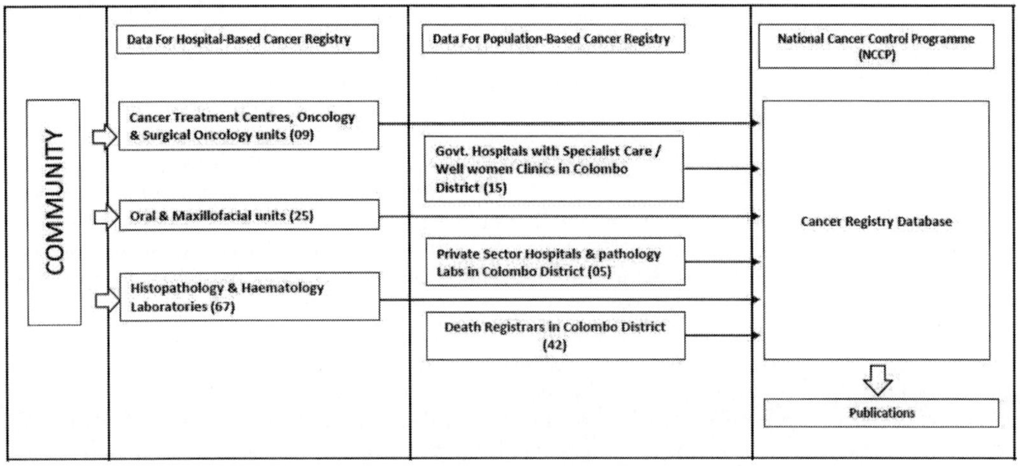

Figure 1. Current Cancer Registry Case Finding Sources and Data Flow

2. MATERIALS AND METHODS

Designing, developing and Implement a web-based cancer registry information management system called Sri Lanka Cancer Registry (SLCR) to maintain all reportable cancer cases is by using web 2.0 technology, open-source software, and mobile technology. Open-source license products are used to decrease the cost for the technical platform and also as a defining factor in terms of profitability. Although there is no clear-cut definition the term web 2.0 suggest an upgraded and updated version of the web (Bear, 2012). The java technology is used as the tool for the web development process. Java is an object-oriented sever-sided programming language and is more suitable for both web and desktop application development and is enabled to write more secure programs (Rabah, et al., n.d.). The server is Linux-based operating system and the Apache is the webserver. The database system for storing information is MySQL. MySQL is multithreaded and multi-user database management system (Bassil, 2012). It is distributed under a GNU General Public License (GPL) and hence has no running cost. The community involvement makes the software development progress to be steady (Stroe, 2011). MySQL can run over 20 operating systems including Linux-based and Windows. Its performance, scalability, and security system have made the system to be used by many large companies world-wide (Stroe, 2011).

The software architecture of the Database Management System (DBMS) is client-server architecture. The client-server architecture is considered because this architecture has several advantages.

a) Centralization: All the information stored in a server. Clients can access the database over an authorized access. Every authorized client is given the opportunity to access information via the desktop or laptop interface and tools like spreadsheets, power point presentations can be used to deal with data with the help of database and application servers resident on the network to produce meaningful information.

b) Scalability: The server where the data are stored can be repaired, replaced, upgraded or relocated while the client or end users remain unaware and unaffected by that change.

c) Availability: As the server is always turned on, data can be accessed at any time throughout the day and throughout the year.

d) Prevent from data losing: As data is stored on a server it is easy to make a backup. In case of data corruption or data loss, it can be recovered easily and efficiently from backup.

e) Security: Data can be better secured from unauthorized access by defining the permissions at the time of the setup of the server.

There are two models of client-server architecture namely, two-tier and three-tier architecture. From these two options, two-tier architecture is selected for the development of DBMS because it has several advantages over three-tier architecture.

a) Two-tier architecture is easy to maintain.
b) Modification of the database is relatively easy.
c) Communication between the end user and the database is faster.
d) Development cost is much less than the three-tier architecture.
e) Two-tier architecture is less complex than the three-tier architecture.

The mobile app is used to gather data from death registrars' offices located in Colombo district, Government hospitals with specialist care, well woman clinics in Colombo district, private sector hospitals and pathology laboratories located in Colombo district. This method of data collection is greatly supported by the fact that more than 23 million mobile connections exist in 2015 (Marketers, 2016). This figure exceeds the population of Sri Lanka (Dept. of Census and Statistics, Sri Lanka, 2016)

The process of data collection procedure begins with abstracting data from case-finding sources to a summary sheet, which is designed based on standard coding and classification procedures and definitions for the population-based cancer registry (Silva, 1999).

A cancer registry of the modern era is a medical registry and fulfil the standards defined by Drolet and Johnson (2008). The standard consists of five features. Data from multiple sources are combined to create an aggregate set of data called mergeable data (M), data such collected are standardized data set (D), in that the same set of characteristics is collected for each patient in the dataset according to the defined set of rules (R). Furthermore, patients in the registry have a unique identifier in order to follow up over time (O), and this helps to assess the occurrence of specific outcomes of interest (k) (Drolet and Johnson, 2008). It is expected that the new system will fulfil the criteria of a modern cancer registry.

As the value of any cancer registry relies on the underlying quality of its data and quality control procedures it has in place, the Cancer Incidence in Five Continents (C15) assesses three dimensions of quality to enable the comparison of cancer incidence rate across different population worldwide (Bray, et al., 2014). They are completeness, comparability, and validity (or accuracy). Completeness is defined as the degree to which all diagnosed neoplasms within a registry's catchment population are included in the registry database (Bray, et al., 2014). Therefore the system is designed to include case-finding sources that are currently available with the facility for future expansion.

Data quality equals to the completeness of data and its accuracy (or validity) (Silva, IDS.,1999). On the context of a cancer registry, it is defined as the proportion of cases in a dataset recorded as having a given characteristics that truly do have that attribute (Bray, et al., 2014). The proposed web-based system accomplish data accuracy in several ways. The coding standards for cancers are maintained using the third edition of the International Classification for Oncology third edition (ICD-O-3) (Fritz, et al., 2000). Errors are notified to the relevant data collection centre by registry staff through the system.

Thus, the system assists in maintaining data consistency as well as timeliness as the data can be viewed as soon as they are included in the system and ability to correct them as soon as possible. The stand-alone databases currently used to collect data in case-finding sources are unable to fulfil such advantages. Once confirmed by the NCCP a record as a completed record, it is shown as confirmed record in the system.

3. CONCLUSIONS

The web-based application will ensure timely availability of information that is needed for better improvement in cancer surveillance. Therefore the system will provide accurate, complete and timely registry data to the stakeholders such as health policy makers, researchers, clinicians, and the general population. The goals expected to be achieved using registry data are: determine the incidence of cancers with respect to geographic, demographic characteristics, monitor trends and pattern of cancer incidence over time, identify high-risk populations, provide data for epidemiological studies and prioritize health resource allocation.

ACKNOWLEDGEMENT

Ministry of Health, Nutrition & Indigenous Medicine, Sri Lanka, International Agency for Research on Cancer (IARC), Health Informatics Society of Sri Lanka (HISSL).

REFERENCES

(IARC), 2015. International Agency for Research on Cancer (IARC). [Online] Available at: http://www.iarc.fr/ [Accessed 16 September 2015].

Association, N. C. R., 2016. Cancer Informatics. [Online] Available at: http://www.ncrausa.org/i4a/pages/index.cfm?pageid=3934 [Accessed 22 September 2016].

Bassil, Y., 2012. A Comparative Study on the Performance of the Top DBMS Systems. Journal of Computer Science & Re-search (JCSCR), 1(1), pp. 20-31.

Bear, D., 2012. Making Friends with Jarvis Cocker: Music Cul-ture in the Context of web 2.0. Culture Sociology, 2(2), pp. 222-241.

Bianconi, F. et al., 2012. Information Technology as Tools for Cancer Registry and Regional Cancer network Integration. IEEE Transactions on Systems, Man, and Cybernetics-Part A:Systems and Humans, 42(6), pp. 1410-1424.

Bray, F. et al., 2012. Global cancer transitions according to the Human Development Index (2008-2030): a population-based study. Lancet Oncol, 13(8), pp. 790-801.

Bray, F. et al., 2014. Planning and Developing Population-Based Cancer Registration in Low-and-Middle-Income Settings, Lyon: International Agency for Research on Cancer.

Bray, F., Kohlar, B. & Ferlay, J., 2014. Data Comparability and Quality. Cancer Incidence in Five Continents Volume X, 7, x(7), pp. 89-97.

Centers for Disease Control and Prevention, 2012. The Role of Public Health Informatics in Enhancing Public Health Surveillance. MMWR, 27 July , pp. 20-24.

Drolet, B. & Johnson, K., 2008. Categorizing the world of registries. Journal of Biomedical Informatics, Volume 41, pp. 1009-1020.

Fritz, A. et al., 2000. International Classification of Disease for Oncology, Third Edition. 3rd ed. Geneva: World Health Organization.

Jason, O. & Whelan, S., 1991. Planning a cancer registry. In:Cancer Registration: Principles and Methods, Lyon: InternationalAgency for Research on Cancer.

Jensen, O. et al., 1991. Purposes and uses of Cancer Registration. In: O. Jensen & H. Storm, eds. Cancer Registration: Principles And Methods. Lyon: International Agency for Research on Cancer, pp. 13-26.

Leal, Y., Fernandez-Garote, L., Mohar-Betancourt, A. & Meneses-Garcia, A., 2016. The importance of registries in Cancer Control. Salud Publica Mex, Volume 58, pp. 309-316.

National Cancer Control Programme, 2013. Cancer Incidence Data, Sri Lanka 2007, Colombo: National Cancer Control Programme.

National Cancer Control Programme, 2015. Cancer Incidence Data Sri Lanka 2015, Colombo: National Cancer Control Programme.

Rabah, S., Li, J., Liu, M. & Lai, Y., n.d. Comparative Studies of 10 Programming Languages within 10 Diverse Criteria -a Team 10 COMP6411-S10 Term Report, Montreal: Department of Computer Science and Software Engineering, Concordia University.

Rajaraman, S., 2012. An evaluation of Survival of Cancer Patients Based on Registry Data from Low or Medium Resource Countries, Finland: University of Tampere.

Silva, IDS., 1999. Cancer Epidemiology: Principles and Methods. Lyon: International Agency for Research on Cancer.

Sri Lanka, Department of Census and Statistics., 2016. [Online] Available at: http://www.statistics.gov.lk/PopHouSat/VitalStatistics/Bulletin/Annual%20bulletin%202015.pdf [Accessed 10 November 2016].

Stroe, I., 2011. MySQL databases as part of the Online Busi-ness, using a platform based on Linux. Database Systems Journal, 2(3/2011), pp. 3-11.

TELE-MICROSCOPY SYSTEM BASED ON MOBILE TELEPHONE FOR DISTANCE DIAGNOSIS OF TUBERCULOSIS IN REMOTE AREAS OF PARAGUAY

Virgilio Cane[1], Margarita Samudio[1], Juan García[1], Natalie Weiler[2] and Xabier Basogain[3]
[1]*Research Institute for Health Sciences - National University of Asunción, Paraguay, Campus Universitario, San Lorenzo*
[2]*Central Laboratory of Public Health. Ministry of Public Health of Paraguay, Asunción, Paraguay*
[3]*School of Engineering - EHU - University of the Basque Country, Campus Universitario, Bilbao, España*

ABSTRACT

Optical microscopy provides a simple and cost-effective method of laboratory diagnosis, which is essential for the diagnosis of infectious and hematological diseases. In many regions of the world, especially in remote areas, there are no health personnel trained for performing a quality diagnosis of these diseases. Therefore, a tele-microscopy system that helps health professionals at national reference centers through remote interpretation of the obtained images will be very important. Tuberculosis remains a major public health problem in Paraguay. Diagnosis and monitoring of treatment are based mainly on bacilloscopy (examination of a sputum smear to verify the presence of specific bacilli). This test can be carried out in a laboratory with minimal resources such as Ziehl Neelsen reagents and a microscope. Most of the primary care centers in Paraguay have these resources. It is a simple, efficient and low-cost test. The objective of this project is to develop a tele-microscopy system for the efficient diagnosis of tuberculosis in remote areas of Paraguay using transmission technology and data storage via mobile networks. This tele-microscopy system will be initially applied to remote interpretation of the cases and evaluation of the smears within the quality control framework of tuberculosis diagnosis carried out at the Central Laboratory of Public Health. Furthermore, its application could be extended to the diagnosis of other diseases of public health importance in Paraguay such as Chagas disease, malaria, leishmaniosis and superficial and deep mycosis.

KEYWORDS

Tele-Microscopy, Tuberculosis, Smartphone, Diagnosis, Internet

1. INTRODUCTION

Light microscopy is an essential instrument in the health service (Breslauer, Maamari, Switz, Lam, & Fletcher, 2009). Historically, microscopic image was taken through a permanently mounted camera unit on a microscope, attaching a camera to an ocular lens or to a port in a trinocular head using a C-mount adapter. Two main types of camera sensors are typically attached to microscopes – charged couple devices (CCD) and complementary metal oxide semiconductor (CMOS) sensors (Park, Pantanowitz, & Parwani, 2012). CCD cameras were preferable for the highest quality image. However, CMOS cameras have narrowed the difference in quality and essentially are equivalent for the majority of photomicrographs. Both types of sensors capture more pixels than is often necessary for acquiring optimal photomicrographs (Park, Pantanowitz, et al., 2012).

Since smartphones were introduced, phones had integrated cameras utilizing CMOS sensors (Park, Parwani, Satyanarayanan, & Pantanowitz, 2012). Cell phone cameras have undergone significant improvements with each new phone release. They are potentially suitable for high quality image acquisition of photomicrographs taken from smears via a microscope.

Many investigators have studied the use and quality of image capture using smartphone cameras from bright-field and fluorescent microscopes (Bellina & Missoni, 2009; Breslauer et al., 2009; McLean, Jury, Bazeos, & Lewis, 2009; Zhu, Yaglidere, Su, Tseng, & Ozcan, 2011). Images were captured either using manual positioning of the phone camera against the microscope eyepiece (Bellina & Missoni, 2009) or by using a modified handheld imaging device for image capture (without using a microscope) (Tseng et al.,

2010; Zhu et al., 2011). Several authors have found that the images acquired via cellphone adapters are comparable with standard captured images (Lehman & Gibson, 2013; Wimmer et al., 2012).

In remote rural areas where people affected by endemic diseases such as tuberculosis live, diagnostic resources are insufficient as well as the means of verifying interpretations or performing quality control of diagnoses. Much of the power of light microscopy, especially fluorescence imaging, the opportunity for remote consultation and electronic record keeping, remains inaccessible in these places due to prohibitive equipment and training costs. This is especially problematic since the diagnosis, screening, and monitoring of treatment for diseases such as tuberculosis depend on light microscopy as a screening tool or a definitive diagnostic test (Dawson, Kim, World Health Organization, Regional Office for the Western Pacific, & Stop TB Special Project, 2003; Steingart et al., 2006). The advent of digital imaging has improved the diagnosis by means of telemedicine, which sends sending images to more technologically advance health centers for consultation, evaluation and quality control. Build tele-microscopy networks may help overcome clinical microbiology infrastructure challenges that are present in resource-limited areas of the world (Rhoads, Mathison, Bishop, da Silva, & Pantanowitz, 2015).

Taking advantage of the increase in the coverage of mobile networks, telephones equipped with digital cameras, 3D printing and web-type applications, the goal of the study was to develop a tele-microscopy system for clinical diagnosis and registration of diseases of public health significance in Paraguay such as tuberculosis. The tele-microscopy system was implemented through the integration of different sustainable technologies that include mobile networks, web applications, cell phone camera and coupling with a 3D printed microscope. Laboratory personnel in remote rural areas was able to capture images of smears using smartphones attached to a light microscope in order that a more detailed analysis could be carried out by medical experts located at the central headquarter. The system will also allow storage, management and analysis of the medical records (images, diagnosis, treatment, and geolocation) to monitor endemic diseases and provide quality control of remote interpretations of the smears.

2. MATERIALS AND METHODS

Phone-microscope adapter: An adapter was designed for various types of telephones and microscopes. It was printed in a 3D printer with plastic PLA due to its rigidity. It was designed in AutoCAD 3D in its 2014 version. Figure 1 shows the assembly and use of the support for the tele-microscopy system.

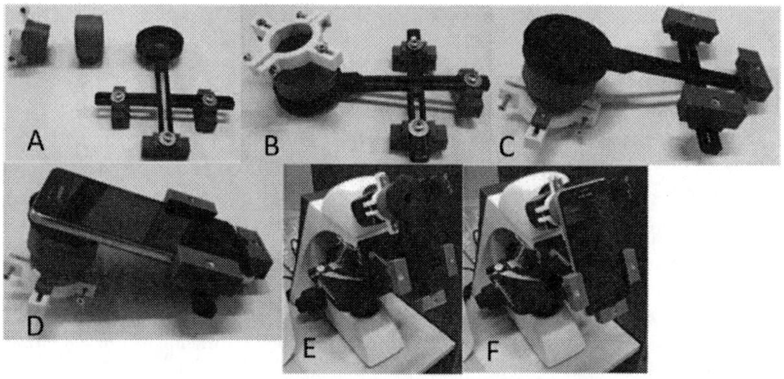

Figure 1. Adapter printed in a 3D Printer: A-C: parts; D: attached to the Phone; E and F: coupled to the Phone and Microscope

Design of the 3D printed adapter allowed a firm attachment to the microscope avoiding the effect of movement in the captured photograph. The quality of the images was evaluated by the laboratory personnel.

WEB Application: A software application was developed to create a computer record as a clinical electronic record for the patient personal data and smear images. The features of the application are as follows : a) it uses HTML5 and JavaScript as an access interface for the client; b) PHP as a server-side programming language; c) The Apache HTTP server as a server of WEB pages, and d) the data is stored in a database management system called MariaDB. To manage the support of the Centos 7 operating system, a DELL R720 server was used. The web access to the system is the URL www.iics.una.py/tm. Figure 2 shows different displays of the web application for the registration and monitoring of clinical electronic records of the tele-microscopy system.

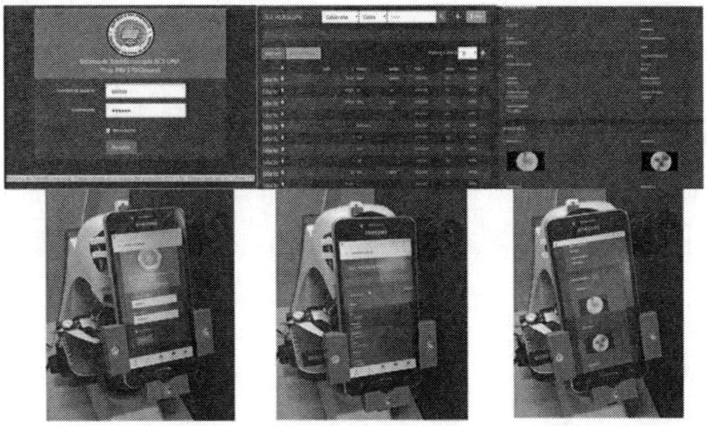

Figure 2. Interface of the WEB Application from the Computer and from the Mobile Phone

Access to the web application by the cell phone of the health personnel in remote rural areas allows the registration of patient data that includes personal identification and images obtained from the microscope by the smartphone camera. Access to previously registered information can be made from any device connected to the internet for later use.

The WEB application transmits information using "Secure Sockets Layer" (SSL) connections, and the data is encrypted in the database for greater security.

Mobile phones equipped with iOS 6 or higher and Android 4.0 "Ice Cream Sandwich" or higher can use an input type = "file" with a capture attribute that has the value "camera". This will launch the camera application and the photo taken by the user will be sent when the HTML form containing the input type = "file" is sent. The incorporation of images to the WEB application file is made during the fillingof this, avoiding the storage of images in the mobile phone.

In addition, the application incorporates a mechanism to prevent simultaneous edition of the registry by different users. All the user actions (Entries / Exits / created registries / modified registries / deleted registries) are stored in a registry of the system.

3. RESULTS

The adapter was successfully tested with several samples under the optical microscope. Once the microscope and the mobile phone were coupled, clear photos were taken without difficulty. In the AutoCAD design, some coupling details were considered in order to be compatible with the majority of microscopes and mobile phones In addition, the ergonomic design aspects was taken into account for the use of the phone coupling.

With half a kilo of PLA plastic, low cost printing of several couplings was achieved, eliminating the need to buy commercial adapters.

The transmission of information in real time over the internet allowed the patient data (personal and images) to be available for diagnosis or control in real time, and from several central and remote points through the WEB application of the tele-microscopy system. The WEB application also allowed listing, and classification of sendings and receptions according to the user and profile assigned.

The WEB application had two profiles for the users; 1) Issuer: It allowed creation of the electronic files with the personal data of the patient and the images captured from the microscope attached to the mobile phone as well as an interpretation of the images. 2) Receiver: It allowed editing of the electronic files created and to issue a diagnosis of the received data and images. These profiles were used according to the type of remote or central point as appropriate. The users were identified at the beginning of the session through their credentials (username and password) that associated them with the institution to which they belonged.

4. CONCLUSION

In this first phase of the project, an adapter for most optical microscopes and mobile telephones, and a web system for the management of clinical electronic files of patients that allows the attachment of images were developed. The electronic file developed allowed the integration of patient data with the images obtained from the mobile phone coupled to the microscope. Access to this file through the Internet helped to interpret the study performed remotely through the diagnosis of specialists at the central level. In addition, quality control was carried out without the physical remission of the smears from the remote laboratory to the reference level.

ACKNOWLEDGEMENTS

Financial support was received from PROCIENCIA-CONACYT within the project 14INV170 framework.

REFERENCES

Bellina, L., & Missoni, E. (2009). Mobile cell-phones (M-phones) in telemicroscopy: increasing connectivity of isolated laboratories. *Diagnostic Pathology*, *4*, 19. https://doi.org/10.1186/1746-1596-4-19

Breslauer, D. N., Maamari, R. N., Switz, N. A., Lam, W. A., & Fletcher, D. A. (2009). Mobile phone based clinical microscopy for global health applications. *PloS One*, *4*(7), e6320. https://doi.org/10.1371/journal.pone.0006320

Dawson, D., Kim, S. J., World Health Organization, Regional Office for the Western Pacific, & Stop TB Special Project. (2003). *Quality assurance of sputum microscopy in DOTS programmes: regional guidelines for countries in the Western Pacific*. Manila, Philippines: World Health Organization, Regional Office for the Western Pacific, Stop TB Special Project.

Lehman, J. S., & Gibson, L. E. (2013). Smart teledermatopathology: a feasibility study of novel, high-value, portable, widely accessible and intuitive telepathology methods using handheld electronic devices. *Journal of Cutaneous Pathology*, *40*(5), 513-518. https://doi.org/10.1111/cup.12108

McLean, R., Jury, C., Bazeos, A., & Lewis, S. M. (2009). Application of camera phones in telehaematology. *Journal of Telemedicine and Telecare*, *15*(7), 339-343. https://doi.org/10.1258/jtt.2009.090114

Park, S., Pantanowitz, L., & Parwani, A. V. (2012). Digital imaging in pathology. *Clinics in Laboratory Medicine*, *32*(4), 557-584. https://doi.org/10.1016/j.cll.2012.07.006

Park, S., Parwani, A., Satyanarayanan, M., & Pantanowitz, L. (2012). Handheld computing in pathology. *Journal of Pathology Informatics*, *3*, 15. https://doi.org/10.4103/2153-3539.95127

Rhoads, D. D., Mathison, B. A., Bishop, H. S., da Silva, A. J., & Pantanowitz, L. (2015). Review of Telemicrobiology. *Archives of Pathology & Laboratory Medicine*, *140*(4), 362-370. https://doi.org/10.5858/arpa.2015-0116-RA

Steingart, K. R., Henry, M., Ng, V., Hopewell, P. C., Ramsay, A., Cunningham, J., … Pai, M. (2006). Fluorescence versus conventional sputum smear microscopy for tuberculosis: a systematic review. *The Lancet. Infectious Diseases*, *6*(9), 570-581. https://doi.org/10.1016/S1473-3099(06)70578-3

Tseng, D., Mudanyali, O., Oztoprak, C., Isikman, S. O., Sencan, I., Yaglidere, O., & Ozcan, A. (2010). Lensfree microscopy on a cellphone. *Lab on a Chip*, *10*(14), 1787-1792. https://doi.org/10.1039/c003477k

Wimmer, J., Dhurandhar, B., Fairley, T., Ramsaywak, J., Scheiber-Pacht, M., Satrum, L., … Thrall, M. (2012). A Novel Smartphone-Microscope Camera Adapter: An Option for Cytology Consultation in Low-Resource Environments. *Journal of the American Society of Cytopathology*, *1*, S124-S125. https://doi.org/10.1016/j.jasc.2012.08.268

Zhu, H., Yaglidere, O., Su, T.-W., Tseng, D., & Ozcan, A. (2011). Cost-effective and compact wide-field fluorescent imaging on a cell-phone. *Lab on a Chip*, *11*(2), 315-322. https://doi.org/10.1039/c0lc00358a

COGGYM GAMES: A STRUCTURED TECHNOLOGY BASED THERAPY TO PREVENT POST INTENSIVE CARE SYNDROME AMONG CRITICALLY ILL PATIENTS

Mona M. El-Hady[1] and Samaa M. Shohieb[2]
[1]*Assistant Professor, Emergency and Critical Care Nursing Dept., Faculty of Nursing, Mansoura University, Egypt;*
Visiting Professor, Emergency and Critical Care Nursing Dept.,
Faculty of Nursing, King AbdAlaziz University, Saudi Arabia
[2]*Assistant Professor, Information Systems Dept., Faculty of Computers and Information, Mansoura University, Egypt*

ABSTRACT

Critically ill patients experience newly acquired long-term cognitive impairment, major physical, mental and functional disabilities. Those patients are given the diagnosis of post-intensive care syndrome. Integrating technology-based activities into occupational therapy sessions means; Intensive Care Unit patients will be given the chance to challenge their cognitive abilities as well as their physical function, and to interact with technology within a physically limiting environment. This paper presents a complete research design (from medical and technical points of views) for a structured technology based therapy to prevent PICS among critically ill patients. Researchers called the new app CogGym (Cognition Gymnastics). CogGym is an android- based devices application with an Arabic User Interface that contains many flexible and easy to use games designed specifically for the Arab critically ill patients. Ditto, CogGym enables them to continue to do therapeutic gymnastics after the therapist leaves the room to keep working toward rehabilitation goals.

KEYWORDS

Technology Based Therapy, Cognition Games, Intensive Care Unit, Post-Intensive Care Syndrome, Critically Ill Patients

1. INTRODUCTION AND MOTIVATION

Yearly, millions of critically ill patients experience newly acquired long-term cognitive impairment and major physical, mental and functional disabilities (Jackson et al., 2012). Those patients are given the diagnosis of post-intensive care syndrome (PICS). PICS is known as a new or worsening impairment in cognitive (e.g., judgment and thinking), physical (e.g., intensive care unit-acquired neuromuscular weakness), or mental functions, developing after acute disorders and continuo beyond the critical care setting discharge (Robinson, 2013).

The main risk factors relevant to PICS include delirium duration in intensive care unit (ICU), acute brain disorders, hypoxic status, hypotension (involving severe sepsis and trauma), glucose deregulation, prolonged mechanical ventilation "MV", use of renal replacement therapy, and history of cognitive impairment, advanced age, preexisting cognitive deficits, or pre-morbid health conditions (Rawal, Yadav & Kumar, 2017).

Patients with PICS can experience different limitations and disabilities that are usually difficult to recognize. These limitations can be apparent in the form of neuromuscular dysfunction, or less evident, like mild cognitive impairments (Robinson, 2013). PICS often persist beyond acute care hospitalization, and may last for months or years that affecting patient's quality of life (Parker, Sricharoenchai & Needham et al., 2013). PICS is recently recognized as a public health burden associated with neuropsychological and functional impairments, however its exact prevalence still vague (Rawal, Yadav & Kumar, 2017).

Cognitive impairment has been documented to occur in more than 75 % of ICU survivors (Needham et al., 2013; Pandharipande et al., 2013; Davydow et al., 2013). Whoever, the risk of developing psychological impairment after ICU discharge, ranges from 1 to 62 % and ICU-acquired neuromuscular weakness is developing in more than 25% of ICU survivors (Fan et al., 2014; Hermans et al., 2014).

Prevent PICS is better than cure. The most important preventive strategies for PICS involve reducing deep sedation used and encouraging patient's early mobility, concomitant with aggressive physical and occupational management (Kress 2013).

Mobile phones are no longer a luxury item. Every person may have one or more mobiles, as well as children. Based on these developments, advanced technology as tablet and smart mobile phones are considered a successful alternative method to assist occupational therapy practitioners for completing therapeutic activities in the ICU setting (Needham, Truong, & Fan, 2009). The tablet is already being used as a tool for management patients in school systems and outpatient rehabilitation facilities through encourage social participation, communication, motor planning, and sensory perception (Hoesterey & Chappelle, 2012; Waite, 2012). However, there are very limited studies all over the world in general, and in Arab world specifically, that investigate the benefits of using this therapy on critically ill patients.

Baysari et. al. (2015) and Erickson and Kelly (2015) implemented a mobile app-based device that was incorporated in the occupational therapy process. In Egypt, there is no research that studies this technology in ICUs. This paper therefore presents CogGym, an Arabic language vocabulary learning game. An android- based devices application with an Arabic User Interface (UI) that contains many flexible and easy use games designed specifically for the Arab critically ill patients to prevent them from PICS. Ditto, CogGym enables them to continue to do therapeutic gymnastics after the therapist leaves the room to keep working toward rehabilitation goals.

The second section of this paper will introduce the aim of the study. Section 3 contains a complete medical plan to deploy the proposed game on real critically ill patients. Deeping into the proposed system will be in section 5. Finally the conclusions and future work will be presented.

2. AIM OF THE STUDY

The aim of the study is to evaluate the effect of implementing a structured technology based therapy on preventing post intensive care syndrome among critically ill patients.

2.1 Hypothesis

1. Technology based therapy will maintain neuro-cognitive capabilities and prevent cognitive impairment among critically ill patients.
2. Technology based therapy will reduce psychological disability among critically ill patients.
3. Technology based therapy will eliminate incidence rate of ICU-acquired neuromuscular weakness.

2.2 Operational Definition

Structured technology based therapy is defined as a mobile-based cognitive training intervention that is used to prevent physical, cognitive and psychological impairments.

3. METHODOLOGY

3.1 Design

A quasi-experimental design will be used to achieve the aim of the current study.

3.2 Setting

The study will be performed in respiratory and postoperative ICU at Mansoura University Hospital, Egypt.

3.3 Subjects

Convenience sampling technique of 120 patients admitting to the predetermined setting will be enrolled in the study. The sample size was estimated by using the EPI info program with confidence coefficient at 90%.

3.4 Inclusion Criteria

Patients who are newly admitted to ICU, alert, mechanically ventilated, hemodynamic stable, normoglycemia and have some form of active range of motion in at least one upper extremity will be enrolled in the study.

3.5 Exclusion Criteria

Patients who have delirium, visual or auditory deficit, spinal disorder, cognitive impairment, short-term expected mortality or admission following cardiac arrest will be excluded from the study.

3.6 Medical Tools

Three tools will be developed by the researchers after reviewing updated literature review (Wergin & Modrykamien, 2012; Jean et al., 2010; Malinowsky et al., 2015).

3.6.1 Tool (I) "Post ICU Syndrome Assessment Sheet"

It's used to collect data about sociodemographic and clinical characteristics and pulmonary function parameters.

3.6.2 Tool (II) "Cognitive Stimulator observation Sheet"

It used to collect data about the received medications affecting patient's neuro-cognitive functions.

3.6.3 Tool (III) "Structured Technology Based Therapy Outcome Follow up Sheet"

It includes twice observations for cognitive capabilities using Johns Hopkins Adapted Cognitive Exam, delirium using CAM-ICU, psychological impairment using Hospital Anxiety and Depression Scale, upper extremities Muscle strength by using Medical Research Council Manual Muscle Testing scale, and vital signs.

3.7 Data Collection

A convenient sample of 120 intensive care unit patients who fulfill inclusion criteria will be included in the study. The subjects will be assigned randomly to a STBT group (30), cognitive stimulation group (30), STBT& cognitive stimulation group (30) & a control group (30). The demographic data will be collected on first day of patient's admission to ICU. Data will be collected on the indicators for PICS during ICU hospitalization period.

For the STBT group, the subjects will receive STBT by the researchers. STBT will involve 10 sessions which perform at a fixed time every day. Each session has 3 levels of complexity and the participant give 7-10 minutes to success in each level to be able to transfer the highest level. All sessions turned on tablet by the researcher and informing the patient about the apps technique as showed in the instruction letter. Cognitive stimulation group will receive pharmacological therapy to prevent PICS by the researchers. STBT& cognitive stimulation group will be given both pharmacological therapy to prevent PICS and STBT by the researchers. The control group will receive the routine care by the ICUs health care providers. The indicators for PICS will be assessed for subjects in the studied groups daily throughout ICU hospitalization period using Structured Technology Based Therapy Outcome Follow up Sheet.

3.8 Ethical Considerations

An ethical approval is obtained from the ethics committee of Faculty of Nursing, Mansoura University. An informed consent will be obtained from patients who will accept to participate in the study after providing detailed description of study. Data confidentiality will be maintained.

4. THE PROPOSED "COGGYM" GAMES

The CogGym game is categorized as an assistive technology that runs on Android-based devices (mobiles or tablets), whose aim is preventing physical, cognitive and psychological impairments among critically ill patients.

4.1 CogGym Design

In the development of our system, authors used the design process for Human Computer Interaction (HCI) science to design CogGym (Figure 1). The explanation of the design methodology as follows:

a. What's wanted & problem identification (User experience research): Actions conducted by researchers in identifying the problem are:

1. Conducting personal interviews with ICU doctors.
2. Conducting field observation (observing critically ill patients' activities in the open field.)
3. Collecting and analysis of related research literature.

b. Analysis Phase: researchers analyzed the findings that in Egypt, as one of the developing countries, it is very difficult (economically) for the ICU to maintain and preserve the critically ill patients form the PICS. Consequently, researchers established their goals in developing a game to solve the mentioned valuable issues.

c. Design Phase: In this process we began the software development and prototyping for the application. That was made in terms of construction models and methods. We began with to sketching the game as a paper-based prototype, then building the algorithm.

d. Implementation and Deployment Phase: The next stage, was defining the major features of the game. On the other hand, researchers designed the prototype interface to be displayed on a full screen of a mobile or a tablet device. Also, a tutorial is used to simulate every feature's functionality and demonstrate the main countenances to both the patient and the caregivers for assuring that countenances satisfied their expectations.

4.2 CogGym Specification and Description

The entire CogGym application interface is in Arabic language. CogGym contains 10 sub-games that are used to enhance the ten different perceptions capabilities for each person, which are (Visual Discrimination, Figure Ground Perception, Form Constancy, Visual Closure, Visual Memory, Visual Sequential Memory, Visual Spatial Relationship, Visual Motor Integration, Visual Tracking, and Visual Focus).

The first level is always a training level to teach the player how to play the game. After completing each level there will be a soft sound of clapping to motivate the player and inform him that he has completed the level successfully. CogGym has not any disturbing noisy sounds or background music to not disturb or/and de-concentrate the patient. The main screen of the game has four buttons (Figure 2) namely:

- "Play" Button: This button is specified for patient to start playing and do the practicing process.
- "Caregiver" Button: This allows therapists and caregivers to follow the patients' improvements and overall performances along the time. The overall information for the caregivers to assess his/her patient is patient's name, last level that the patient arrived, number of levels played, number of times to win and loose, number of icons, best score in the time based level, time of starting last play and time of ending, total time for the last level, total time in playing the game, the percentage of winning times, longest and shortest time spent in a level. Also, the caregiver can change the patient's name / choose the types of icons that he needs his patient to learn (selected form the ten types) from their control panel after some verification.
- "About Us" Button: that contains general information and data about the game's designers.
- "Exit" Button: To exit from the game.

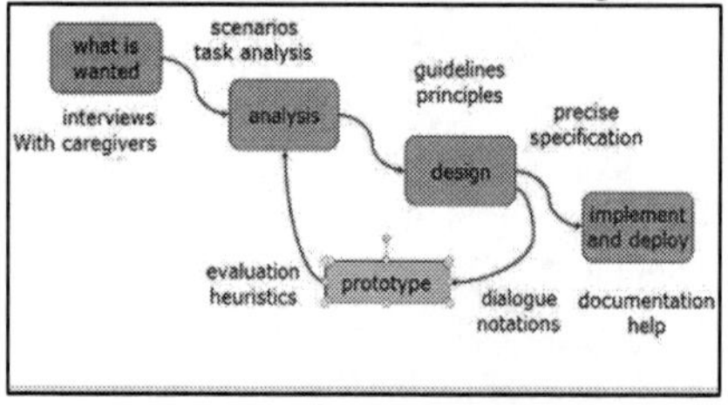

Figure 1. HCI- Based Design Process

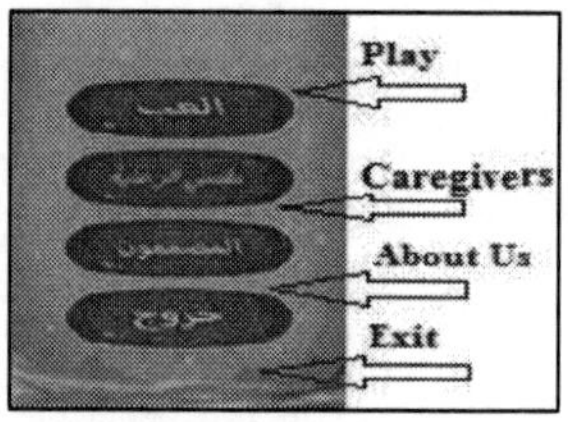

Figure 2. Main screen of CogGym

5. CONCLUSION & FUTURE WORK

Critically ill patients usually suffer some degree of decline in their cognitive capacities. Those patients are given the diagnosis of post-intensive care syndrome (PICS). Technology based therapy provides new training chances to improve the decrease in cognitive functions. This paper detailed the design of CogGym. CogGym contains serious games with Arabic UI helping to train planning, memory, attention and reasoning. CogGym has been designed as a structured technology based therapy to prevent PICS among Arab critically ill patients. This paper also contains a complete medical proposal to deploy CogGym on patients in Mansoura University hospitals in Egypt. The application will be applied for the critical ill patients and feedbacks will be collected to enhance the application usability consequently achieving the most benefits for this type of patients.

REFERENCES

Baysari, K. Adams, E. C. Lehnbom, J. I. Westbrook, R. O. Day October 2014. iPad use at the bedside: a tool for engaging patients in care processes during ward rounds?. *Internal Medicine Journal,* Vol. 44. Issue 10, pp. 986–990 DOI: 10.1111/imj.12518

Davydow DS, Zatzick D, Hough CL, Katon WJ., 2013. In-hospital acute stress symptoms are associated with impairment in cognition 1 year after intensive care unit admission. *Ann Am Thorac Soc.* Vol. 10. pp. 450–57.

Erickson Kelly, 2015. Evidence Considerations for Mobile Devices in the Occupational Therapy Process. *The Open Journal of Occupational Therapy*: Vol. 3: Issue 2, Article 7. : https://doi.org/10.15453/2168-6408.1132

Fan E, Cheek F, Chlan L, Gosselink R, Hart N, Herridge MS, 2014. Physical complications in acute lung injury survivors: a two-year longitudinal prospective study. *Crit Care Med.* Vol. 42. Issue 2. pp. 849–59.

Formal Game Elements, [Online] Available: http://www.acagamic.com/courses/infr1330-2014/theformalsystems of-games-and-game-design-atoms/

Jackson JC1, Ely EW, Morey MC, Anderson VM, Denne LB, Clune J, 2012. Cognitive and physical rehabilitation of intensive care unit survivors: results of the RETURN randomized controlled pilot investigation. *Crit Care Med.* Vol. 40. pp. 1088–1097

Hermans G, Van Mechelen H, Clerckx B, Vanhullebusch T, Mesotten D, Wilmer A, 2014. Acute outcomes and 1-year mortality of intensive care unit-acquired weakness: A cohort study and propensity-matched analysis. *Am J Respir Crit Care Med.* Vol. 190. pp. 410–420.

Hoesterey C & Chappelle C., 2012. Touch the future: Using iPads as a therapeutic tool. *OT Practice.* Vol. 17. Issue 13. pp. 7–9.

Jean L, Bergeron ME, Thivierge S, Simard M., 2010. Cognitive intervention programs for individuals with mild cognitive impairment: systematic review of the literature. *Am J Geriatr Psychiatry.* Vol. 18.pp. 281–296.

Kress JP., 2013. Sedation and mobility: changing the paradigm. *Crit Care Clin.* Vol. 29. pp. 67–75.

Malinowsky Camilla, Anders Kottorp , Rumi Tanemura, Eric Asaba , Toru Nagao, Kazue Noda, Jiro Sagara, December 2015. "Validation of the Everyday Technology Use Questionnaire in a Japanese context", *Hong Kong Journal of Occupational Therapy*, Volume 26, pp. 1-8, https://doi.org/10.1016/j.hkjot.2015.08.002

Needham DM, Truong AD & Fan E., 2009. Technology to enhance physical rehabilitation of critically ill patients. *Critical Care Medicine.*Vol. 37. pp. S436–S441.

Needham DM, Dinglas VD, Morris PE, Jackson JC, Hough CL, Mendez-Tellez PA. 2013. Physical and cognitive performance of patients with acute lung injury 1 year after initial trophic versus full enteral feeding. EDEN trial follow-up. Am J Respir. *Crit Care Med.* Vol. 188. pp. 567–576.

Parker AM, Sricharoenchai T & Needham DM., 2013. Early rehabilitation in the intensive care unit: Preventing impairment of physical and mental health. *Current Physical Medicine and Rehabilitation Reports.* pp. 307–314.

Pandharipande PP, Girard TD, Jackson JC, Morandi A, Thompson JL, Pun BT., 2013. Long-term cognitive impairment after critical illness. *N Engl J Med.* Vol. 369. pp. 1306–1316.

Robinson MR., 2013. Post intensive care syndrome: The important implications for occupational therapy practitioners. *OT Practice.* Vol. 18. Issue 13. pp. 7–9.

Rawal G., Yadav S., Kumar R., 2017. Post-intensive Care Syndrome: an Overview. *J Transl Int Med.*Vol. 5. Issue. 2. pp. 90–92.

Waite A. , 2012. App"titude: Smart gadget applications showing their worth in practice. *OT Practice.* Vol. 17. Issue. 12. pp. 9–12.

Wergin R., Modrykamien A. 2012. Cognitive impairment in ICU survivors: Assessment and therapy. *Cleveland Clinic Journal of Medicine.*Vol. 79. Issue 10. pp. 705-712.

SHARING AND COLLABORATING WITH THE END-USERS AROUND THE ROBOT EXTENTION PARADIGM. USING ROBOTS AND COMPUTERS BY OLDER PEOPLE WITH DEMENTIA

Dimitri Delacroix[1], Rénald Gaboriau[2], Sophie Sakka[2], Laurent Billonnet[3] and Nicolas Couégnas[1]
[1]Université de Limoges – CeReS, EA 3648, 39 rue Camille Guerin 87000 Limoges- France
[2]Centrale Nantes – Laboratory of Numerical Sciences of Nantes (LS2N, UMR 6004)
1 rue de la Noe – BP 92101 – 44321 Nantes cedex 03 - France
[3]Université de Limoges- XLIM, UMR CNRS n°7252, 123 avenue Albert Thomas – 87060 Limoges cedex -France

ABSTRACT

Rob'Zheimer proposes an approach to improve communication skills of people with dementia. The program focuses on the people's active position and checks the participation effects. Rob'Zheimer is divided into 20 sessions of 1 hour each, alternating non-robotic and robotic sessions. During non-robotic sessions people register the voices reading story and create the scenery. During the robotic sessions, people program the robot gestures according to the registered voices. In this program, the robots are used as an extension, and not as a companion as traditionally performed in other research programs. Consequently, people use it as a tool in order to communicate with others showing improvements of their communication. This article review a theorical approach to set up a use of robots by older people with dementia and preliminary observations of Rob'Zheimer.

KEYWORDS

Robot Extension, Elderly People, Alzheimer Disease, Autonomy, Active Participation, Mutual Shaping

1. INTRODUCTION

Robotic services and therapies know a strong development and interest for older people and especially for people with dementia like Alzheimer diseases (AD). People present difficulties that be managed by non-pharmacological therapies; uses of robot integrate this categorization. According to studies they show benefits like reducing stress and foster communication. In this context robots are pre-determined and used in a second step by users. Scenarios are created by the stakeholders: medical staff, robotic industrials and politics. The paradigm developed is the companion robot that splits in two sections: the robot as a caregiver or the older people as caregiver. Although studies show positives effects – the results have to be nuanced by the methods and evaluations – this technical determinism confines people in a passive position and generates negative effects. In contrast to this technical determinism this paper goes on to explore the use of mutual shaping of people and robots as a framework for social robot design. Sharing and collaborating in a service design involve the active people's participation in the process from the beginning: it is necessary to maintain people with Alzheimer disease as a subject. According to the design service the risk of reification is strong and in this way both stakeholders and people are losing because none mutual shaping exists between humans and non-human elements. To illustrate this process this paper describes a preliminary work aimed at setting a communication support for Alzheimer's disease using three humanoid robot NAO (SoftBank Robotics). The approach is based on master puppet concept: the subjects program the robots, and use them an extension for communication. Twenty sessions are organized, alterning ten preparatory sessions and ten robotics programming sessions. During the preparatory sessions, the subject prepares the story: voice recording, scenery creation, etc. During the robot programming session, the subjects program the motion to be realized to make the robot tell the story. The program is concluded by a public performance. The experiment involves six AD aged to 60-85. As first observations, a dynamic of the group a presence and action of everybody and very positive reactions are observed. The approach allows the subject to shift a passive position to an active one, and therefore helped them remaining human being.

2. ALZHEIMER DISEASE AND ROBOTS

2.1 Background

Alzheimer's diseases (AD) was identified in 1906 by Aloïs Alzheimer as an irreversible, progressive brain disorder. In last 30 years, researches have focused into symptoms, causes, risk factors and treatments. There are lots of elements that they are still unknown. About symptoms, AD affects people in different ways during a long process that be divides in stages. So, the AD is qualified as early, moderate or advanced. This categorization is established by the mini mental state (MMS) scores: it consists of 30 questions; if the score is more than 20 the disease is considerate as early, between 10 and 20 as moderate and less than 10 as advanced. But every generalization about AD is dangerous because every patient is different, and the situations are very complexes; so different steps can be multiplied. For all stages of the dementia and overall for the last one the pathological process affects cognition, functioning, behavior. The following are common symptoms of AD: memory loss, poor judgment, wandering and getting lost, mood and personality changes, increased anxiety and/or aggression, losing things or misplacing them in odd places, difficulties completing familiar tasks, etc. According to stages the intensity of symptoms are different. For example, in the final stage patients seem to loss every means of communication. In the degenerative process communication is much altered. In generalizing, AD can ben intended by the illness of the link, of the relationship. Links to the previous world and to people are vanishing: links between people are attacked and are deleted. The "relation enter" is getting more complicated.

2.2 From Dependent People to a "Collectif Actant"

AD people become *dependent people* involving stakeholders: informal helpers, caregivers and medical structures. All stakeholders must work together to facilitate AD people's life course and to ensure their dignity. An AD person becomes an *individu collectif* (Thomas, 2015) e.g. a *collective person*, because several human beings work together and AD people belong to this collective. In semiotic terms, this notion refers to an "actant collectif". "Actant" means that a human or non-human element gets a role, and "collective" underlines the fact that they share the same goals, particularly the quality of life (QoL) of AD people. Standard indicators of QoL include lots of data on wealth or employment, and also on the whole environment. AD people's way of establishing classic judgment is altered by a diminution of the recognition of spacial and temporal units. As a consequence, enunciations are different and it is difficult to establish the QoL degree. For service design, QoL is seen as a basic human definition: AD people have to act, create and listen to influence their environment. In this complex environment, there is a mutual shaping between service design and the actant collectif. According to service design, the actant collectif does not exist or get another path that may not answer to AD people's QoL. According to service design and synergies created, the actant collectif could shift into an "actant social" as stated by Couégnas & Halary (2000, p.33). So, dependency is not seen as a negative term but as a necessary link to the "existing together". *(Couégnas & Halary, 2000, p.45).*

2.3 Non-pharmacological Therapies and Robot

Robots create phantasms, confusion and ambiguity because science fiction and industrial revolution precede technologies. So, an unconscious collective thinking exists and influences our imaginaries. Robots are described simultaneously as dangerous technologies or benefic companion. The use of the term "robot" is very large, and it must be outlined. A unified definition is lacking but common features are the fact that robots can interact with people in a socio-emotional way. (Wrobel *et al.*, 2014 ; Wu *et al.*, 2014). Robots meet informatics, control, electronics and mechanics fields. The categorizations done by Heerink (2010), on the one hand, and by Glende *et al.,* on the second hand, allow to see a large part of the robots and robotic devices.The robots used for AD people enter in the the assistive social robots (service or companion) Heerink's categorization. The main existing experimentations in order to support aging or Alzheimer environment are the giraff+, Care-O-Bot, Hobbit, Mobiserv, RobAlz, Accompagny, Domeo, Robot-Era,

Paro, Babyloid, CompanionAble Hector, Aibo, Nao. They show positive effects like diminution of stress and foster of interaction (Kachouie *et al.*, 2014). But some limits remain about methods. The first one is the fuzziness of the clarity and validity of results on current studies. The second limit points out the way people are integrated in the process and how they are able to influence it. Here, the rise of robotics imposes questioning the methods of development including how stakeholders manage AD people, care and robots, in order to constitute an actant social with AD people as actors.

Most experimentations currently follow to the companion robot paradigm – versus the extension robot paradigm that will illustrated in what follows – and AD people are excluded of the design process: they are kept in a passive position in this companion design process. The scenarios include the fact that the robot must be tailored before for each AD people. The technological – and medical – determinism is total; matters of concern for social robots and older people are underlined by Britt (2014) and as well in the RobAlz study. So AD people are excluded and it puts them in a passive position initiating a reification process. At this point of the analysis the passive or active position in the use of the robot determines the paradigm – companion or extension – that implies consequences for the protocol and, in fine, people. In the companion paradigm, the robot aims at replacing a human being in the interaction with AD people. The robot is pre-programmed with scenarios at the beginning of the design process without the end-users. They become an element among others but to avoid the total exclusion they are "consulted". Are they really integrated in the different processes? Consequences of a not interactive design are to not foster interactions between people. Thinking interactions before they take place eliminates creation of the life and it categorizes people into boxes and processes. This approach develops a process of *reification*. Reification can be defined when persons' judgment, consideration, activities or simple presence are deleted in a creation and decision process by humans or non-human elements. So, AD people first are excluded from the design process and they get a passive position. They do not act and when they do it, they answer to a stimulation done by few stakeholders: the actant collectif does not work. When stakeholders act in the robot companion paradigm way, they have too much power compared to AD persons. They decide about everything – even in an inconspicuous way – for themselves and not for the elderly people who become a mean to reach their goals. A robotic system acceptability does not work in a companion paradigm and presents issues in terms of ethics, use worthiness, usability. A service design with a robotic system for AD people needs a more intense construction and collaboration with the AD people and between stakeholders because without sharing and collaborating the system does not work. Moreover, it creates negative effects like infantilization, deception and reduction of human contact.To respond to the previous issues, a real integration process has to be developed with a dialogic link between human and non-human elements in limited adaptation and adoption – *concretization*, in Simondon's terms. Such approach exists in the robot extension paradigm.

The robot extension paradigm means that people use robots as tools to communicate, to take place and exist in the world. In this way, people act by themselves and they are becoming creators (of their life). They are not in a passive position like in the companion paradigm, but in an active one. They are not alone. Indeed, they use the robot in order collaborate with other persons while being protected in its rigid body. The robot, in this perspective, attracts people and actions. It is a collective adventure between all, the people exist together, and the robot allows a co-enunciation: the actant social is established. In AD situations, communication between people and caregivers is often very difficult. So, this approach can create a bend to generate communication, share affective and actions together.

This collective adventure correspond to a real design service conception with an effective integration and observable benefic effects on QoL: AD people remain subjects with their own enunciation. The different environments – human and non-human elements, robotic – meet and create a new environment: the mutual shaping provides a third environment, the actant social acts. This new environment does not result from the elimination of the previous one; but on the contrary they provide each other elements: there is a mutual shaping. This process can be associated to the notion of hyperthelie developed by Simondon (1958/p.62/64).

To conclude, in the companion paradigm, AD people are not implicated in a decision making process: they may be considered as a data for stakeholders. In other terms, people are a non-subject functional in Coquet's terms (2007). They act in a structure without getting goals and issues. They are marginalized as a person although they are in the center of stakeholders' interests. So, they live in a heteronomy environment: there is a whole objectivation up to become a functional being in Coquet 's perspective (2007, p.68). On the opposite side, in the extension paradigm, AD people do not play their parts, but a part imposed by stakeholders; they are "subjects who act" and they are put in a dynamic situation: they create. Overall, they decide, so they are in an autonomy sphere where they live in a heteronomy sphere. In the process they act with stakeholders. Power has to be shared by people for themselves.

In order to illustrate this process and how to involve people from the beginning to sharing and collaborating, the paper presents the Rob'Zheimer project. It is a communication support for AD using humanoid robots; it is inspired by Rob'Autism project that is a therapeutic support for autistic teenagers using humanoid robots (Gaboriau & Sakka (2016); Sakka & Gaboriau (2017)). It results from the collaboration between a nonprofit organization (*Robots !, dedicated to robotics and arts), a* nursing home (*Les Eglantines*) and the engineer school *Centale Nantes*. The project was born in 2017; the first-year program is applying in 2018. Six elderly people aged between 60 to 85 years old are participating in this experiment, 3 females and 3 males. They suffer from degenerative ills and they have some ability to read and write. The sessions take place in the retirement house (Les Eglantines - France). Two different rooms are necessary: one for non-robotic sessions and another one for programming sessions but the same during the experiment. The staff is composed of 5 persons: 2 animators and 3 caregivers. The program uses 3 humanoid robots NAO from SoftBank Robotics. The robots are programmed by the subjects using the software interface *Choregraphe :* it is the classical programming interface sold with the robots (i.e., no software is used for the programming). We may that the programming interface is in English and the subjects do not speak English. Nevertheless, this language specificity doesn't affect their ill to make the robot talk or move. Rob'Zheimer consist in 20 sessions of 1 hour each, once in a week. The sessions alternate 10 non-robotics and 10 robotics programming work groups. A public show concludes the program. Each session is linked to the preparation of the public show. The robotics sessions are dedicated to making the AD people program the robot. They are organized as follows: two AD people per robot (Nao, SoftBank Robotics) using the same computer. The working tables are in the center of the room in such disposition that each group can see the two other ones. Each group is assisted by one caregiver trained on using Choregraphe (Softbank Robotics software necessary to program the robot). An animator and a robotic specialist are also present in the room to help with specific programming requests and specific supervision. When AD people enter in the room, they find the robots and computers always at the same place. The robot programming sessions take place at Les Eglantines the retirement house where AD people are living. For the non-robotic sessions AD people are sitting in arc form. It is structure in four distinct moments: (1) speaking slot, (2) reading and/or recording the story, (3) activities and (4) singing time; and it must add the enter and the exit.

3. CONCLUSION

This paper explains the 20 weeks Rob'Zheimer protocol aimed at improving AD people creation and communication capabilities in an actor position within a protocol. People with Alzheimer's disease – due to the neuro-degenerative illness – can be considered as collective persons: persons with several stakeholders. They have to act all together: it is a complex and dynamic phenomenon that creates either autonomy or heteronomy environment. Robots as a mediation tool can foster actions and interactions between people and their environment. In this way robotic services and therapies know a strong development and interest for older people and especially for people with dementia like Alzheimer's disease. According to studies, the use of robots shows benefits like reducing stress and foster voluntary communication and interactions. In this context robots are pre-determined, pre-programmed and used in a second step by users as companion robots. This service design avoids sharing and collaborating with the end-users and creates a technical determinism. This one confines people in a passive position and generates negative effects, such as dependency to external solicitation. In opposition to this technical determinism, the robot can be viewed as an extension like every tool and may be used as a mediator. This paradigm creates a real use of mutual shaping of people and robots – and stakeholders. In this perspective, the whole of the service has to be challenged as a framework for social robot service design that involves AD people. The paper demonstrates that sharing and collaborating in a service design involving an active people's participation in the process from the beginning, keeps AD people as acting subjects, authors and creators of their life and avoids the reification process. Observed results showed fast changes in communication behavior, spontaneous interactions and presence feeling. Thanks to the use of the robots as an extension tool resident create situations with a mix of personal and imaginary elements.

This collective adventure tries to put first steps in the approaches and uses of the robot extension paradigm. The next step should be understanding in a deep level – after the fist experimentation- the other questions on the program effects.

REFERENCES

Banks, M.R. (2008) Animal-assited therapy and loneliness in nursing homes: use of robotic versus living dog. *JAMDA* -9-73.

Coquet, J-C. (2007) *Phusis et Logos. Une phénoménologie du langage.* (*A phenomenology of language*)Paris, France: Presses universitaires de Vincennes La philosophie hors de soi.

Couégnas, N. and Halary, M-P. (2000) De l'actant collectif à l'actant social : la communication identitaire d'une confédération syndicale.in Recherches socio-sémiotiques : l'actant collectif. (from the collective agent to the social agent : the identity communcation of the trade union confederation.) *Actes sémiotiques 71 ; 72.*

Ennuyer, B. (2014) *Repenser le maintien à domicile. Enjeux, acteurs, organisation.* (*Reconsider the aging in place. Issues, Actors, organisation.)* Paris, France : Dunod 2nd edition.

Escaffre, J-P. (2016) *Autonomie, dépendance et stratégie.*(*Autonomy, dependecy and strategies*)Paris, France : Nuvis.

Glende et al. (2015) Increasing the acceptance of assistive robots for older people through marketing strategies based on stakeholders needs. *International Journal of social Robotics.*

Greimas A.J. and Courtès.J., (1979) *Sémiotique. Dictionnaire raisonné de la théorie du langage. (Semiotic. Dictionnary of the language theory)* Paris, France : Hachette.

Heerink, M. (2010) Assessing acceptance of assistive social robots by aging adults. *Thesis completed* University of Amsterdam.

Heerink, M. et al. (2010) Assessing acceptance of assistive social agent technology by older adults: the Almere model. *International Journal of Social Robotics.*

Moyle, W. et al. (2013) Exploring the effects of companion robot on emotional expression in older adults with dementia: a pilot randomized controlled trial. *J Gerontol Nurs* 39:46-53.

Ostlund, B. (2014) Review: Seven Matters of Concern of Social Robots and Older people. *International Journal of Social Robotics* 6(2), 299-310.

Ricoeur, P. (1990) *Soi-même comme un autre.* (*Oneself as another*)Paris, France : Seuil.

Sakka, S. & Gaboriau, R. (2017) A robotic « Puppet Master » application to ASD therapeutic support. *World Academy of sicence, Engineering and technonoly.*

Salichs et al. (2015) Study of scenarios and technical requirements of a social assistive robot for Alzheimer's Disease patients and their caregivers. *International journal of social Robotics.*

Simondon, G. (1958) *Du mode d'existence des objets techniques.* (*On the mode of existence of technicla objects*) Paris, France : Aubier (2012).

Sim, D. & Loo, C. (2015*)* Extensive assessment and evaluation methodologies on assistive social robots for modelling human-robot interaction – A review. *Information Sciences.*

Thomas, H. (2005) Le métier de vieillard. Institutionnalisation de la dépendance et processus de désindividualisation dans la grande vieillesse. (*The eldery job. The intitutionalisation of the dependency and the desindividualisation process in the aging*) *Politix. n°72 De Boeck Supérieur.*

Wada, K. and Shibata, T. (2007) Living with seal robots – its sociopsychological and psychological influenceson the eldery at care house. *IEEE Trans robo*t; 23:972-80.

Wada, K. and Shibata, T. et al. (2005) Effects of robot therapy for demented people evaluated by EGG. *International Conference on intelligent robots and systems.*

Wada, K. and Takasawa, Y et al. (2013) Robot therapy at facilities for the eldery in Kanagawa prefecture – A report on the experimental result of the first week. *IEEE*757-61.

Wrobel, J et al. (2014) Robots and virtual agents to assist older adults: A review of present day trends in gerontechnology. *Neurologie-Psychiatrie-Gériatrie Vol14 Issue 82.*

WU, Y-H et al (2014) Emotional robots for institutionalized people with dementia. *Neurologie-Psychiatrie-Gériatrie Vol14 Issue 82.*

UNDERSTANDING CHILDREN AND PARENTS' USE OF ICT FOR CHRONIC DISEASE MANAGEMENT: A REVIEW OF THE LITERATURE

Wala'a Barifah[1,2], Bill Davey[1] and Vince Bruno[1]
[1]*School of Business IT & Logistics, RMIT University, Melbourne, Australia*
[2]*Management Information Systems, TAIF University, Taif, Saudi Arabia.*

ABSTRACT

This paper reviews existing literature on the role and types of information and communication technology (ICT) in managing chronic diseases (CD) among children and adolescent (C&A) populations. With the rise in population levels globally and the growing prevalence of CDs, many individuals including C&As are likely to experience the adverse effects associated with CD, including decreased quality of life, decreased physical performance, frequent hospitalisations and long periods of sick leave. Interestingly, many studies have reported that the use of various ICT tools can play a role in the treatment and management of CDs. ICTs can help patients and their parents access and retrieve valuable information about CDs. They can also help make self-care or self-management more effective. This review indicates that the most common ICT tools that parents and patients can use in the management of CD are video conferencing, integrated self-management systems and mobile phone applications. This review also reveals significant limitations in the existing research such as the trust in health information sources, challenges in education, use of multiple ICT tools to manage chronic disease, nature of awareness, and improve the effectiveness and the use of ICT tools.

KEYWORDS

Parent, Child, ICT Use, Review, Chronic Disease, Management

1. INTRODUCTION

In the last few decades, the prevalence of chronic diseases (CDs) has increased with the rise in global population (Scalvini et al. 2014). The demand for healthcare services targeted at CDs has also grown proportionally. Those suffering from CDs face numerous challenges, including progressive decrease in physical performance and quality of life, long periods of sick leave and frequent hospitalisations. Consequently, strategies for effective management of CDs have become a key priority among researchers.

A recurring theme in the literature on the management of CDs is the important role of the collaboration between patients and care providers. According to Celler et al. (2003), the successful management of CDs relies on a close partnership between healthcare providers and patients. While professional caregivers are responsible for administering treatment, patients with CDs too have a crucial role in ensuring their day-to-day care. This is because the patients are in the best position to estimate the severity of their symptoms and the efficacy of treatments (Celler,.et, al 2003). Wildevuur et al. (2017) agree that the management of CDs is a demanding task that requires patients' commitment on a day-to-day basis, in addition to help and support from healthcare providers and family members. Thus, the effective management of CDs does not entirely rely on the expertise of professional caregivers.

With the rapid development of information and communication technology (ICT) in the recent years, scholars have identified various modes in which technology can be utilised in the management of CDs. For instance, Wildevuur et al. (2017) argue that ICT plays a crucial role in the collaboration between patients and caregivers by connecting the two parties at any time and from any distance. ICTs are especially helpful in ensuring effective communication that is paramount for the management of CDs as well as in addressing compliance issues that often emerge (Celler et al. 2003). CD patients are often asked to comply with numerous instructions: reduce stress levels, report any emerging symptoms, monitor blood sugar and intake of sugar or take drugs at certain a dosage. Many patients are often unable to follow all the instructions, and their compliance rates become poor. However, researchers have shown that ICT tools can help enhance patient compliance.

This paper seeks to review existing literature on the role that ICTs play in managing CDs especially among children and adolescent (C&A) populations. Stinson et al. (2009) highlight that CDs among C&A patients can adversely influence all aspects of their quality of life. This literature review focuses on two aspects of CD management. First, it seeks to identify research on patients, specifically adolescents and children, and caregivers who have used ICT for the management of CDs. Second, it explores the effectiveness of various ICT tools in reducing the impact of chronic illnesses. This review identifies various gaps in the existing literature.

2. REVIEW

2.1 The Role of ICT in Managing CDs

Recently, there has been an increase in the adoption of ICTs towards the prevention and management of CDs. Adler-Milstein et al. (2014) evaluated the adoption and utilisation of ICTs for enhancing the care given to chronically ill populations. By sampling data from four countries, the researchers established the opportunities and challenges that modern-day, high-income countries are facing in the use of ICTs for CD management. For instance, they found that although most countries have a national-level strategy for the adoption of health-based ICTs, inequalities in the utilisation of the ICTs are common, and some regions derive more benefits than others. The researchers also found that most countries are struggling to ensure that ICT health information is available to patients across all settings. This study is a clear indication that although many high-income countries have established strategies for the adoption of e-health in the management of CDs, specific populations are yet to fully and equitably benefit from the advantages that these ICTs can offer. Traynor et al. (2016) support the above findings with their study aimed at assessing the effectiveness of psychological interventions delivered through ICTs for C&A with chronic pain. They found that unlike with adults – for whom ICT is a critical mode of treatment delivery – the utilisation of ICTs for the treatment and management of CDs in children has yielded mixed results.

Nevertheless, a number of studies have explored routes through which ICTs can be used to manage chronic illnesses in C&A and facilitate the collection of relevant information. For instance, Zhu and Winkel (2014) conducted a study to investigate the importance of ICT tools in meeting the social and educational needs of adolescents with long-term illnesses and CDs. Their findings showed that utilising ICT helps C&As keep up with their schoolwork and maintain social contacts with teachers and classmates. The results encouraged parents and teachers to embrace ICTs and integrate them into the lives of children with CDs to help them in their day-to-day activities.

An analysis of the available literature demonstrates that ICTs can play a crucial role in meeting the information needs of children with CD. Hicks et al. (2013) conducted a study on the experiences of children with cancer who use virtual communities and found that computer systems help these children in accessing medical information, which in turn leads to an enhanced understanding of their healthcare conditions. Nordfedt et al. (2013) concur that ICTs and the internet assist parents of adolescents with diabetes in accessing and retrieving information about the illness. However, Nordfedt et al. (2013) also found through their study that a parent's choice of information source depends on many factors including their level of trust in the source, their previous experience with the source and the present situation. Schiffman et al. (2008) studied adolescents and young individuals with cancer to establish how ICTs help in the management of their conditions in as far as information retrieval is concerned. They found that adolescents and young individuals with cancer often prefer to visit cancer websites to retrieve information related to their illness and interact with other young individuals living with cancer. Thus, the papers discussed above indicate that although ICTs provide crucial health-related information to children with CDs and their parents, not all sources meet their information needs. Hence, healthcare professionals should be strongly involved in helping individuals living with CDs filter the most useful information.

Another way in which ICTs can contribute to the management of CDs among C&A patients is by making self-care or self-management more effective. For instance, Stinson et al. (2009) state that self-management interventions provided through the internet can enhance outcomes such as the quality of life among young ones living with CD. Huygens et al. (2016) further explained that that primary care professionals consider self-management or self-care as a critical component of chronic care. Through their study that targeted all populations of CD patients, they established that eHealth is a paramount enabler of chronically ill patients' self-management. For this reason, other than information provision, healthcare providers should emphasise the utilisation of ICTs as way of enhancing self-care among individuals with CDs.

2.2 Types of ICTs Used in CD Management Research

Literature also discusses various ICT tools that can aid in CD management. Froisland and Arsand (2015) showed that mobile phone-based self-care management applications integrated with visual dietary communication can be helpful in assisting adolescents with type 1 diabetes understand the basics of their illnesses. Similarly, Gustafson et al. (2012) conducted a randomised controlled trial to analyse the ICT tools that can be utilised in the management of paediatric asthma. They established that a combination of mobile applications and web-based eHealth platforms can enhance information retrieval about the illness and self-management. However, they also found that these tools cannot improve adherence to medication. According to Lindberg et al. (2013), many ICT tools, including video technology, health monitoring and text messages, are effective in enabling efficient homecare for individuals living with CDs. A'Bear (2014) reported that chronically ill students often face significant challenges in their education because of their intermittent and prolonged absences from school. Various researchers have conducted studies on the technological tools that can assist in addressing these challenges. For instance, Ellis et al. (2013) observed that videoconferencing technologies can be used to connect paediatric cancer patients from hospitals to their schools. Interestingly, all the studies discussed above suggest that ICTs can only be beneficial in the management of CDs when two or more ICT tools are combined. That is, the use of a single tool cannot be entirely effective.

3. CONCLUSION

Rapid developments in ICT in the recent past have benefited many fields including healthcare. The application of new ICT tools has played a significant role in addressing the challenges that have characterised healthcare systems around the world for many years. One of these challenges is the treatment and management of CDs across populations. Studies have shown that CDs can have severely negative effects on the quality of life of the patients. Moreover, they disrupt the day-to-day lives of the patients by decreasing their physical performance and, consequently, impacting their economic and social lives. As is evident in the literature review above, C&As also experience the adverse effects of CDs. Researchers have therefore attempted to identify how the use of various ICT tools can be helpful in managing CDs. A review of the relevant available literature shows that ICTs could play various roles in the management and treatment of CDs. For instance, ICTs can help C&As and their parents in accessing and retrieving crucial information about their CDs. The emergence of virtual communities and the internet has promoted many platforms, such as websites, on which individuals with chronic illnesses can access medical information. This information subsequently results in an enhanced understanding of the individual's conditions. Another role that ICTs play in CD management is the enhancement of self-care and self-management. As stated earlier, for effective management and treatment of CD, patients should play a key role in managing their conditions, and ICTs can help in accomplishing this goal. To facilitate the roles identified above, diverse ICT tools are available. They include videoconferencing, mobile applications and web-based eHealth platforms among others. In summary, the available literature is relatively adequate in addressing the roles that ICT can play in managing CDs, especially among adolescents and children. However, an analysis of the literature also revealed the need for further research, especially on issues such as the trust in health information sources, challenges in education, use of multiple ICT tools to manage chronic disease and nature of awareness. The literature has shown that while ICT has contributed positively towards the management of CDs by caregivers, further research is needed to improve the effectiveness and the use of ICT tools.

REFERENCES

A'Bear, D., 2014. Supporting the Learning of Children with Chronic Illness. *The Canadian Journal of Action Research*, Vol. 15, No. 1, pp. 22-39.

Adler-Milstein, J. et al, 2014. A Comparison of How Four Countries Use Health IT to Support Care for People with Chronic Conditions. *Health Conditions*, Vol. 33, No. 9, pp. 1559-1566. Doi.org/10.1377/hlthaff.2014.0424

Celler, B. et al, 2003. Using Information Technology to Improve the Management of Chronic Disease. *The Medical Journal of Australia*, Vol. 1, No. 179, pp. 242-246.

Ellis, S. J. et al, 2013. Results of Nurse-Led Intervention: Connecting Pediatric Cancer Patients from the Hospital to the School Using Videoconferencing Technologies. *Journal of Pediatric Oncology Nursing*, Vol. 30, No. 6, pp. 333-341. doi/abs/10.1177/1043454213514633

Froisland, D. H. and Arsand, E., 2015. Integrating Visual Dietary Documentation in Mobile-Phone-Based Self-Management Application for Adolescents with Type 1 Diabetes. *Journal of Diabetes Science and Technology*, Vol. 9, No. 3, pp. 541-548. doi/full/10.1177/1932296815576956

Gustafson, D. et al, 2014. The Effects of Combining Web-Based eHealth with Telephone Nurse Case Management for Pediatric Asthma Control: A Randomized Controlled Trial. *Journal of Medical Internet Research*, Vol. 14, No. 4.

Hicks, P. et al, 2013. With a Little Help from My Friends: Experiences of Building a Virtual Community for Children with Cancer. *The Journal of Community Informatics*, Vol. 9, No. 2. Retrieved 18 March 2018 from http://ci-journal.net/index.php/ciej/article/view/840/1009

Huygens, M. et al, 2016. Expectations and Needs of Patients with a Chronic Disease Toward Self-Management and eHealth for Self-Management Purposes. BMC Health Services Research, Vol. 16, 232. Retrieved from https://doi.org/10.1186/s12913-016-1484-5

Lindberg, B. et al, 2013. Using Information and Communication Technology in Home Care for Communication between Patients, Family Members, and Healthcare Professionals: A Systematic Review. *International Journal of Telemedicine and Applications*, Vol. 2013, pp. 1-31.

Norfeldt, S. et al, 2013. Parents of Adolescents with Type 1 Diabetes–Their Views on Information and Communication Needs and Internet Use. A Qualitative Study. *PLOS One*, Vol. 8, No. 4, e62096.

Scalvini, S. et al, 2014. Information and Communication Technology in Chronic Diseases: A Patient's Opportunity. *Journal of Medicine and the Person*, Vol. 12, No. 3, pp. 91-95.

Schiffman, J. et al, 2008. Internet Use among Adolescent and Young Adults (AYA) with Cancer. *Journal of Pediatric Blood Cancer*, Vol. 51, No. 3, pp. 410-415.

Stinson, J. et al, 2009. A Systematic Review of Internet-based Self-Management Interventions for Youth with Health Conditions. *Journal of Pediatric Psychology*, Vol. 34, No. 1, pp. 495-510.

Traynor, A. et al, 2016. The Effectiveness of Information and Communication Technology-Based Psychological Interventions for Pediatric Chronic Pain: Protocol for a Systematic Review, Meta-Analysis and Intervention Content Analysis. Systematic Reviews, Vol. 5, 175. Retrieved from http://www.jmir.org/2012/4/e101/

Wildevuur, S. E. et al, 2017. Information and Communication Technologies to Support Chronic Disease elf-Management: Preconditions for Enhancing the Partnership in Person-Centered Care. *Journal of Participatory Medicine*, Vol. 9, No. 1, e12.

Zhu, C. and Winkel, L. V., 2013. Using an ICT Tool as a Solution for the Educational and Social Needs of Long-Term Sick Adolescents. *Journal of Technology, Pedagogy and Education*, Vol. 24, No. 2, pp. 231-245.

UNOBTRUSIVE SYSTEM FOR THE DETECTION OF MENTAL FOCUS DEPLETION

John Gialelis[1], Panos Pavlou[1], Chris Panagiotou[1], Stavros Koubias[1], Nuno Pombo[2] and Nuno Garcia[2]

[1]Applied Electronics Lab, Electrical & Computer Engineering Department, University of Patras, Patras, Greece
[2]Instituto de Telecomunicações and ALLab Assisted Living Computing and Telecommunications Lab, Department of Informatics, Universidade da Beira Interior, Covilhã, Portugal and Universidade Lusófona de Humanidades e Tecnologias, Lisbon, Portugal

ABSTRACT

Advances in Wearable Devices and Artificial Intelligence upgrade the use of ICT aiming both at improving humans' life quality as well as safety. This paper presents a system for the detection of mental focus depletion comprising a robust embedded system used for aggregating EEG data and transferring them in real time to a central server where specific purpose software is used for further processing. The performance of the system is tested using experimental data, as well as its accuracy is depicted against a well-known commercial device.

KEYWORDS

Intelligent Health /Brain Informatics, Brain Signal Processing

1. INTRODUCTION

The advancements taking place in the domain of pervasive health and wellbeing assisted by the evolution of Internet of Things (IoT) technologies have led to the hyper growth of the wearable devices and applications market. In fact, the respective market is expected to increase rapidly from US$750 million in 2012 to US$5.8 billion in 2018 (Wei 2014). Moreover, significant work has been published, regarding systems employing data mining techniques tailored on the detection of specific pathological anomalies such as apnea, heart diseases, epilepsy, and mental disorders (Shukla et al. 2014, Pombo et al. 2016). On the other hand, the acquisition of multi-modal and context data using wearable devices gains great interest among researchers. These data along with computerized algorithms are expected to help researchers to extract meaningful patterns that will help them monitor and assess human's routines that jeopardize health and deteriorates quality of life either in short or long term. The electroencephalogram (EEG) is an electrophysiological monitoring method to record electrical activity of the brain, utilizing noninvasive electrodes placed along the scalp. The analysis of the EEG provides vital information regarding the diagnosis of mental illnesses and psychiatric disorders as well as the assessment of mental state in various contexts (i.e., sleeping, resting, driving, working, etc). The improvement of human safety during work activities on the one hand, along with the wellbeing and the quality of life on the other hand challenge researchers and practitioners to develop novel techniques that will enable accurate diagnosis and treatments based on data obtained from the EEG. The study of Alpha waves, for example, has acquired a lot of attention with respect to quality of sleep assessment (Wang et al. 2014). In addition, state of the art unobtrusive, low cost, and ubiquitous systems complemented with personalized and structured presentations of the aggregated data provide physicians and healthcare personnel with more effective decision support tools, thus making feasible a broader adoption of the EEG for the detection of mental illnesses and psychiatric disorders as well as the assessment of mental state (Welch 1967).

This paper presents the design and development of an activity tracking platform that intends to deliver unobtrusively a fine-grained activity analysis. This paper extends the primary work presented in (Gialelis et al. 2014) by focusing on the detection in the frequency band of the Alpha waves, constituting the core part of an embedded system that could prevent workers from losing their mental focus which could result to less mistakes and accidents.

The rest of the paper is organized as follows. Section 2 briefly describes the background in regard to EEG signals characteristics and related work about their utilization. Section 3 depicts our approach and decomposes our proposed system in its components while Section 4 depicts the intelligence of the proposed system that is responsible for the Alpha waves detection and the mental focus assessment. Section 5 depicts the conclusions and summarizes the progress made so far along with the future steps to follow.

2. BACKGROUND AND RELATED WORK

Polysomnography (PSG) is a test conducted to study sleep and to diagnose a variety of sleep disorders utilizing – among others- EEG signals. Unlike polysomnography (PSG), which requires a great variety of sensors, tools aiming to detect the loss of attention by merely using unobtrusive commercial devices that aggregate and process brain data and draw conclusions about the focus level of a subject are in need. In the forthcoming Table **1**, EEG signals are presented, depicting signals' amplitude and frequency spectra for each band. Research has shown that the appearance of Alpha waves is the trigger event for a subject to transit from a wake-up phase to a relaxing phase eventually preparing for sleep or more generally speaking for a shutdown process (Rosenberg et al. 2013). In a workplace context, this event while appearing to people that handle machinery notes the lack of attention and concentration [Indoriko et al. 2013). As the importance of attention and concentration in such contexts is undisputed, its lack may cause catastrophic or even fatal results for humans and environment with wide economical and societal effects

Table 1. EEG Signals Characteristics

Wave	**Signal Amplitude**	**Frequency band**	**Occurrence**
Alpha	<=50μV	8-12Hz	Loosy attention
Beta	<30μV	13-60Hz	Subject awake processing information
Theta	50-100μV	4-8Hz	Correlated with memory and feelings
Delta	100-200μV	0.5-4	Subject in deep sleep or in comma

Authors in (Huang et al. 2014) present a unobtrusive approach by utilizing smartphone's sensors (microphone and inertial measurement unit) for the detection of body movements, snoring, heartbeat and respiration while authors in (Ronzhina et al. 2012) detect sleep patterns by studying smartphone's usage patterns (e.g., the time and length of smartphone usage) and ambient observations (e.g., prolonged silence and darkness). As boredom is being a significant issue in education, a system for the assessment of students' concentration or depletion level in a context of a classroom during the teaching process, so when focus depletion is detected the teachers are notified to alter their teaching strategy in order to regain the interest of their students is also welcome [Rosegard et al. 2013, Rebolledo et al. 2009). Therefore, there is an imperative need for low cost unobtrusive wearable self-assessment tools aggregating EEG signals for the detection of the attention and the concentration levels of people while being in various contexts to either address safety or upgrade quality issues.

3. SYSTEM DESCRIPTION

The solutions presented in the referenced literature highlight the single channel EEG as the most prominent device to provide unobtrusively all the necessary information regarding brain activity during sleep fullness and awake fullness. In line with that, our proposed system comprises three main subsystems: the aggregation subsystem, the gateway subsystem and the backend subsystem. According to our approach, for the signals' aggregation a single channel commercial EEG device with Bluetooth communication is utilized. An Android smartphone is used as a gateway node to preprocess, store and forward EEG data. A backend server is used as a backend subsystem for further processing, analysis and visualization.

3.1 Aggregation Subsystem

Wearable devices is the cornerstone for health monitoring and ambient assisted living (AAL) systems, as they constitute an unobtrusive raw data source [Mporas et al. 2014). In our use case we use components of the self (COTS) that employ widely adopted and popular wireless standards, such as Bluetooth or Wi-F and incorporate built- in flexible and interoperable communication interfaces while at the same time the devices address the needs for comfort and unobtrusiveness. As far as the primary source of data, the EEG, the source used is based on the MyndPlay BrainBandXL, based on the Neurosky platform, with two dry sensors and an ear clip as a reference point. The Surface Mount Device (SMD) board is attached on a headband made of a soft and comfortable fabric. The wireless connectivity offered by the MyndPlay device is based on Bluetooth. The same applies, comfort and unobtrusiveness wise, with respect to the other lesser importance sources of data (Electrocardiograph (ECG), Pulse Oximeter (PO) and Blood Pressure Gauge (BPG)).

3.2 Gateway Subsystem

The smartphone platform could be any Android device with Bluetooth and Wi-Fi connectivity while its contribution is twofold. Its primary role is derived by its capability to provide interconnection between the wearable devices and the backend system, thus playing the role of a gateway. Communication drivers have been implemented and deployed at the application level for the interaction with the wearable devices. The required connectivity is implemented as individual android services that run in the background. As long as a wearable device is detected in the vicinity of the smartphone's wireless interfaces, a connection is established and the smartphone receives data streams from the EEG (or the ECG) and single values from the PO and the BPG. All received values are stored in the smartphone's storage unit until they are transmitted to the backend system for further processing. The second functionality that the smartphone could provide is derived by the native services that any Android device implements. For example, accelerator's values are also gathered providing indication of the person's lying activity.

3.3 Backend Subsystem

The backend server is responsible for storing, processing and displaying data that have been collected and transmitted by the wearable devices, during users' sessions, via the Android services deployed within the smartphone. Apart from simple data gathering, the backend system also incorporates data processing components for high level information creation and storage. Overall, the backend server consists of four main components as described below:

Web services: this component is mainly responsible for gathering, handling and further processing of transmitted data through the smartphone services. Since continuous recording of EEG can easily derive huge amount of raw data a special treatment has been applied as a custom communication protocol between the backend server and the smartphone's services towards minimization of data volumes that have to be transmitted. Such a technique results in major advantages in processing and storage resources in the backend as well as in the mobile pillar. Finally this component is responsible for feeding the visualization component through the proper web interface (Web UI). Data processing: the software modules responsible for processing the bio signals are integrated in the backend server as standalone applications, executing signal analysis utilizing a specific dataset. An extensive description of the processing approach conducted on the EEG signals for the Alpha waves detection is presented in section 4.Data storage: raw data and extracted high level information from the data processing component are stored in a backend database which is deployed with a typical MySQL server. Access to these datasets is available through the Web UI. Web UI: the Web application allows the users to access the stored datasets as well as conduct signal analysis using the standalone applications.

4. ALPHA WAVES ASSESSMENT

The signal processing unit handles the EEG signal characterized as non-stationary, nonlinear and noisy. EEG is an error-prone signal since it coexists with other stronger bio signals that interfere and introduce artifacts.

These artifacts originate mainly from ECG, EMG and eye blinks. Furthermore, noise is introduced by the power line inference and the body movements. The methodology followed for the identification of Alpha waves comprises three sequential steps: signal preprocessing (artifacts/noise elimination), segmentation of signal in parametric time intervals, signal filtering (frequency & time domain analysis) and finally energy computation and extraction of the actual percentage of the Alpha waves presence in the signals segment. To maximize the noise elimination without eliminating valuable brain data frequencies a 2nd order Butterworth filter with low and high cutoff frequencies set at 0.3 and 35Hz is applied. Since the EEG is a non-stationary signal, it is processed in epochs of dependable parametric discrete time intervals in points where stationarity is assumed. The goal is to get the filtered signal with frequency components not over 35Hz and to distinguish the specific frequency band of the Alpha waves. To realize this goal the basic principle of signal processing was used by calculating the energy of the filtered signal. Subsequently it was passed through a band stop filter with cut off frequencies at 8 and 12 Hz respectively to eliminate the contribution of these frequency components in signal's energy. The next step was to calculate the percentage of the existence of Alpha waves in the signal by the metric:

Equation 1 Alpha wave percentage=100-Efiltered / Einitial

Where Efiltered stands for the energy of the signal after having adapted the signal portion in two sequential filters and Einitial stands for the pure signal energy all derived from the corresponding type:

Equation 2 $E[n]=\sum x[n]$

After the length of the discrete time intervals (in secs) is preselected, the signal processing unit yields the percentages of the Alpha waves presence in the EEG signal. Specifically, the signal is segmented in epochs of 15 or 30 seconds each for better accuracy. In the time domain, filters are applied in order to eliminate specific frequency components from the signal. In the frequency domain, Fourier transformation is used on the signal before and after in order to eliminate the contribution of specific frequencies in the signal's spectra. Then energy computation provides the best metric of how much portion of energy was eliminated from the filtering. The final step is the calculation of the Alpha waves percentage in the signals segment through the reverse procedure. The validation is carried out through a comparison between the outcome of the proposed method vs Neurosky's device native outcome. Neurosky device encapsulates in its protocol a proprietary field called "Meditation" updated once a second. This field indicates the focus level of a subject and its value varies from 0 to 100 having 100 indicating total lack of focus. To validate the proposed method and to conclude about the quality of the extracted results an epoch of 15 seconds is used. With a dataset of four minutes having the subject not to think during the first minute of the experiment while having eyes closed, Figure 1 depicts both results having with the red line the results of the proposed method and with the blue line the average percentage of the "Meditation" field. The comparison is lucid and reveals that the proposed method can accurately detect focus depletion.

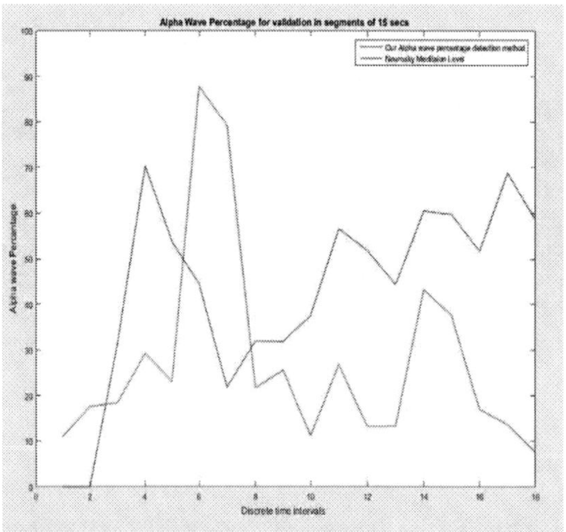

Figure 1. Experimental Results

5. CONCLUSION

In this paper the main architectural and implementation aspects of a system that detects subjects' focus depletion were presented. Design as well as development details of the final system were provided along with the presentation of the measurements that secure the reliability of our results. Focus attempts would lead to low percentage of Alpha waves while relaxing or losing attention would lead to high percentage of Alpha waves. The preliminary studies revealed the feasibility as well as the accuracy of the proposed system for focus depletion detection. Further studies should be contacted in order to upgrade this prototype to a close to market system.

ACKNOWLEDGEMENT

This work is currently supported by the ERASMUS+ Program through the project 591945-EPP-1-2017-1-DE-EPPKA2-SSA "Digital Skills on Computational Biology for Health Professionals (BIOS) and by FCT project UID/EEA/50 0 08/2013. The authors also acknowledge the contribution of the COST Action IC1303 – AAPELE – Architectures, Algorithms and Protocols for Enhanced Living Environments.

REFERENCES

Gialelis, John, Chris Panagiotou, Petros Chondros, Dimitris Karadimas and Elefteria Batzi.2014. "End-to-End System for the Relationship Modeling of Physical Exercise with Sleep Quality." 8th Multi Conference on Computer Science and Information Systems: 367 – 372.

Huang, Chih-Sheng, CL Lin, LW Ko, SY Liu, TP Su, CT Lin. 2014. "Knowledge-based identification of sleep stages based on two forehead electroencephalogram channels." Frontiers in Neuroscience (8): 263.

Indoriko Shin, Bambang Setiawan, Febriliyan Samopa , Hatma Suryotrisongko, Faizal Johan, Atletiko, Radityo P. Wibowo. 2013. "Development of Drowsiness Detection System with Analyzing Attention and Meditation Wave using Support Vector Machine Method." Information Systems International Conference: 2 – 4.

Mporas, Iosif, Vasiliki Tsirka, Evangelia I. Zacharaki, Michalis Koutroumanidis, Vasileios Megalooikonomou. 2014. "Online seizure detection from eeg and ecg signals for monitoring of epileptic patients." Artificial Intelligence- Methods and Applications:442-447.

Pombo, Nuno, Nuno M. Garcia. 2016. "ubiSleep: An ubiquitous sensor system for sleep monitoring." IEEE 12th International Conference on Wireless and Mobile Computing, Networking and Communications:1–4. https://doi.org/10.1109/7763192.

Rebolledo-Mendez, Genaro, Ian Dunwell, Erika A. Martínez-Mirón, María Dolores Vargas-Cerdán, Sara de Freitas, Fotis Liarokapis, Alma R. García-Gaona. 2009. "Assessing NeuroSky's Usability to Detect Attention Levels in an Assessment Exercise." Human-Computer Interaction. New Trends: pp 149-158.

Ronzhina, Marina, Oto Janoušek, Jana Janoušek, Jana Kolářová, Ivo Provaznik. 2012. "Sleep scoring using artificial neural networks." Sleep Medicine Reviews 16.3: 251-263.

Rosegard, Erik, Jackson Wilson. 2013. "Capturing students' attention: An empirical study." Journal of the Scholarship of Teaching and Learning, Vol. 13(5):1-20.

Rosenberg, Richard S., and Steven Van Hout.2013. "The American Academy of Sleep Medicine inter-scorer reliability program: sleep stage scoring." Journal of Clinical Sleep Medicine 9(1): 81–87.

Shukla, D. P., Shamsher Bahadur Patel, and Ashish Kumar Sen.2014. "A literature review in health informatics using data mining techniques" International Journal of Software & Hardware Research in Engineering. vol 2(2):123-129.

Singh, Mandeep, and Mahak Narang. 2014. "Changes in Brain wave rhythms during tasks involving Attention and Working memory." International journal of information technology and knowledge management 7.2: 71-97.

Wang, Xuewen, and Shawn D. Youngstedt. 2014. "Sleep quality improved following a single session of aerobic exercise in older women: Results from a pilot study." Journal of Sport and Health Science, vol.3(4): 338-342.

Wei, Joseph.2014. "How Wearables Intersect with the Cloud and the Internet of Things: Considerations for the developers of wearables." IEEE Consumer Electronics Magazine, vol.3(3): 53-56.

Welch, Peter D. 1967. "The Use of FFT for the Estimation of Power Spectra: A Method Based on Time Averaging Over Short, Modified Periodograms", IEEE Transactions on Audio Electroacoustics, AU-15: 70–73.

M-HEALTH APPLICATION INTERFACE DESIGN FOR SYMPTOM CHECKING

Christopher Copeland, Patricia Morreale and J. Jenny Li
School of Computer Science, Kean University, Union, NJ 07083 USA

ABSTRACT

The demand for mobile health (mHealth) information by today's wireless users has never been greater. Apart from proprietary applications, linked to pharmacies and other medical and wellness service providers, a widely-used simple health and wellness symptom checker is not yet available. A mHealth symptom checker, envisioned with basic diagnostics and using location information, was designed to provide first-level diagnosis and direction for users seeking information regarding their self-reported health and wellness symptoms. This research project used human computer interaction (HCI) principles to develop a user-centered mHealth symptom checker to narrow the user inquiries and provide prompt information based on the symptoms reported by the users. A user evaluation, based on the System Usability Scale (SUS) and the Usability Metric for User Experience LITE (UMUX-LITE), was conducted. The results show the mHealth application was well-regarded by users, who appreciated the accessible interface design and location-specific information provided. Plans are for further development of the mHealth application and integration with heath care sites, including urgent care clinics and hospitals to provide users with responsive, personalized care from the mHealth application.

KEYWORDS

Mobile Health, mHealth, Application Interface Design, Symptom Checker, System Usability Scale, UMUX-LITE

1. INTRODUCTION

While the global population is increasing, the number and availability of medical doctors is not keeping pace (Carroll 2016). The public perception in many communities is that a doctor is not available when needed. When illness strikes, scheduling a doctor's appointment may be difficult, with the next closest appointment several weeks or months away. Often, a quick medical appraisal is all that is needed, to determine what might be done immediately to provide comfort and begin healing. By the time the patient can see the doctor, an individual's symptoms may have become more serious or, if contagious, others may have been impacted. One approach to providing more widely available medical care is the creation of a symptom checker application, resident on a mobile phone, in the hand of the prospective patient.

Given today's mobile users, a mobile application for symptom checking is an ideal solution. Mobile users would be able to communicate with doctors or other health care providers and receive information appropriate for their condition. Immediate information could be received, and users could be directed to the nearest available medical provider, if appropriate. While a mobile application is not a substitute for a medical diagnosis, sufficient patient information can be gathered to appropriately triage or prepare for the patient's later appointment or further medical care, if needed.

2. PRIOR WORK

Prior work in mobile application design has focused on personal health services (Lähteenmäki, et al., 2017), specifically personalized service delivery to support self-care. The mobile user is envisioned as the patient, and the focus of the delivered information is on meeting the needs of the patient. The focus of the model presented was on making sure that developed prototypes could be successfully integrated into the existing health IT infrastructure, rather than on meeting the needs of the user. While some work has involved end-users in the development of web-based services (Bruinessen, et al. 2017; Locatelli, et al. 2017), the focus has been on elderly users, which are a portion of the target market. Moving to a younger user community, mobile apps have been developed to manage dental anxiety in children (Romero-Ternero, et al, 2017) and the adherence to asthma treatments in active patients (Huckvale, et al., 2015; Jácome, et al., 2017). Wearable devices have also been studied, particularly for mobile health predictive data (Bocaj, et al., 2017), again with both an emphasis on the data analysis environment and on elderly.

The need for privacy protection has also been investigated as any health IT usability must also preserve privacy (Cooly, et al., 2013), while offering personalized service. The data gathering, analysis, and delivery of information back to the user is commonly thought of as a round-trip process which must maintain data integrity while protecting patient privacy. Investigations of usability and healthcare data breaches (Anthony, et al., 2013) frequently focus on the limitations of electronic systems. Mobile health presents special challenges. Using the definition of mobile health from the literature (Anthony, et al., 2013) that mobile health refers to the use of mobile computing and communications technology in the delivery of healthcare or collection of health care information, mHealth is distinguished further (Anthony, et al., 2013) by the sensitive nature of the data, the amount and variance of data that can be collected, and the personal and immediate impact that any security problems can have on the patient.

Survey articles looking at methods to connect patients, drugs, and clinicians (Gan, et al., 2016) highlighted the important role the smartphone can play in connecting the clinician and patient, effectively closing the gap in personalized care. However, the mental models of clinicians do not agree with the physical reality of patients (Smith and Koppel, 2013) and this must still be addressed by healthcare IT. One approach to bringing clinicians closer to patients and enabling more effective dialog is through symptom checkers.

Symptom checkers have been used for self-diagnosis and triage, with evaluation summaries (Semigran, et al., 2015) finding that the correct diagnosis was provided first in 34% of the standardized patient evaluations, which increased to 58% within the top 20 diagnoses, and the appropriate triage advice in 57% of standardized patient evaluations. Performance ranged from 33% to 78% on appropriate triage advice. The conclusion of the authors was that while symptom checkers had deficiencies in both triage and diagnosis, the advice from symptom checkers is usually not harmful and does encourage users to seek care for conditions in an appropriate manner. Essentially, symptom checkers do more good than harm in their current form and will continue to evolve. The beneficial use of symptom checkers is highlighted (Morita et al., 2017) for doctors, as well as providing access to health care for individuals in rural areas or areas without medical facilities. Given the need, a convergence with artificial intelligence can be used to improve the uncertainty of imprecise reporting and potential ambiguity from various symptoms (Huang, et al., 2018), allowing greater accuracy in the results returned to the user.

Previously developed symptom checker apps are often text based. Design concepts from human computer interaction (HCI) recommend less text (Preece, 2015; Shneiderman, 2017) as many mobile users just glance at their smartphones, preferring not to read, and want their tasks to be clear and simple. Examples of symptom checkers are WebMD's mobile app (Gillingham, 2012) and Steps2Care app (Paramount, 2018) (Figure 1).

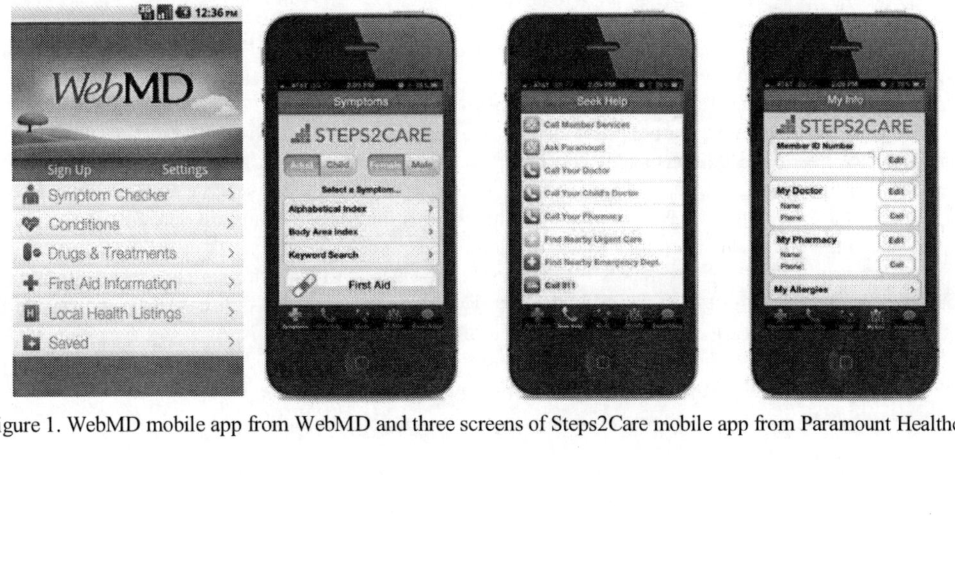

Figure 1. WebMD mobile app from WebMD and three screens of Steps2Care mobile app from Paramount Healthcare

While the WebMD mobile application is text based, the transition between sections is smoothed using animations. The colors are simple, serve to mark sections on the app and do not detract from the task. In addition to symptom checking, the first option on a list, other options are presented to the user.

Steps2Care by Paramount Healthcare (Figure 1) also offers a simple interface. The Steps2Care interface is primarily text driven with recognizable icons. This design is ideal for users who understand the purpose of the app. The interface clearly displays choices for the users to make while also providing readable text.

Another app design provided for comparison is from Schmitt-Thompson Clinical (Figure 2). The SymptomMD app (SelfCare 2018) is entirely text based and the user interface (UI) is cluttered. A large amount of information is displayed but the text is too small for the user to read. The goal should be to reduce the cognitive load of the user and provide users with the requested information.

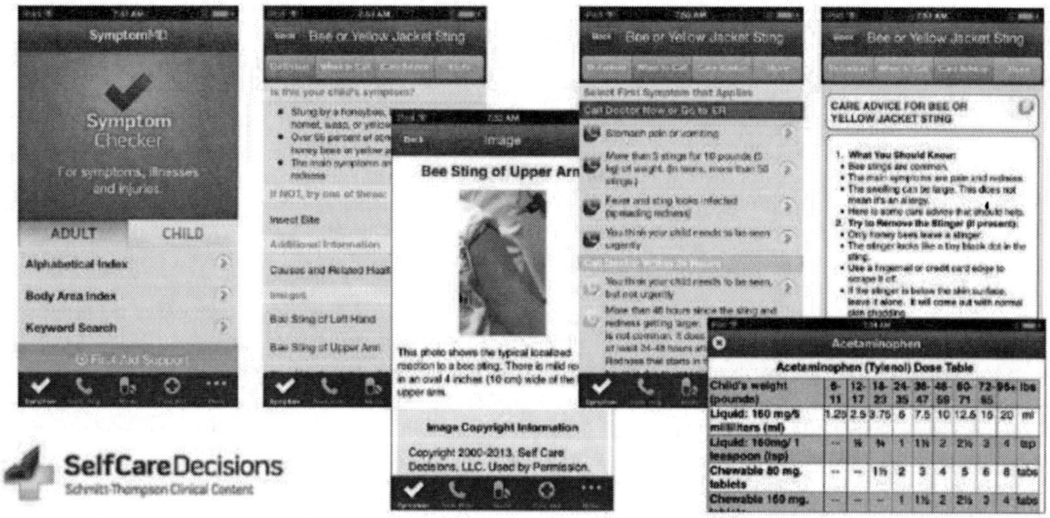

Figure 2. SymptomMD Mobile app from Self Care Decisions and Schmitt-Thompson Clinical Content

The figures presented show a range of symptom checker apps, with user interface design limitations. The objective of the research was to improve the UI, while reducing the cognitive load to improve the user experience with a mobile health symptom checker. Once designed, user feedback was solicited.

3. M-HEALTH METHOLOGY AND USER INTERFACE DESIGN

The human computer interaction (HCI) principles followed for user interface design were Tolerance, Simplicity, Visibility, Affordance, Consistency, Structure and Feedback (Preece, 2015; Shneiderman, 2017). The focus principles were Simplicity, Consistency and Feedback.

The goal was to create a simple interface that did not distract the user from completing the designated task. The interface needed to be consistent otherwise the cognitive load would be too heavy for the user. Popular apps such as Instagram, Twitter and Netflix have a consistent layout that users are familiar with. Feedback was supported as the users must clearly understand the choices they made using the app, due to the health aspect. Upon selecting a menu, the user must know where they are and what they are doing.

The concept which follows makes the best use of the space given. The concept was inspired by features found in the mobile apps of Netflix and Instagram. Representative screens of the mHealth app are in Figure 3.

- The colors are purposely light to draw attention to the text.
- A menu is added to the bottom for navigation
- If the user doesn't want to type, a voice option is provided
- If the user doesn't know how they are feeling a link to the database is provided

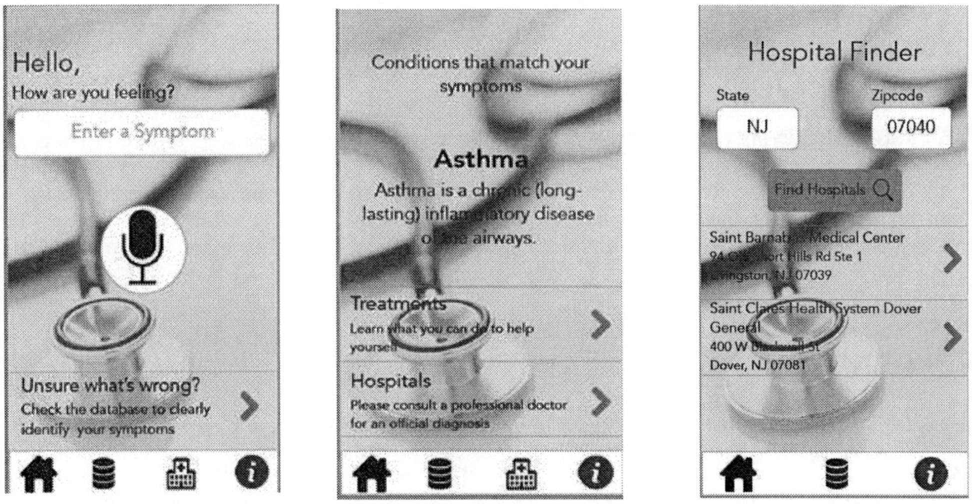

Figure 3. The mHealth app Welcome, Diagnosis, and Hospital Finder Screens (left to right)

The mobile health screens are distinctive for their clarity, with an icon-driven footnote on each page which identifies home, symptom database, hospital locater, and information screens. The clear, set-apart text boxes for symptom entry, and hospital searches support low cognitive load with high feedback and recognition.

4. M-HEALTH DESIGN EVALUATION

After the concept was developed, a survey, based on the System Usability Scale (SUS) (Brook, 1996; Shneiderman, 2017) and the Usability Metric for User Experience (UMUX) (Finstad, 2010), particularly UMUX-LITE (Lewis, et.al. 2013), was sent out asking general questions on symptom checkers and about the concept itself. For the concept four questions were asked.

1. Can you understand the purpose of the prototype?
2. Are the colors pleasing to the eye?
3. Would you be comfortable with this on your phone?
4. In their opinion which feature is more important?

User responses to the first question were all affirmative, as all users could understand the purpose of the prototype screen shots presented. Most users (69.2%) liked the colors used in the prototype, with only 30.8% replying negatively. Users were comfortable having an app like the prototype on their phone, with affirmative answers (58.3%) in the majority over those that replied maybe (38.5%) or no (3.2%). When asked what feature is more important for the mobile health application, both personalized health tracker (46.2%) and emergency SOS (46.2%) were highly in demand, with hospital locations a distant choice (7.7%). Based on the user responses (n=13) the mHealth app concept yielded both positive and negative responses. The positive responses were that users can easily identify tasks and majority of mobile application users feel comfortable with a symptom checker on their phone. Some users did not find the colors appealing. Overall the mHealth symptom checker was well-received, and enhancement will continue. The symptom checker works well on a modest range of common ailments, and all users are presented with a statement that the information provided by the mHealth app is not a substitute for actual medical advice or information. The goal of reducing cognitive load and text reading was met using icons, less text per screen than comparable apps, and consistent placement of input areas on the UI.

5. FUTURE WORK

The project is being further refined to improve the HCI and symptom-checking. Overall, a user-friendly interface remains the goal, with one possibility being a UI that adapts to the user, particularly users with chronic disease monitoring. Continued work on symptom-checking is proceeding as well. Depending on the areas where the mHealth application is adopted, appropriate medical information can be included, such as information on localize epidemics or flu outbreaks, seasonal allergies, and other geographically specific information. Additionally, once symptoms are reported and diagnosed, timely follow-up, such as medication reminders or physician visits can be provided, further increasing the utility of mHealth beyond symptom checking.

REFERENCES

Anthony, D., et al. 2013. Securing IT in Healthcare. *In IEEE Security and Privacy*, Vol. 11, pp. 25-33.

Bocaj, E., et al. 2017. MHealth Predictive Data Analysis of Daily Activities and Physiological Conditions. *In Proceedings of the International Conference on e-Health*, pp. 45-52.

Brook, J. 1996. SUS: A quick and dirty usability scale. *In Usability Evaluation in Industry*, Taylor and Francis Pub., UK.

Bruinessen, I., et al. 2017. Involving End-Users in the Development and Implementation of a Web-based, Physcial Activity Platform for Elderly. *In Proceedings of the International Conference on e-Health*, pp. 145-149.

Carroll, A. 2016. We're Thinking About Doctor Shortages the Wrong Way, in *The New York Times*, Nov. 8, 2016, page A3.

Cooley, J. and Smith, S. 2013. Privacy-preserving screen capture: Towards closing the loop for health IT usability. *In Journal of Biomedical Informatics*, Vol. 46, pp. 721-733.

Finstad, K. 2010. The Usability Metric for User Experience. *In Interacting with Computers*, Vol. 22, No. 5, pp. 323-327.

Gan, S.K. et al. 2016. An overview of clinically and healthcare related apps in Google and Apple app stores: connecting patients, drugs, and clinicians. *In Scientific Phone Apps and Mobile Devices*, Vol. 2, pp. 1-9.

Gillingham, F. 2012. Online. http://www.travelwellworldwide.com/app-reviews/webmd/ Accessed on June 25, 2018.

Huang, Y., et al. 2018. Health Symptom Checking System for Elderly People Using Fuzzy Analytic Hierarchy Process. *In Applied System Innovation*, Vol. 1, Issue 10, pp. 1-16.

Huckvale, K. et al. 2015. The evolution of mobile apps for asthma: an updated systematic assessment of content and tools, *In BioMed Central Medicine*, Vol. 13, pp. 1-15.

Jácome, C. et al. 2017. Inspirers: An App to Measure and Improve Adherence to Inhaled Treatment. *In Proceedings of the International Conference on eHealth*, pp. 135-139.

Lähteenmäki, J. et al. 2017. Pre-commercial Procurement of Personal Health Services – Interoperability Considerations. *In Proceedings of the International Conference on e-Health*, pp. 79-86.

Lewis, J., et.al. 2013. UMUX-LITE-When There's No Time for the SUS. *In Proceedings of the SIGCHI Conference on Human Factors in Computing Systems*, pp. 2099-2102.

Locatelli, P., et al. 2017. Progressively Developing a Business Model to Assist Elderly Patients with Cognitive Impairment through a Digital Ecosystem: A Methodical Approach. *In Proceedings of the International Conference on e-Health*, pp. 19-26.

Morita, T., et al. 2017. The Potential Possibility of Symptom Checker. *In International Journal of Health Policy Management*, Vol. 6, No. 10, pp. 615-616.

Paramount. 2018. Online http://www.paramounthealthcare.com/productguide/img/slideassets/8.8%20SymptomChecker.pdf Accessed on June 25, 2018.

Preece, J. et al. 2015. *Interaction Design: Beyond Human Computer Interaction*, 4th edition. Wiley, Sussex, UK.

Romero-Ternero, M.C., et al. 2017. A Mobile App to Manage Children's Dental Anxiety: Context and Approach, *In Proceedings of the International Conference on e-Health*, pp. 131-134.

SelfCare Decisions. 2018. Online http://www.selfcare.info/pdf/SCD_brief_hplan.pdf Accessed on June 25, 2018.

Semigran, H., et al. 2015. Evaluation of symptom checkers for self-diagnosis and triage: audit study, *In BJM: British Medical Journal* 351, pp. 1-9.

Shneiderman, B, et al. 2017. *Designing the User Interface: Strategies for Effective Human-Computer Interaction*, 6th edition. Pearson Publishers, Hoboken, NJ, USA.

Smith, S. and Koppel, R. 2014. Healthcase information technology's relativity problems: a typology of how patients' physcial reality, clinicians' mental models, and healthcare information technology differ. *In Journal of American Medical Informatics Association*, Vol. 21, Issue 1, pp. 117-131.

A CROWDSOURCE BASED HOSPITAL BI SYSTEM FOR THE NATIONAL HEALTH SERVICE

Duarte Sequeira and Bruno Trigo
SPMS – Serviços Partilhados do Ministério da Saúde, Av. da República, nº61, 1050-099, Lisboa, Portugal

ABSTRACT

Healthcare professionals are facing a growing demand for both clinical and administrative information to improve their practice and comply with top-management decisions.
In the past, each hospital tried to address this need in different ways, duplicating efforts and creating heterogeneity in terms of indicators. As a Shared Services organization working for the National Health Service (NHS), a project was put in place to envision and develop a clinical indicator Business intelligence (BI) platform that could provide health professionals with safe and efficient access to activity data in a crowdsource model for more than 50 hospitals. With such a model, it was intended to create a living system that could include several hospitals in its development and respond to the need for continuous improvement, research, national standardization of catalogues and indicators and information security with the goal of improving care practice and providing patient cantered care.
More than 200 metrics are available today with the contribution of healthcare professionals that can address their demand of having an easy access BI platform. By opposition to other projects, this platform targets primarily health professionals, instead of top management. The preliminary results show a considerable interest in using this platform due to its good user experience in accessing several indicators, as well as actionable insights for the daily practice enabling patient care improvement.

KEYWORDS

Health Professionals, Hospital, Clinical Indicators, Shared Services, Business Intelligence, Crowdsource

1. INTRODUCTION

Over the years, hospitals and central administration had the need to produce indicators concerned with production/activity KPI related with monitoring and management control of the national health service financing model.

Nowadays, coexistence with clinical record software is a reality for most healthcare professionals in Europe. In a country where core hospital software is mostly uniform and provided by the NHS, with a good level of informatization and whose activity and records generates enormous data, it's clear that it's possible to produce highly detailed measures with very good uniformization. The opportunity for secondary use of data becomes increasingly relevant and accessible with decision makers in the healthcare sector facing a growing demand for both clinical and administrative information to comply with both legal and customer-specific requirements (Mettler and Vimarlund, 2009).

But this data is generally not translated into actionable insights to the health professional's actual practice. In this sense, we are developing a project which aims to collect, analyze and make clinical information available to its producers – physicians and nurses - in a bottom-up logic, that is, first for health professionals and only later for the managers/boards.

This paper describes an innovative crowdsource model, with a true concept of shared services, and development of this platform, as well its first results and challenges.

2. BODY OF PAPER

2.1 Background

SClínico Hospitalar is an information system developed by Serviços Partilhados do Ministério da Saúde (SPMS), a state public enterprise, used by thousands of doctors, nurses and other health professionals. This software is available in more than 50 hospital centres, and other Institutes, accounting for most of the NHS. SClínico is part of the strategy set by the Ministry of Health for the area of clinical digitalization of the Portuguese NHS, which provides for the standardization of procedures of medical records to ensure information standardization.

However, the platform had some issues regarding information management to give its users some feedback that could inform on their clinical practice. So, one year ago, a small team was set up to envision and apply a governance model of a clinical indicator platform that could provide health professionals with safe access to activity data (GDPR compliant), based on information generated daily by these professionals while using the clinical records software. The governance model should address several challenges: a) promote efficiency and savings by developing one common solution for all NHS hospitals; b) enable individual and team focus on process improvement and outcomes; c) enable research; d) standardization of parameters and indicators at the national level, leading to hospital benchmark opportunities and better compliance with electronic health records; d) deal with local databases and e) different proprietary software application.

2.2 Analysis and Design

To tackle these challenges, we decided to use Business Intelligence technologies. Data warehouses, a crucial tool in the data preparation phase of a Business Intelligence system, provide a practical solution to deal with integration of data from these silos into a central repository ensuring that the reporting and dashboarding that follows is possible and easier to implement and maintain.

The solution is based on a hybrid architecture where source/operational systems (SClínico) are hosted in each hospital or institution and this platform is centralized on the cloud. This architecture allows the flexibility needed to create more components for the solution and evolve it according to new business needs in a faster way, as well as features like scalability in terms of computational resources or data warehousing and powerful security services, that were harder to achieve other way.

As a Business Intelligence system, BI SClinico is composed of 3 layers of architecture. In all three layers, data is submitted to masking and pseudoanonimization processes.

In terms of project governance, it was decided to adopt a shared model of implementation with hospitals in both designing and achieving consensus on metrics and indicators. The first approach was to challenge some hospitals to participate, decide and design some metrics. Once engaged, they wanted to contribute and to even share human and financial resources in developing new indicators. The second approach was the methodology and infrastructures in which this platform should be used, with rules and central governance, ensuring that one hospital contributes for all of them, instead of the previous models where indicators had local definition and low central control and monitoring. This insures that all indicators are uniform allowing easy benchmarking. This approach constitutes the crowdsource model.

2.3 Implementation

As stated, BI SClínico is composed of 3 layers of architecture. The first layer is integration, information collection, and standardization, with data collection being accomplished through integration of data from clinical software application (Sources).

Then, the second layer is the Data Zone, that has all the information stored and combined in a star schema, providing simpler queries and business reporting logic, query performance gains, fast aggregations and feeding cubes. The latter being a relevant advantage since BI SClínico has a multidimensional model that is the foundation of the reporting model and of the ad-hoc tool designed for power users that allows these users to query the multidimensional database and extract data with no technical knowledge of querying language.

The third layer is the Information Zone. The multidimensional attributes and facts and its data is accessed with DAX scripts. The objective is to provide answers quickly and the multidimensional data model is designed to solve complex queries in real time. Additionally, the structure of the cube makes it much easier to compare historical data, or to create cumulative values such as year-to-date production. The multidimensional data model is composed of logical cubes, measures, dimensions, hierarchies, levels, and attributes and the information is refreshed 1 time a day during the night period in order of shielding the clinical and operational platform.

Furthermore, we have the reporting model. The reporting platform allows users to access tabular and graphic data. It also allows users to export this information and work data in local spreadsheets files. It is useful for high level decision making because it gives very easy access to data in a structured form. Lastly, we have dashboards based on Microsoft Power BI ® that were recently publicly presented and early access is being given to key users, physicians and nurses.

Figure 2 outlines that currently around 2 years of data are stored in the Data Warehouse, accounting to about 12M medical appointments, 0.75M hospitalizations, 5.5 emergency dep. admissions and 0.5M surgeries. This data includes, for example, information on 26M lab and exams, 1.6M ambulatory prescriptions and 0.4M diagnostics.

2.4 Evaluation

This initiative began with the creation of few indicators concerning physicians and that were sent weekly and individually in a PDF report by email. These included comparisons with the previous week in various areas such as activity indicators, clinical records compliance (e.g., coding percentage of ICD-10 diagnoses), and analysis of the population covered by the professional in terms of scope (consultation, hospitalization, urgency, surgery), sex, age, origin, type of procedures, among others.

Twelve hospitals were invited to try out the solution, and each of the approximately initial 500 covered physicians were asked to provide feedback on the quality and relevance of the report received, needs for new indicators and priorities that they would like to see answered by the project. The feedback provided by around 150 doctors was collected and analysed, and a roadmap was prepared that could respond to the requested indicators, as well as a platform was developed to enable dynamic exploration of intuitive dashboards. In six months, this project is available for 2500 users.

A similar process is being applied for nursing professionals, which prioritize as indicators those based on metrics that translate results and gains in patient centred care, namely at the level of complications prevention and resolution of nursing diagnoses, and in indicators that translate the evolution of health conditions from admission to the discharge. (Hackl et al., 2015) In addition to these, the platform also focuses on epidemiological indicators, conditions inherent in the provision of care, as well as the care process itself.

This project enabled several findings on EHR compliance, data quality and maturity. SNOMED adoption, for example, revealed itself one of the priorities for catalog standardization and mapping at the national level.

Initial users feedback outlines the easiness of access (inside or outside hospital facilities, web and mobile application), the ability to get insights that could improve clinical practice and the importance of direct exploration of data using tools such as Excel ®.

2.5 Figures and Tables

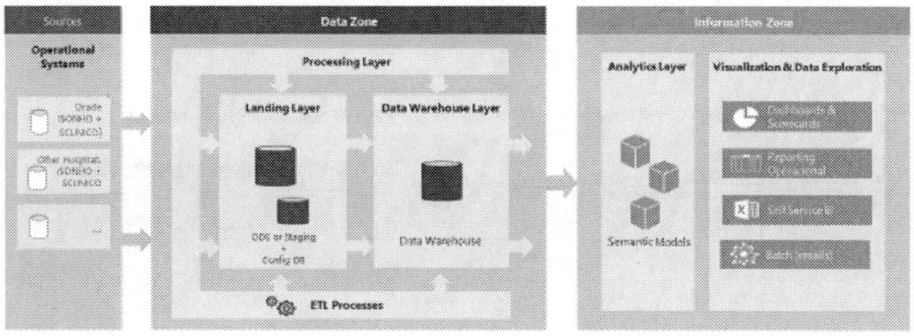

Figure 1a. Solution Architecture

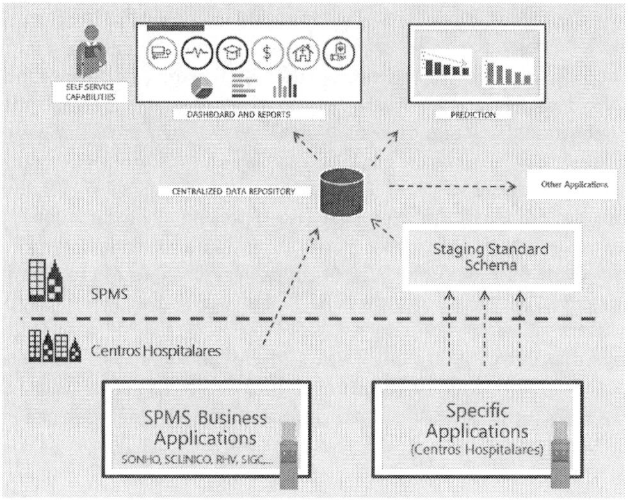

Figure 2b. Solution Architecture

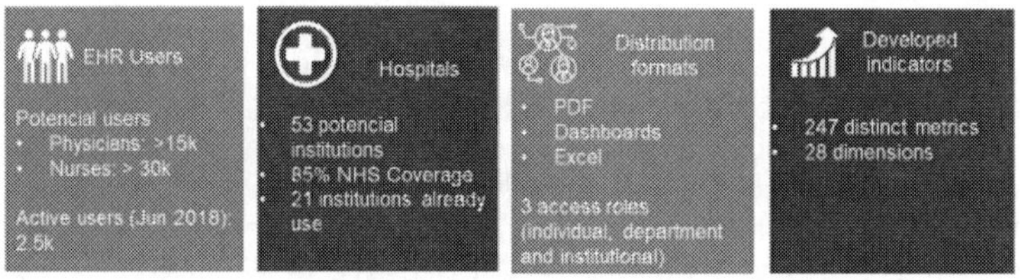

Figure 2. Some Project Figures

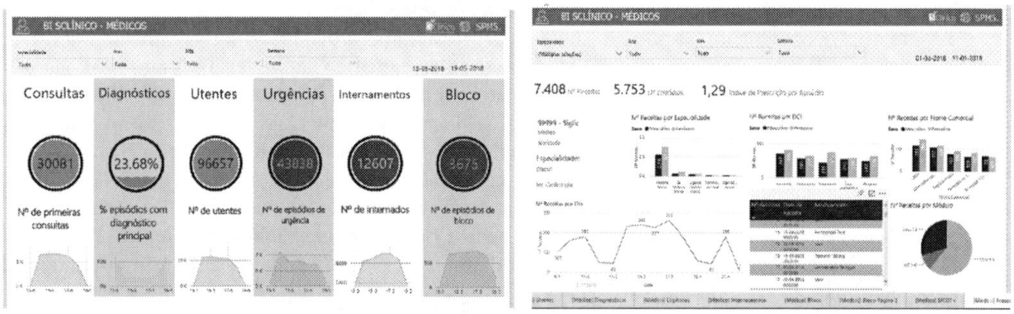

Figure 3. Application Examples for the Physicians' Dashboards

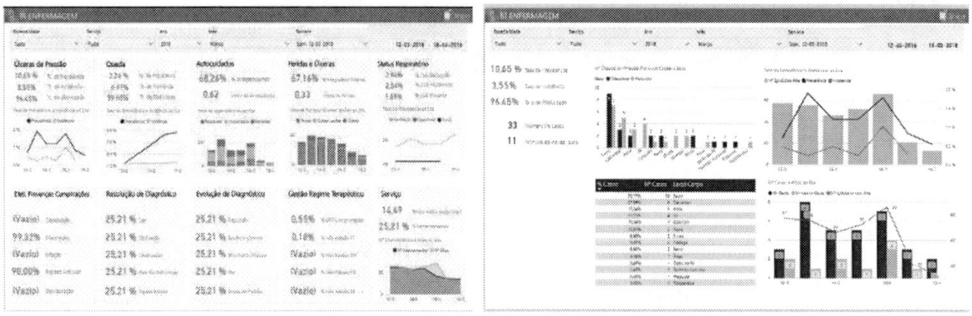

Figure 4. Application Examples for the Nurses' Dashboards

3. CONCLUSION

This innovative platform - BI SClínico Hospitalar - relies on business intelligence technologies and allows users to explore indicators with security and reliability.

The involvement of the health professionals is key to this project, however, should persist beyond the definition of requirements. (Hackl et al., 2015) It is necessary to have all processes and relations between the defined business areas to ensure greater cooperation and understanding. To ensure the engagement with the system, it must be responsive in the sense that the response time must be short, ensuring nevertheless a high effectiveness of the produced information. Moreover, it is important that users are free to access, explore and combine information, as relying on others to make changes in dimensions of analysis, for example, is one of the main factors leading to the search for alternative solutions.

When compared to other business sectors, the healthcare sector presents a few key differences that add to the intricacy of the system (Bennett and Evelson, 2013). Ensuring the data quality are also of great importance. The adoption of robust data cleaning and data quality frameworks is imperative.

The main conclusions are:

- Standardization of practices between the different health care units can benefit from the adoption of normalized terminology to facilitate the process of integrating data from different sources;
- Actionable insights can promote process improvement and quality of care;
- The crowdsource model is an interesting approach to promote consensus on indicators for easier benchmarking; also, to achieve a culture of shared services, so hospitals can contribute with financial resources, avoiding the multiplication of individual silos;
- The platform provides health indicators in various areas of health care to support healthcare units and the development of professional skills, being an important aid in the production of scientific papers, various activity reports, among other needs felt by the services or professionals.

So, there is, nonetheless, work to be done. From the organizational standpoint, an integrated and holistic vision that supports managerial decision making, both clinical and administrative, coupled with change management and transformation strategies, while bearing in mind that BI is a complex and an ever-changing area but nevertheless a priority for the future, is fundamental. Scalability, high performance, adaptiveness to different realities, and more sophisticated approaches to deriving information from data, such as advanced analytics, should also be driving future BI initiatives (Ferranti et al., 2010). The challenge of managing the burden of disease and dealing with an ageing population is also shaping healthcare. To tackle these challenges, the integration of health, environmental, and social care data should also be a key driver in the future initiatives of integration between the Portuguese primary and hospital care systems.

ACKNOWLEDGEMENT

All teams from Information Systems Department involved in the project as well as NHS hospitals and its users.

REFERENCES

B. J. Kolowitz, R. B. Shresth, 2011, 200 Enabling Business Intelligence, Knowledge Management and Clinical Workflow with SingleView, *Issues in Information Systems, Vol. XII, No. 1*,; p. 70-77, 201.

Bennett, M., Evelson, B., 2013, Best Practices: Maximize Your Chances Of Business Intelligence Success, Forrester Research, Inc., http://www.primenumerics.com/White Papers/Maximize Your Chances of Business Intelligence Success.pdf

Ferranti JM, Langman MK, Tanaka D, McCall J, Ahmad A., 2010, Bridging the gap: leveraging business intelligence tools in support of patient safety and financial effectiveness. Journal of the American Medical Informatics Association: JAMIA.;17(2):136-143. doi:10.1136/jamia.2009.002220.

Hackl WO, Rauchegger F, Ammenwerth E, 2015, A Nursing Intelligence System to Support Secondary Use of Nursing Routine Data. Appl Clin Inform; 6:418–428 http://dx.doi.org/10.4338/ACI2015-04-RA-0037

Mettler, T., Vimarlund, V., 2009, Understanding business intelligence in the context of healthcare. Health Informatics Journal 2009; 15(3), p. 254-264.

MANAGING ICT PROFESSIONALS' WORK STRESS THROUGH PSYCHOLOGICAL AWARENESS

Eugene Kaluzniacky
Applied Computer Science, University of Winnipeg, 515 Portage Avenue, Winnipeg, Manitoba, Canada

ABSTRACT

With the increasing reliance of organizations on the work of information systems developers, there has been increasing pressure on ICT professionals to create impactful systems which are very precise, need to satisfy a broad spectrum of users, and must be delivered within strict time deadlines. Such a climate has created significant work stress among ICT professionals worldwide. While traditional stress management techniques such as exercise, diet, relaxing music and meditation have been mentioned as factors for ICT stress management, this paper promotes psychological self-awareness to enable stressed workers to relate to common work stressor in a different psychological state that would mitigate the effect of the stressors on the workers. Two personality typing systems, the Myers-Briggs Type Indicator and the Enneagram and also generative and adaptive creativity styles are outlined along with insight as to how increased awareness and application of such factors in a work environment could decrease undesirable stress factors.

KEYWORDS

Stress Factors, Personality Types, Myers-Briggs, Enneagram, Creativity Styles

1. INTRODUCTION: DIMENSIONS FOR PSYCHOLOGICAL AWARENESS

Information and Communication Technology (ICT) professionals encompass a variety of persons in an ever-expanding variety of positions. The term comprises, for example, systems analysts, business information analysts, database designers, telecommunications engineers, software engineers, knowledge engineers and many others. Such personnel may found in the category producers of specific deliverables, often in ambiguous circumstances. As well, ICT personnel would include, at least, middle management, planning and overseeing the work of subordinates in the previous category. In a majority of circumstances, however varied the ICT personnel can be, work stress is a common and often debilitating denominator. With rapid organizational and technological change, increased expectations from computers and often declining budgets, today's ICT worker is often required to be creative yet precise and within strict time limits. Factors such as rapid change, unrealistic expectations of users and skill obsolescence are being reported as causing significant, ongoing stress among ICT professionals. Moreover, ICT burnout is becoming a pervasive reality. In a short Internet poll of 250 IT workers, 48% responded being "close to burnout", answering at least 5 on a 7-point scale (Kaluzniacky, 1998).

While at least a certain fraction of ICT management and their subordinates is concerned about impending large-scale burnout, proposed remedies have often followed traditional stress management techniques such as healthier diet, physical activity in job design (Shanthi, 2016) and consciously developing work-life balance (Dash et al. 2012). Yoga (Gupta, 2016) and meditation (Sreekumar, 2016) for stressed IT personnel have also been advocated. While such prescriptions are undoubtedly beneficial, they do have limitations. The stress-managing ICT worker is still likely to have the same attitudes and emotional reactions to his/her work stresses, but he may have more physical and perhaps emotional stamina to deal with the inner effects of these reactions.

For example, we can recognize an ICT worker experiencing disappointment, frustration and insecurity with accompanying headaches, indigestion, poor circulation and lessened motivation as a result of not being able to meet deadlines and satisfy user expectation. After changing his diet, working out regularly and intentionally taking time for recreational family activity, this same worker will still likely experience disappointment, frustration and insecurity in the same work environment, only now the effects of these experiences on his person will be lessened.

However, what can be done so that this ICT worker will no longer react in disappointment, frustration or insecurity in the same work circumstances? Such a change in inner reactions to the same work environment can indeed be achieved through a change in personal consciousness, or psychological self-awareness. Below are described several dimensions for personal psychological awareness i.e., two personality systems and cognitive style are outlined. Potential benefit of each dimension for ICT workers is then discussed in the context of alleviating stress.

2. MOTIVATION FOR INNER CONSCIOUSNESS

Over thirty-five years ago, researchers Couger and Zawacki (1980) reported that, while IS professionals (systems analysts and programmers), had the lowest needs for social interaction on the job in a group of surveyed professionals, they reported much higher "growth needs". While at the time, "growth" may have been largely understood as greater development of professional competencies, today the scope of growth among ICT professionals can be expanded to include a deeper awareness of the dynamics of one's psyche in terms of, for example, personality functioning, cognitive style and spiritual, core-energy connectedness. In a survey on stress among Canadian IS professionals, the most frequently mentioned desired coping resource was "personal development seminars", followed closely by "conflict resolution workshops" (Kaluzniacky, 1998).

Thus, ICT professionals, long recognized for their intellectual, analytic orientation may indeed be ready to adopt a more wholistic approach to their work and their life. However, a blueprint is needed for journeying into what for many will likely be uncharted inner territory. ICT workers, accustomed to controlled, specific logic and easily assemble feedback may largely be uncomfortable with what they perceive as "that touchy, feely stuff". As a result, many may be ignoring valuable self-development avenues which could indeed affect profoundly their emotional, intellectual and physiological reactions to a work environment that may not be likely to change regarding stressful challenges.

The truth is that growth in awareness of various dimensions of one's psyche need not be ambiguous, "poetic" and "without a roadmap". Is there another way of explaining and promoting growth and awareness of the "inner person" that may be more palatable to the ICT mentality? Can such increased inner awareness be concretely related to specific ICT work situations?

Just as an Olympic athlete and his coaches will delve into a broad spectrum of knowledge in physiology, kinesiology, nutrition, biochemistry, physics, and psychology so as to be optimally prepared for "the big test", so, in this era, can ICT professional engage in cross-disciplinary growth so as to be optimally prepared for ongoing work challenges. .

3. PERSONALITY DIMENSIONS

3.1 Myers-Briggs Types

Personality can be defined as "a complex set of relatively stable behavioral and emotional characteristics of a person" (Hohmann, 1997). It can be viewed, classified, and assessed according to different aspects. A number of personality classifications have arisen over time with each focusing on specific characteristics, possibly for specific purposes. One of the most widely-used systems within the last 40 years is the Myers-Briggs Type Indicator (MBTI) system. This system was developed privately by a mother and daughter team (Katherine Briggs and Isabel Briggs Myers) in the US in the first part of the twentieth century. It reflects the work of noted psychologist C.G. Jung. Myers-Briggs describes a personality in terms of the following four dimensions:

(1) **Introversion/Extraversion** (I/E): where a personal focuses more easily, within oneself or on other people and the surrounding environment.

(2) **Sensing/Intuition** (S/N): how a person takes in information, a sensing (practical) person perceives through the senses and focuses on the real, factual and tangible; an intuitive person perceives through the "sixth sense" and focuses more on the imaginary, on what is "behind the scenes".

(3) **Thinking/Feeling** (T/F): how a person comes to conclusions/makes decisions; a thinking person employs logical analysis, using objective and impersonal criteria, while a feeling type uses person-centered values and motives to arrive at a conclusion.

(4) **Judging/Perceiving** (J/P): how much structure and predictability one prefers in one's life; a judging (structured) person focuses more on results rather than process; he prefers to make judgments about what he notices, and a perceiving person focuses on the process rather than on results - she prefers to notice one's outer environment without making judgments.

Although each person will have abilities in both components of each dimension, he/she will exhibit a preference for one side in each case. Since there are four dimensions and two possibilities for each dimension, the Myers-Briggs system identifies 16 different types. The Association for Psychological type is the overseer of the MBTI. Its trained members administer the official MBTI questionnaire to assist in identifying one's type. The well-known book *Please Understand Me* by Keirsey and Bates (1998) is an excellent primer for the uninitiated and contains a shorter "approximator" questionnaire.

Psychologist David Keirsey has collapsed the 16 types into four temperaments and each type corresponds with one and only one temperament, being the NT (Promethean), NF (Apollonian), ST (Guardian) and SP (Dionysian). Each temperament has a basic operating characteristic and an essential need. For example, the NT is the planner and visionary and has a strong need for competence. The NF is the motivator and the conciliator and has a strong need for authenticity.

MBTI has been mentioned reasonably often in IT circles. Key applications have been in system development team members' awareness of each other's types and strengths and in the fit of different types with different development tasks. An MBTI consciousness in the ICT organization could allow ICT workers to express openly to co-workers and supervisors why specific tasks are easy or difficult for them and where they would appreciate the assistance of another specific complementary type. ICT workers could also monitor their energy level, noting which aspects of their work seem to cause a drainage of energy - they could then try to relate this phenomenon specifically to Myers-Briggs personality considerations. Those in IT management could become more explicitly aware of how different types are energized and stressed. Such awareness could then become a regular part of their management skills and strategies.

Thus, increased open communication about MBTI and stress in ICT and an availability of personality-related strategies to decrease stress levels in commonly occurring work situations can indeed contribute to the war on burnout in ICT. As well, specific, applied research in this area will need to be intensified. .

3.2 Enneagram Types

While the Myers-Briggs personality classification system has its definite contribution, no one system can hope to address all aspects of personality. Thus, both individual ICT workers engaged in personal growth and ICT management interested in furthering their effectiveness can also look to the Enneagram personality system. The essence of the Enneagram is said to have descended from the ancient Sufis and modern adaptations have been made by a variety of contemporary authors. While MBTI attempts to explain *how* we function, the Enneagram focuses more on *why* we function in a particular way - what is likely our underlying emotion that guides the way we act.

The Enneagram system proposes nine personality types, each type being assigned a specific number, one to nine. Each type views himself and his style to be "OK" when a certain condition is met. The following is a summary:

(1) Type 1, the **Perfectionist** needs to be performing nearly perfectly, to be doing the "right" thing; he has a fear of being condemned.

(2) Type 2, the **Helper**, needs to be helping someone and receiving appreciation for the help; she has a fear of being unloved.

(3) Type 3, the **Status Seeker**, needs to be seen by others a successful and important; he has a fear of being rejected.

(4) Type 4, the **Artist**, needs to be free to be unique, often to extremes; she has a fear of being defective.

(5) Type 5, the **Knowledge Seeker**, needs to observe, analyze and learn as much as possible about his environment; he has a fear of being overwhelmed by the world.

(6) Type 6, The **Loyalist**, needs to "hang on" to external rules or dictates, often hesitating to act out of personal initiative; she fears being abandoned.

(7) Type 7, the **Fun Seeker**, needs to be "having fun" no matter what the circumstances; he fears emotional deprivation.

(8) Type 8, the **Power Seeker**, needs to be a "mover and a shaker"; she has a fear of submitting to others.

(9) Type 9, the **Peace Maker**, needs to experience no conflict and hopefully harmony, often at any **cost**; he has a fear of separation.

Both the needs and the fears of each type may be largely subconscious, yet they drive a person's behaviors and reactions, including those on the job. Each type is identified with a particular "compulsion" (reflecting the imperfection of humanity). However, the 9-point diagram of this system, the actual "Enneagram", offers what are called directions of integration for each type. Conscious progression for each type along the indicated path allows the type to become broader, deeper and more balanced. Moreover, it is shown that under stress and without explicit consciousness for Enneagram growth, each type will move in his direction of disintegration, that is, in the opposite way to his intended growth path on the diagram. To illustrate, type 6, the loyalist, wants to "do it right", to be loyal to external rules. Yet when circumstances prevent him from doing so, he will proceed in the direction of type 3, the image seeker. Thus he will do all he can so that *it looks like* he is doing it right. This will drain him of considerable energy. Yet the direction of integration for a 6, on the Enneagram, is towards 9, the peace maker. When it is impossible to do something "the way it should be", the type 6 should adopt the attitude of the 9, saying "I won't be excessively concerned here, I want to have inner peace above all". This liberating inner shift will allow the type 6 to handle the difficult situation with much less distress. Each type, of course, has different directions of integration and disintegration. Conscious personal growth can help one to recognize and alter appropriately what have so far been natural reactions to stress. An informative primer on the Enneagram is *The Enneagram Spectrum of Personality Styles: An Introductory Guide* by Jerome Wagner. The Enneagram has been applied to management, notably by Michael Goldberg in his book, *Know Your Boss's Number.*

4. COGNITION AND CREATIVITY CONSCIOUSNESS

In addition to a deeper consciousness related to one's personality characteristics, it can be beneficial for an ICT professional to become more aware of how he/she thinks, creates and learns. Cognitive style has been defined as a person's preferred way of gathering, processing and evaluating information. Although there are numerous ways of classifying cognitive style, one significant approach relates to analytic, sequential functioning versus intuitive, wholistic processing. Some psychologists have referred to the former as "left-brain" and the latter as "right-brain" thinking.

The analytic person abstracts, analyzes, structures, organizes and plans systematically. He can articulate clearly and may focus on details. The intuitive person integrates many perspectives, finds problems, discovers opportunities and generates new visions. She is sensitive to both logical and emotional issues, viewing them as one. The analytic is like a tax lawyer, the intuitive, like a criminal lawyer, the analytic like a chemist or accountant, the intuitive like an athlete or artist; the analytic prides himself on intellectual rigor, the intuitive on imagination.

It is easy to see that people of both cognitive styles are valuable to the ICT profession. However, if people are not specifically aware of their styles and those of co-workers or users, needless stress can result. If development team members are to co-operate effectively each should be aware of what he can contribute and when he can use assistance. It remains for further research to estimate productivity losses in system development work due to task and cognitive style mismatch. In addition, the employee's professional self-confidence may be eroded when, for example a promising Intuitive is misassigned and then poorly evaluated by a highly sequential manager.

Differences in thinking style also show up in ICT tasks requiring creativity. Robert Glass, in his book *Software Creativity* (1995) states that "creativity is an essential part of the work of software development" and that "analysis and design are arguably the most intellectual and creative of the life cycle phases". Parallel to the two basic cognitive styles, we recognize differences between adaptive creativity and generative creativity. The former, practiced more by analytics, accepts a current paradigm but extends approaches found within it. For example, ERP Systems can be thought of as resulting from adaptive creativity. They are essentially a large-scale extension of existing basic, integrated business systems. Generative creativity, belonging more to intuitives, challenges a current paradigm and creates an entirely new one. The object-oriented paradigm for system development is likely a result of such creativity. The adaptor focuses on doing things *better*; the innovator (with generative creativity) focuses on doing things *differently*.

With expansion of the scope and influence of business information systems and with the increasing integration of technologies, the need for both types of creativity in IS development will undoubtedly increase. An enhanced consciousness in this area can indeed increase effectiveness and decrease stress.

5. CONCLUSION

It may indeed take some time for personality and cognition consciousness to become part of the everyday work world of an ICT professional. Initial steps have already been taken in a number of ICT settings. To move into "another gear" however, will require a concerted effort on the part of ICT employees, their professional associations, MIS and interdisciplinary academic researchers and human resources departments of the ICT workers' employers. Perhaps, then, "personal re-engineering" will parallel process and system re-engineering within ICT. With a gradual but determined growth in individual and collective self-awareness, the looming threat of a burnout epidemic in ICT can be significantly addressed by empowering the affected to turn initially perceived stressors into edifying and enriching experiences.

REFERENCES

Couger, J.D., and Zawacki, R., (1980), Motivating and Managing Computer Personnel, *John Wiley & Sons, Inc., New York, NY. USA.*

Dash, M., Anand, V., & Gangadharan, A (2012) Perceptions of Work-Life Balance Among IT Professionals *IUP Journal of Organizational Behavior, 11, 51-65.*

Glass, R., (1995), Software Creativity, *Prentice-Hall, USA.*

Goldberg, M., (1996), Know Your Boss's Number, *Harper Business, USA.*

Gupta, S., (2016), Occupational Stress in IT Professionals and Coping Strategies, *International Journal of Home Science, 2(3) 29-32*

Hohmann, L., (1997) Journey of the Software Professional, *,Prentice-Hall, USA.*

Kaluzniacky, E., (1998) An Assessment of Stress Factors Among Information Systems Professionals in Manitoba. *Proceedings of the ACM SIGCPR Conference on Computer Personnel Research*,, pp. 254-257.

Keirsey, D. and Bates, M., (1978), Please Understand Me, *Prometheus, USA*

Shanthi, R., (2016), A Study on Stress Management and Coping Strategies with Reference to IT Companies, *International Journal of Management, 7(2), 636-645.*

Sreekumar, V., (2016), Occupational Stress Management Among Information Technology Professionals, *International Journal of Information Research and Review, 03(11), 3074-3076.*

Wagner, J., (1996), The Enneagram Spectrum of Personality Styles: An Introductory Guide, *Metamorphous Press, USA.*

THE EFFECT OF ATTITUDE TOWARD CHANGE AND COMPUTER SELF-EFFICACY ON THE ERP ADOPTION

Kee-Young Kwahk
Kookmin University, 77 Jeongneung-ro, Seongbuk-gu, Seoul, Korea

ABSTRACT

To rapidly respond to uncertainties in the business environment whilst remaining competitive, every organization needs to be able to successfully introduce and manage organizational change. Cognizant of the role of information systems (IS) as an enabler of organizational change, many organizations have paid attention to Enterprise Resource Planning (ERP) systems for successful organizational change primarily because of their change-driving forces across organizations. This study introduces a theoretical model which views ERP systems as both an organizational change driver and a sophisticated information system in order to better explain phenomenon of ERP systems adoption. Based on the literature review and a proposed research model, the potential contributions of the study are discussed.

KEYWORDS

Intention to Adopt ERP Systems, Attitude Toward Change, Computer Self-Efficacy, Perceived Usefulness

1. INTRODUCTION

Because of today's increasingly dynamic environments, organizations are continually faced with the need to change their structures, objectives, processes, and technologies. Therefore, every organization needs to be able to successfully introduce and manage organizational change to achieve and sustain competitive advantage. Cognizant of the role of IS as an enabler of organizational change, many organizations have paid attention to Enterprise Resource Planning (ERP) systems for successful organizational change primarily because of their change-driving forces across organizations. ERP systems promise to improve organizational performance and competitiveness by streamlining business processes and eliminating duplication of effort and data. Despite the promised benefits, many companies have been plagued by a high failure rate which is estimated to be between 60 and 90 percent. Some plausible reasons for their failure can be derived from the inherent characteristics of ERP systems which may be interpreted in the context of socio-technical systems.

On the one hand, from a social or organizational system perspective, ERP systems are considered having a change-driving force over the adopting companies, which leads to enterprise-wide organizational changes. Previous studies (Lapointe and Rivard 2005) have indicated that one of the most critical failure reasons for ERP systems adoption is potential user resistance to change. It is widely reported that the sources of ERP systems failures are not limited to technical issues. Rather, there are various causes arising from the interactions among people, task, environment and technology. Although an ERP system could be developed successfully from the technical point of view, if users are not willing to use it because of their negative attitude toward change, the ERP system would not bring the expected benefits to the company. On the other hand, from a technical system perspective, ERP systems are considered technologically sophisticated information systems (IS), which often require more technical knowledge about the adopting systems compared to the traditional transaction-processing systems. Technical knowledge, which is usually learned through education and training, is likely to be acquired more effectively and efficiently when users' self-efficacy for using computers and information technologies (IT) is high than when users are reluctant to use them because of the lack of computer self-efficacy. It is widely argued that computer self-efficacy can influence the successful IS adoption by positively influencing individual expectancies about the job performance when using the specific IS for performing the job.

This study introduces a theoretical model which views ERP systems as both an organizational change driver and a sophisticated IS in order to better explain phenomenon of ERP systems adoption. For this purpose, the proposed model includes attitude toward change and computer self-efficacy, two individual factors from the socio-technical systems perspective, which may affect ERP systems adoption behavior through perceived usefulness for the systems.

2. THEORETICAL BACKGROUND AND RESEARCH MODEL

Change is a fundamental theme in human life and organizational behavior, which individuals generally resist (Joshi 1991). Organizational change is defined as an attempt, or series of attempts, to modify an organization's structure, goals, technology or work tasks. A range of perspectives on organizational change have developed over the past decades. Among them, three paradigms have influenced studies of technology-based organizational change: planned change, technological imperative and punctuated equilibrium (Orlikowski, 1996). Transforming organizations with IT is an important business objective since traditional structures and processes are often ineffective in producing desired levels of productivity and customer service. As each new generation of technology and each major technological advance have emerged, organizations will be radically and fundamentally altered.

Attitude toward change refers to the extent to which individuals hold positive views about the need for organizational change, as well as the extent to which they believe that such changes are likely to have positive implications for themselves and the organization. Thus, an individual's attitude toward change can play an important role in determining whether the individual chooses to support or resist a change. Attitude toward change, in general, consists of a person's affective reactions to change, cognitions about change, and behavioral tendency toward change. Consistent with this, individuals' or groups' responses to organizational change are classified into three types. First, affective responses are a greater or lesser feeling of being linked to, satisfied with, or anxious about change. Second, cognitive responses are the opinions one has about the advantages, disadvantages, usefulness and necessity of the change, and about the knowledge and information required to handle it. Third, behavioral responses are the actions one has already taken or may take in the future for or against the change.

Previous research has attributed many IS failures to user propensity to resist change (e.g., Jiang et al. 2000). Organizational resistance to change has been identified as a top critical success factor for ERP adoption (Lapointe and Rivard 2005), with user resistance blamed for many ERP project failures. In addition, the absence of an adequate organizational change attitude can result in the failure of an ERP initiative, regardless of how competent the organization is technically. Creating the belief that organizational change is needed implies that there is a performance gap between the current state and some desired end-state. In general, an ERP system as a means of organizational change is introduced into a company to improve organizational effectiveness and thus fill up the performance gap. Organizational members who have favorable perceptions of organizational transformation are likely to proactively participate in any organizational change situations such as an ERP system introduction, and possibly look forward to changes in work patterns while expecting enhanced performance. In addition, when people have positive attitude toward change, they tend to be more interested in new features of the ERP system and expect more benefits from them. Individuals with such traits are more likely to focus on positive outcomes and pay less attention to negative outcomes when using the ERP system. Thus, this leads to higher perceived usefulness for the ERP system. In contrast, when people have negative attitude toward change, they tend to be rather passive to the new features of the ERP system and expect nothing or little beneficial from them. Individuals with such traits are more likely to focus on negative outcomes and disregard positive outcomes when using the ERP system, which results in lower perceived usefulness for the ERP system. Therefore, we hypothesize:

H1: Attitude toward change will have a significant effect on perceived usefulness.

Self-efficacy is the belief that one has the capability to perform a particular job and is reflective of the confidence in one's ability (Bandura 1997). In the context of using computers and IT, computer self-efficacy, then, is defined as a judgment of one's capability to use a computer (Compeau and Higgins 1995). Computer self-efficacy judgment consists of three distinct, but interrelated, dimensions: magnitude, strength, and generalizability (Compeau and Higgins 1995). First, the magnitude of computer self-efficacy reflects the level of capability expected. Second, the strength of computer self-efficacy refers to the level of conviction about the judgment. Third, the generalizability of computer self-efficacy refers to the extent to which the judgment is limited to a particular domain of activity. Computer self-efficacy has been proposed and has accumulated empirical support as an important antecedent of perceived usefulness (Agarwal and Karahanna 2000, Compeau and Higgins 1995). Based on the social cognitive theory (Bandura 1991), it has been argued that computer self-efficacy has a positive influence on both personal and performance-related outcome expectations about the consequences of performing a specific behavior, because it is often difficult for individuals to separate the anticipated consequences of the behavior from their expectations of performance achievements (Compeau and Higgins 1995). Such outcome expectations can be interpreted as the notion of perceived usefulness (Agarwal and Karahanna 2000). When individuals believe that they will be able to use computers and IT with great skill, they are more likely to expect beneficial outcomes from using computers and IT compared to when they doubt their computer related capabilities. In other words, computer self-efficacy can be established as an additional important predictor of perceived usefulness. Hence, we hypothesize:

H2: Computer self-efficacy will have a significant effect on perceived usefulness.

Many previous studies based on the Technology Acceptance Model (e.g., Venkatesh et al. 2003) have indicated that individuals' behavioral intention to use a new system is influenced by perceived usefulness for the system. The more useful the system is in enabling employees to accomplish their tasks, the more it will be used. Hence, we hypothesize:

H3: Perceived usefulness will have a significant effect on intention to adopt an ERP system.

3. CONCLUSION

This research is expected to make some potential contributions to IS adoption literature, in particular, ERP systems adoption literature. It proposes a theoretical model to account for ERP systems adoption through perceived usefulness for the system by introducing two perceptual factors, an organizational factor (attitude toward change) and a technological factor (computer self-efficacy). From the perspective of practice, there are some additional implications. Despite the promised benefits, the ERP systems are considered to be inherently risky because they require significant organizational resources; and organizations often adjust slowly to complex enterprise system packages. Thus, the ERP systems are viewed as a completely different class of IT application compared with traditional IT systems. Cognizant of the different features of the ERP systems, therefore, it is imperative that organizations understand the particular conditions under which they will be adopted and utilized. Investigating the role of employees' attitudes toward change might explain aspects of why many organizations installing and launching the same ERP system show the mixed results with their acceptance. This study sheds some light on this issue by showing that attitude toward change influences the expected usefulness about the system, which in turn affects the behavioral intention to use the system. Acknowledging that more than two thirds of ERP systems projects result in failure and ERP systems continue to grow with promising potential benefits, this study thus will have value for theoretical advancement for IS researchers as well as practical development for ERP systems practitioners, while several avenues for future research remain.

REFERENCES

Agarwal, R. and Karahanna, E. 2000. Time Flies When You're Having Fun: Cognitive Absorption and Beliefs about Information Technology Usage, *MIS Quarterly*, Vol.24, No.4, pp.665-694.

Bandura, A., 1991. Social Cognitive Theory of Self-regulation. *Organizational Behavior and Human Decision Processes*, Vol. 50, No. 2, pp. 248-287.

Bandura, A., 1997. *Self-Efficacy: The Exercise of Control*. W.H. Freeman, New York.

Compeau, D.R. and Higgins, C.A., 1995. Computer Self-Efficacy: Development of a Measure and Initial Test. *MIS Quarterly*, Vol. 19, No. 2, pp. 189-211.

Jiang, J.J., et al., 2000. User Resistance and Strategies for Promoting Acceptance across System Types. *Information & Management*, Vol. 37, No. 1, pp. 25-36.

Joshi, K. 1991. A Model of Users' Perspective on Change: The Case of Information Systems Technology Implementation. *MIS Quarterly*, Vol. 15, No. 2, pp. 229-242.

Lapointe, L. and Rivard, S., 2005. A Multilevel Model of Resistance to Information Technology Implementation. *MIS Quarterly*, Vol. 29, No. 3, pp. 461-491.

Orlikowski, W.J. 1996. Improvising Organizational Transformation over Time: A Situated Change Perspective. *Information Systems Research*, Vol.7, No.1, pp. 63-92.

Venkatesh, V., et al., 2003. User Acceptance of Information Technology: Toward a Unified View. *MIS Quarterly*, Vol. 27, No. 3, pp. 425-478.

INFLUENCE OF ORGANIZATIONAL CULTURE IN INFORMATION SYSTEMS PROJECTS IN A SOUTH AFRICAN BANK

Ray M Kekwaletswe[1] and Tsholofelo Ncube[2]
[1]*University of the Witwatersrand, Johannesburg, South Africa*
[2]*Tshwane University of Technology, Pretoria, South Africa*

ABSTRACT

Although information systems projects are social activities which require interactions amongst those involved, literature provides little understanding of how organizational culture plays a role during the activities. This paper argues that organizational culture is one of the contributing factors which influence the outcome of IS projects. Failure and success of information systems projects is an over-studied area but notwithstanding this, how other facets affect these outcomes is still inadequately addressed in literature. To this point, this paper aims to bridge this knowledge gap by describing how organizational culture may be influencing information systems projects, especially in the context of a South African bank. The paper subsequently suggests ways to improve the outcome by giving insights into how organizational culture manifests and its influence on information systems projects.

KEYWORDS

Cultural Dimension Theory, Information Systems Project, Organizational Culture, Project Management, South African Bank

1. INTRODUCTION

Traditional project management methodologies tend to focus on the schedule, scope, schedule, budget, quality and control, risk, communication, and procurement but less on the organization culture. The focus of this paper is, therefore, on the people-centric aspect of information systems projects. Although information systems projects are social activities which require interactions amongst those involved, there is still little understanding of how organizational culture plays a role in the outcome. The research argument driving this paper is that organizational culture is one of the contributing factors to the IS project outcome. Thus, this paper determines the influence of culture and subsequently suggest ways to improve the outcome of the IS projects. From the surveyed scholarship, it is evident that there is still not enough literature which addresses the influence of organizational culture in information systems projects and, therefore, there is still little understanding of its role. The contribution of this paper is on the extended insight on how culture may influence IS project outcome.

The rest of this paper is outlined as follows: the field studied is first introduced as well as the theoretical foundations; this is followed by the research methods used. The paper then interprets the findings and the implications thereof, which leads to the conclusion of the paper.

2. SCHOLARSHIP AND THEORETICAL FOUNDATIONS

This section introduces the scholarship by briefly discussing the terms in the research topic. The theoretical frameworks underpinning the study are then discussed.

Polyaninova (2011) asserts that a *project* is a combination of resources pulled together to create something that did not previously exist. He further explains that organisations implement various projects in order to produce a specific outcome and to reach specific goals. There are various factors like global competition, new product development, customer focus, innovative technologies and compression of product life cycle that influence organisations to start implementing project management practices to conduct their business related activities.

Information Systems Project Management: Berggreen and Kampf (2015) mentioned that Project Management has tools and methodologies that Project managers should comply to however it's also result-oriented, it's management style builds on collaborative relationship amongst various stakeholders involved in Projects. This implies that managing a project calls for an understanding of the various factors that may influence and affect information system projects. In the paper, organizational culture is argued to be one of the factors/facets whose influence to information systems projects needs to be studied and understood.

Organizational Culture in the Context of Information Systems Project: Organisational culture may be defined as "the way things are done around and in a work place" and this definition can be applied to projects (Du Plessis and Hoole, 2006). They further advised that the following conclusions can be made about this definition : "The way" – may refer to how things are done in projects i.e processes, "We" – refers to the project stakeholders and the project team, "Do things" – refers to the project management methodologies applied to projects, "Around here" – refers to the environment where the projects are delivered.

Eberlein (2008) explains that the underlying concepts of culture and culture itself are a matter of dispute where almost 160 definitions of culture are discussed. The definition of culture chosen for this paper and one which can be applied to project management is stated as "Culture is a way in which a group of people solves problems and reconciles drama (*ibid.*, 2008)". Types of organisational culture as listed by Mawu (2015) include achievement, role and power of culture and support culture. These types all have different implications on the overall organisational success, including projects.

Project Management is the application of knowledge, skills, tools, and techniques to project activities to meet project requirements (PMI, 2013). Du Plessis & Hoole (2006) maintain that using project management as a lens to view organisational culture refers to looking at how projects are run in an organisation by interdependent project team members. According to Du Plessis & Hoole (2006), project failure still happens despite the extensive research conducted on the technical aspects of project management, the theory and methodology written on how to organise project teams in order to execute and deliver projects successfully.

The project manager's role focuses on functional issues such as recruitment, selection, development,evaluation and compensation (Rodrigues & Sbragia, 2011). Ensuring the team is effective is one of the fundamental responsibilities of a project manager (Ranf, 2010).

Berggreen & Kampf (2015) posit that Project Management is more than just a set of tools, it is a result-oriented management style that builds on collaborative relationship amongst many stakeholders. Park & Lee (2014) explain that it is critical for project team members to have the ability to learn and the ability to use technology-related knowledge as well as domain knowledge effectively as this is a very critical factor for the success implementation of information systems projects. The PMBOK (2013) states that projects are composed of five process groups and these groups are categorised into nine knowledge areas. However, cultural management is not explicitly mentioned in the PMBOK guide. There is no attention paid to culture as a central element of Human resource management or communication management in the PMBOK (Eberlein, 2008).

2.1 Theoretical Frameworks

The aim of the paper is to determine the influence of organizational culture in information systems projects, and subsequently suggest ways to improve the outcome. For one to understand the role of culture, this study was underpinned by the Cultural dimensions theory and the Organizational culture model. In this section, an overview and description of both theories are provided.

2.1.1 Hofstede's Cultural Dimensions Theory

Culture can be defined as the collective programming of the human mind that distinguishes one group or category of people from another (Hofstede, 2010). Organisational culture can be defined as the manner in which members of the same organization relate to each other, to their work and to the outside world that distinguishes them from other organisations. He further suggests that when assessing how members of an organisation are related, culture can be described as the way superior relates to direct reports, the way direct reports relate to their direct superior, and the way colleagues on the same hierarchical level relate to each other (Waisfisz, 2015). Hofstede (1983) states that culture has various patterns, and these patterns may vary according to countries. He also mentions that there are generally five dimensions of culture. The five dimensions of culture include Individualism – Collectivism, Power distance, Uncertainty avoidance, Masculinity – Femininity, and Long Term Orientation (Hofstede, 2010)

2.1.2 Organizational Culture Model

The second theory underpinning the study was the Organisational Culture Model (Harrison, 1993). In organizations, management practices which are suitable for one environment may not be suitable for the other, thus said managers need to understand the core concept of culture specific to the organization they are at currently (Podrug et al., 2006). Organisational culture can be diagnosed in four cultural dimensions namely: Role, Achievement, Support and Power. He further explains that the model creates awareness of the culture gap between the existing and preferred cultures in an organization (Harrison, 1993). Organizational culture model implies that the four cultural dimensions are measured in two modes of operation which are formalisation and centralization; these modes can be measured on a scale of high and low levels (*ibid.* 1993). A weak organisational culture can be detected by its low formalisation of rules and regulations (Martins and Martins, 2003). High formality creates predictability, consistency and order in organizations and this means a strong culture in an organization may serve as a replacement for formalisation. They further explain that the organization's processes, rules and regulations can be internalized by members of an organization when they adapt to the organization's culture. (*ibid.*, 2003).

2.2 Conceptual Research Framework

Having discussed the elements of the underpinning theories, the following section discusses the conceptual research framework followed in the study. The framework shows that an IS project environment is characterized by four culture facets.

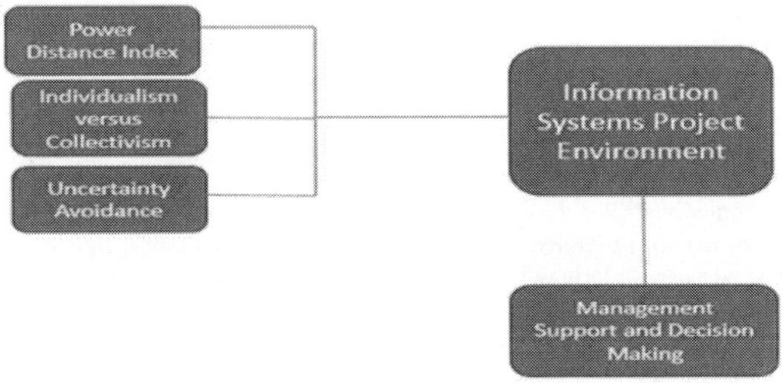

Figure 1. Conceptual research framework for the influence of organizational culture in IS Projects

The study looked at Power Distance, Individualism-Collectivism, Uncertainty Avoidance and Masculinity- Femininity as well as Management support and decision making. Power distance index is related to the level at which the less powerful team members of the IS projects, in the bank, accept or expect power to be distributed equally. Uncertainty avoidance looks at how project members deal with anxiety. Individualism versus Collectivism focuses on the ways in which members either work as individuals or are integrated into groups. Management support and decision making looks at the mutual trust between the organization and individuals, which is the people orientation characteristic of organizational culture. People orientation culture is defined as the extent to which management decisions affects people (Martins and Martins, 2003). The study analyzed how each of the facets manifested or played a role during an IS project life cycle.

3. FINDINGS AND INTERPRETATIONS

This section interprets the study findings. The interpretation is done against literature as well as the study context.

Agnantounkpatin & Zhang (2011) maintain that organisational culture can be viewed as principles, shared values and traditional ways of doing things that influence how employees of a certain organisation should act. The study participants defined it as *organizational culture talks to people's beliefs and values as the component drivers of business... organizational culture is a set of values that are shaped by the staff in the organization.* These responses are an indication that the participants are aware of the concept of organizational culture and its value to the organization. Organisational culture impacts employees' capabilities and the manner in which they view, define, analyse and resolve issues and problems Agnantounkpatin & Zhang (2011). The study participants posit that ... *organizational culture gives the uniformity and common standards and understanding of what is required in an organization and that translate into project success... it can only improve if the organization can share similar values across. Some of the values can be trained in an organization so that every employee would understand how their role fits in the role of the strategist.* It is also evident that the employees understand the role that organizational culture plays in the success of IS projects in the organization. Du Plessis and Hoole (2006) posit that one of the main causes of project failure is that organisational culture in which projects have to be delivered within is not supportive of projects.

3.1 Power Distance Index

In information systems projects, knowledge sharing encourages project team members to maintain social capital (Park and Lee, 2014). Participants said *that ... if the organization has good communication platforms there is success, because in project execution there is a need for common understanding which would align all the team members to be focused and have the zeal to achieve the organizational objectives...sharing of knowledge and information is done on a one on one basis and also through SharePoint.... there is a top down approach to information delivery.* Good communication platforms are important to allow for efficient and smooth communication. The response above is an indication that team members believe it is important for the bank to offer efficient platforms for communication and should encourage a culture of continuous sharing of knowledge amongst employees. The participants indicate that the top down approach to information delivery is not efficient as it tends to delay some of the activities. Knowledge sharing encourages project team members to become more innovative and more creative and also encourages project team members to have increased performance (Park & Lee, 2014). The study participants said... *it is difficult for junior managers to communicate with senior managers and this makes projects difficult to accomplish ...I do not see open communication between junior management and senior management because junior management cannot pass on opinions; they are only given instructions which have to be met without fail.* There's a culture of passing directives from senior management to junior members thus junior members are not free to voice their opinions and using position of power forcefully by senior management tend to lead to resistance from staff. The responses above indicate that there is no open communication between junior members and senior members.

3.2 Individualism vs Collectivism

Ranf (2010) explains that threatening global competition has prompted a need for a flexible and highly competitive operation and this is therefore the reason why effective teams are required in organizations. He mentions that an effective team is a key to success of projects and this applies to all phases of the project lifecycle. The study participants said... *collaboration allows sharing of knowledge and development of ideas driving the project to progress easily... collaboration also develops a sense of community.* Collaboration encourages team work and improves level of productivity in a team. It also encourages team members to share ideas which may lead to success of IS Projects. An individual may not be successful while working alone. The study participants agree with this finding and said... *I do not agree that working individually would contribute to project success... with projects, it is not easy to achieve the organizational goals alone.* According to the participants, teamwork is a key to success in IS Projects, they also agree that collaboration encourages team work and improves the level of productivity amongst project teams. Team members will not succeed if they work individually, therefore, the bank should always encourage team members to collaborate and not work individually.

3.3 Uncertainty Avoidance

A learning organization promotes systematic thinking, extended individual and organisational capacity for an organization to change their structure, their culture, the way they do their work and also change their assumptions about the way things are (Soltani et al., 2011). A participant remarked *...if one is not sure of the scope of the project one should always ask to avoid risks ...doubts and uncertainties can be dealt with by keeping in touch with the managers and experts to establish solutions to problems.* Doubts and uncertainties bring team members together, since they then need to consult with each other, sharing what they think is what is expected of them or what will help them unravel the problem situation *...task duration, time and stakeholder accountability are some of the doubts and uncertainties of the project.* Participants showed awareness of what is expected from them in terms of deliverables and they all concurred that if they have uncertainties they need to consult with the relevant stakeholders to get clarity i.e. (superiors, project manager or other technical experts) ...change of timelines and the scope of the projects are issues that bring about uncertainties. Project scope should be clearly defined and signed-off for team members to be clear on scope and timelines. This would minimize uncertainties amongst members.

3.4 Management Support and Decision Making

Du Plessis and Hoole (2006) posit that one of the main causes of project failure is that organisational culture and the environment in which projects have to be delivered within does not support projects. Literature shows that organization culture is an important factor of project success... *the executives provide moral support, financial support and social support.* Senior management supports the team members by being involved throughout the project lifecycle and their involvement has a positive influence on the projects. Du Plessis and Hoole (2006) maintain that a project management culture which is non-supportive is the main cause for project failure in organizations. They further mentioned that numerous organizations are not used to running projects and they don't comply with available project management principles and practices and this is what contributes more to failure*...the way the sponsors and executives influence the team members trigger the success of the project.... they can trigger that by motivating the team members.* ..When management is not consulted, they are often then left out. Not being kept in the loop then means they can't make informed decisions or give the necessary support*....management do not want to provide resources on time and management sometimes is aggressive in dealing with financial matters...the organization needs a proper project plan before starting projects.* Based on the preceding it is evident that support from executives and senior management is key. This then means that for management to give their on-demand support. There should be open communication between management and the members throughout the project lifecycle. Management should avoid initiating projects without adequate planning. This lack of proper planning often leads to project delays.

4. CONCLUSION

This paper argued that organizational culture is one of the contributing factors influencing the outcome of IS projects. It showed how the varied facets of culture manifested during a project life cycle and how each influences the information systems project environment. The insights may help project managers and project executives with ways to nourish cultural aspects that consequently minimize project delays and other undesirable outcomes.

REFERENCES

Agnantounkpatin, A. R. & Zhang, L. Y., 2011. Framework For Culture Management In Sino-African International Construction Projects. China, IEEE.

Andersen, E. S., 2001. Understanding Your IT Project Organization's Character: Exploring the Differences Between the Cultures of an IT Project and its Base Organisation. Hawaii, IEEE.

Berggreen, L. & Kampf, E. C., 2015. Project Management Communication 2.0. *The Socio Technical Design of P M for Professional Communicators,* Denmark: IEEE.

Du Plessis, Y. & Hoole, C., 2006. An Operational ' Project Management Culture' Framework. *SA Journal of Human Resource Management,* 4(1), pp. 36-43.

Eberlein, M., 2008. Culture as a critical success factor for successful global project management in multi-national IT Service Projects. *Journal of Information Technology Management*, 19(3), pp. 27-42.

Fletcher, D., Massis, A. D. & Nordqvist, M., 2016. Qualitative research practices and family business scholarship: A review and future research agends. *Journal of Family Business Strategy*, 7(1), pp. 8-25.

Harrison, R., 1993. Diagnosing organizational culture: Trainer's manual. Amsterdam, Pfeiffer.

Hofstede, G., 1983. Cultural Dimensions for Project Management. *International Journal for Project Management*, 1(1), pp. 41-48.

Hofstede, G., 2011. Dimensionalizing Cultures: The Hofstede Model in Context. *Conceptual Issues in Psychology and Culture,* 1 (12).

Hofstede, G. & De Mooij, M., 2011. Cross-Cultural Consumer Behavior: A Review of Research Findings. Journal of International Consumer Marketing,, pp. 181-192.

Hofstede, G. H. G., 2010. Cultures and Organisations : Intercultural Cooperation and Its Importance for Survival. 3rd ed. s.l.:McGraw Hill.Polyaninova, T., 2011. Knowledge Management in a Project Environment: *Organizational CT and Project Influences.* 41(3).

Institute, P. M., 2013. Guide to the Project Management Body of Knowledge. 5th ed. Pennsylvania: Project Management Institute, Inc.

Kendra, K. & Taplin, L. J., 2004. Project Success:A Cultural Framework. *Project Management Journal*, 35(1), pp. 30-45

Martins, N & Martins, E. (2003). Organisational culture. In *Organisational behaviour: Global and South African perspectives.* Edited by SP Robbins, A Odendaal and G Roodt: Pearson Education Cape Town (pp. 379- 400).

Mawu, A., 2015. A comparative study for organisational culture in the Banking Sector in India. 4(1), pp. 10-26

Montequin, V., Cousillas, S., F, O. & J, V., 2014. Analysis of the success factors and failure causes in Information and Communication Technology (ICT) projects in Spain. Spain, Elsevier Ltd., pp. 992 - 999.

Park, J.-G. & Lee, J., 2014. Knowledge Sharing in information systems development projects: Explicating the role of dependence and trust. *International Journal of Project Management*, Volume 32, pp. 153-165.

Podrug, N., Pavicis, J. & Bratic, V., 2006. Cross-Cultural Comparison of Hofstede's Dimensions and Decision Making Style within CEE Context. School of Economics and Business, University of Sarajevco , 12/13(10), pp. 339-343

Polyaninova, T., 2011. Knowledge Management in a Project Environment:Organizational CT and Project Influences. 2011.41(3).

Ranf, D. E., 2010. Cultural Differences in Project management. *Series Oeconomica*: Annales Universitatis Apulensis, 12(2), pp. 657 - 662.

Rodrigues, I. & Sbragia, R., 2011. The Cultural Challenges of managing Global Project Teams: a Study of Brazilian Multinationals. *Journal of Technology Management and Innovation*, 8(Special Issue ALTEC), pp. 38 – 52

Soltani, E., Liao, Y.-Y. & Wang, w.-Y., 2011. *A learning organisation approach to managing service quality.* UK, IEEE.

Waisfisz, B., 2015. An organisational cultural perspective. *ITIM International.* Netherlands:.

Watson D, R., 2007. *Information Systems.* Global Text Project. Georgia.

EXPLORING THE IMPLICATIONS OF DIGITIZATION IN THE POST-DIGITAL ERA

Arturo Serrano-Santoyo[1] and Mayer Rainiero Cabrera-Flores[2]
[1]*CICESE Research Center, Carretera Ensenada-Tijuana 3918, Zona Playitas, Ensenada, Baja California, Mexico*
[2]*CETYS Universidad, Camino a Microondas Trinidad, Km 1, Las Palmas 3ra. Seccion, Ensenada, Baja California, Mexico*

ABSTRACT

There is no doubt that the exponential growth of digital technologies has affected the modus operandi of the individual, communities and institutions. We pose that the dynamics of change is of such magnitude and depth that many of the phenomena generated in the process of adopting these technologies have not been studied with sufficient breadth and critical analysis. On the one hand, the seductive and glamorous nature of digital devices, systems and applications has managed to capture the attention of users who acquire an increasing dependence on the functionalities that the digital world offers in relation to efficiency, agility and immediacy for developing an increasing number of human activities. On the other hand, the intense competition between the dominant incumbents of the digital world, makes the pace of innovation unstoppable. Given the overwhelming and fast-paced nature of this scenario, it is essential to understand and reflect on the implications of technology-society interactions in order to take full advantage of the potential benefits of digitization for social progress and community wellbeing. The objective of this contribution is to identify some of the implications of greater influence and relevance, which in turn require immediate and urgent attention. We recommend preliminary steps to structure comprehensive frameworks to understand such implications in order to develop strategies and policies for the adoption of digitization in a balanced and comprehensive manner.

KEYWORDS

Digitization, Innovation, Digital Inclusion, Socio-Technical Systems

1. INTRODUCTION

The development and adoption of digital technologies have brought profound transformations in all fields of human knowledge. However, these technologies do not evolve in isolation, they interact with society (individuals, communities and institutions). The intertwined operation of globalization and digitization have shaped the dynamics of the social fabric affecting the way in which we communicate, learn, work and access information. Furthermore, it is generally acknowledged that with the increase of wireless and wired connectivity capabilities, the emergence of virtualization of network functions, cloud computing, and above all, the explosion of mobile communications; we have at our disposal significant tools with potential to improve the socioeconomic conditions of the world. ITU (2017), Nomino J, M and Carriere (2016).

Since the final decades of the last century, digitization, driven by accelerated technological change, managed to permeate virtually all aspects of human activity. By spreading and penetrating "in all corners", digitization is embedded in communication, culture, design and many other disciplines and activities, becoming ubiquitous and transparent. Under these circumstances, Negroponte (1998) and Pardo-Kulinski (2010) consider that the evolution of the digitization processes from something new to something of common use, acceptable; and at the same time imperceptible, gives rise to the use of the term post digital. This term is not only used to connote patterns of a society that evolves towards another condition even more influenced by ICT, but also the post digital is characterized by the omnipresence and ubiquity of digital information flows in social interactions. So when using the term post digital, it is recognized that the technology-society symbiosis has advanced to a more complex condition where the interactions between all agents of the digital ecosystem take place.

In summary, the current post-digital context is characterized by information exchanges where connectivity and ubiquity give rise to new patterns of human interaction. From a business perspective, the company Deloitte (2013) identified key elements of the post-digital society when referring to the current convergence environment characterized by the confluence of: 1. Data analytics, 2. Mobility, 3. The importance of the social, 4. Cloud Computing and 5. Other complementary Cybernetic applications (Artificial Intelligence, Cybersecurity, Internet of Things (IoT), Big Data, etc.). These five elements or "forces" give rise to new rules of operation, performance and competence reflecting the "new normal" of business dynamics.

2. IMPLICATIONS OF THE TECHNO-SOCIAL INTERACTIONS OF THE POST DIGITAL ERA

The implications of techno-social interactions of the post digital era are enormous. The pace of innovation and technological change is of such magnitude and acceleration that their consequences are not easily discernible. We have to acknowledge that in the post digital era there are available significant applications of great benefit and impact on health, commerce, industry (for example: Industry 4.0), culture and so on. We argue that attention to aspects of education, regulation, and ethical and moral development is crucial to cope with the exponential technology growth and its associated social changes. These three agents of the post digital ecosystem play a key role to accomplish a balanced and beneficial adoption of digitization providing resources and strategies for compensating and mitigating the undesired effects of the current techno-centric environment.

As a first attempt, we identified a set of implications strongly related and circumscribed to education, regulation and ethical and moral development. We consider that policies and strategies regarding these three fundamental areas may support the emergence of adoption patterns of digitization for the benefit and socioeconomic advancement of society, both at local and national levels. We acknowledge that there are other implications that scape the scope of this first stage of our research, however, we suggest that our analysis might provide an initial platform for formal discussions and conversations on the balanced adoption and adaptation of digitization by society. The implications we integrate in our proposal are: Cybersecurity and Privacy, Health, Economic Control of Interconnected Systems, Human Behavior and Social Interaction, Technology and Innovation Pace, Work of the Future/The Future of Work. Figure 1 shows the interplay of education, regulation and ethical and moral development and the selected implications from a connectivity perspective. This list is not exhaustive, there are other implications that might be included in our initial proposal, nevertheless, such list provides a first approximation for a more in-depth analysis. Similarly, the weights assigned to the techno-social interactions depicted in Figure 1, do not describe a concrete case. Figure 1 is just a basic graphical representation to highlight the levels of interaction involved. We acknowledge that the task of defining the most relevant implications regarding digitization is a more complex exercise. Conversations and consultations on this matter with the general public and other actors, such as public servants in charge of developing policies, technology practitioners and academia need to start in order to define and evaluate more precisely these implications and to increase the awareness about the effects, benefits and consequences of adopting digitization by society. Whether society is ready or not to digest and understand the potentially serious; and probably not intentional consequences; it is suggested to urgently develop epistemic, conceptual and action frameworks to support society in the understanding and balanced adoption of digitization. In-depth analysis, reflection and attention regarding the implications of the techno-social interactions of the post digital era can't be postponed or neglected if we want to turn digitization into a true vehicle of integral human progress.

3. WHO IS WRITING THE FUTURE OF THE POST DIGITAL ERA?

In the current post digital era, new educational, economic, technological, socio-political and human behavior models emerge. The relevance and degree of digitization interventions must respond to real and necessary demands of individuals, communities and institutions; and not just to the impulse of modernization and pervasive consumerism patterns. The incumbent powerful agents of the current post digital ecosystem exert an overwhelming influence given their financial, innovation, and marketing capabilities; and above all due to their increasing number of users.

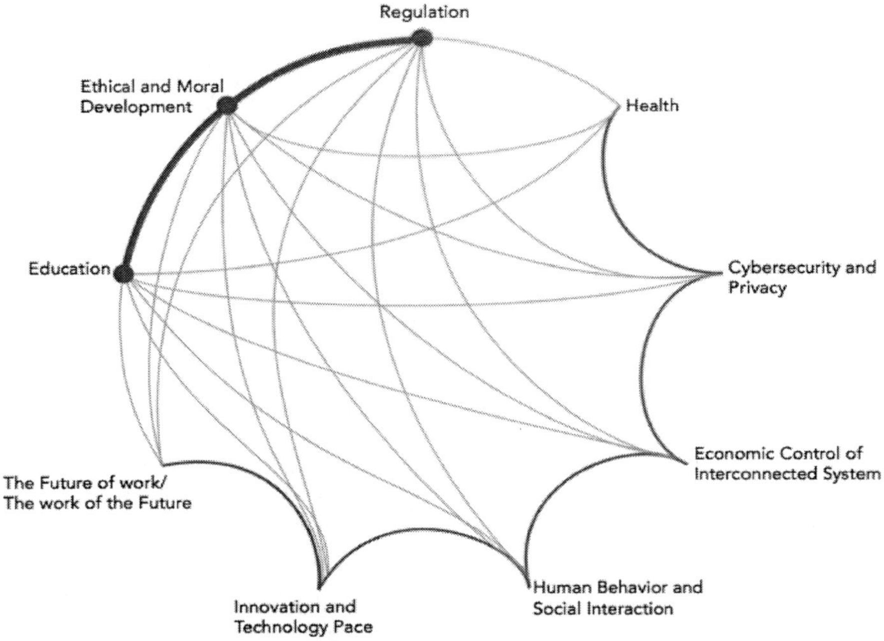

Figure 1. A first Approach to Study the Implications of Techno-Social Interactions in the Post Digital Era

Since the dawn of the "Information Revolution" in the middle of the last century, the activities of research and Innovation in ICT and related areas (OECD (2017)), were emerging as fundamental ingredients of the economic positioning of industrialized countries. The strengthening of the so-called Triple Helix (Etzkowitz (2008)) and the emergence of innovation poles with a high concentration of talent and availability of investment, led to an accelerated development of ICT, which in turn gave rise to the current post-digital scenario. Technological change fueled by digitization brought, on the one hand unprecedented advances for great benefit to society in various disciplines, on the other hand, significant innovation gaps were generated between developed and developing countries with a marked concentration of knowledge and know-how regarding the tasks of transport, processing and storage of information and the creation of advanced systems (hardware and software). The generation and concentration of talent in the academic nuclei of the most important universities and research centers of the world, the systematization of the processes of technological transfer from the academic sector to the business world, the proliferation of the entrepreneurial impulse and the availability of venture capital, - including cultural and contextual aspects- have been key factors in the emergence of regions of global impact with marked global influence and economic gains .

The innovation gap is overwhelming, and at a certain extent, worrisome. Emerging economies, with perhaps the exception of India and China, experience important challenges to cope with the advances and pace of digitization. The case for the least developed economies is even more drastic. These asymmetries and inequalities in the condition of the global innovation context are not restricted to academic environments. The general public, distracted by the seductive nature of digital devices and systems, is unaware; and in many cases indifferent to understanding the implications, nature and pervasiveness of the techno- social interactions of the post digital era. This scenario, with a marked dominance of a handful of powerful global technology players, is becoming increasingly complex (The Economist (2018)). It describes an intricate panorama where in many instances the general tendency and priorities of technology providers and developers are focused on increasing their influence in convergent markets with applications of great technological power to ensure the "engagement" of users and strengthening their position as global platforms of social interaction.

4. CONCLUSION

The exploration offered in this contribution suggests that an in-depth analysis of the implications of society's digitization adoption is necessary with intervention of government, academia, industry and civil society. We propose that it is urgent to developed strategies and their associated lines of action, adapted to the requirements of particular contexts, to face the challenges of the post digital era. In view of this situation, and as a general conclusion, the following actions are suggested:

1. Organize and foster fora and other outreach mechanisms to increase the awareness and basic understanding of the implications of adoption of digitization by society. Parents and education practitioners are key players in this endeavor.
2. Explore mechanisms for the strengthening of national regulatory contexts to foster competitive environments centered on the benefits of digitization, as well as on the protection of the population from consequences and risks derived from adopting emergent digital technologies, and more importantly, it is also advisable the development of policies and guidelines to safeguard the rights of digital consumers under the current hyper-connectivity scenario of the post digital era.
3. Explore mechanisms for the development and application of ethical and moral frameworks to promote and nurture the integral and balanced adoption of digitization with service to society in mind.
4. Explore mechanisms of cooperation and knowledge transfer to mitigate the growing ICT innovation gap, in particular on the generation and deployment of applications in Artificial Intelligence and related branches and other emergent technologies such as Genomics, Nanotechnology and Synthetic Biology, to mention the most important.
5. Accelerate the deployment of training programs to develop digital skills and entrepreneurial capabilities to promote employability and sustainable development initiatives with emphasis on social action and service to society. This is particularly relevant in underserved regions facing socio-economic challenges that historically have been approached with the promise of advancing their progress and quality of life through the application of techno-centric formulas or technological packages.

These actions are just a sample of initiatives, nor are they necessarily the most significant. Furthermore, it is more likely that some of these actions are already under development or in the planning process of governments or other organizations of the civil society. Nevertheless, urgent and concentrated efforts are required. Our objective has been to delineate and emphasize aspects that, in the opinion of the authors, should be taken into account so that technology change and innovation can be fully exploited in favor to society; and to explore the implications that demand urgent actions conducive to the "humanization" of digitization.

REFERENCES

Deloitte, 2013 "*Tech Trends, Elements of postdigital*", available at: https://www2.deloitte.com/content/dam/Deloitte/global/Documents/Technology/gx-cons-tech-trends-2013-elements-postdigital.pdf

Etzkowitz, H. 2008, *The Triple Helix: University-Industry-Government Innovation in action*. Routledge, London.

ITU 2017, *Measuring the Information Society*, ITU, Geneva, Switzerland.

Momino, J. M, Carrere, J (2016). Model for Obtaining ICT indicators in Education, *UNESCO Working Papers on Education Policy N° 3*

Negroponte, N. 1998, *"Beyond Digital"*, Wired Magazine, available at: https://www.wired.com/1998/12/negroponte-55/

OECD 2017, *Science, Technology and Industry Scoreboard 2017: The digital transformation*, OECD Publishing, Paris http://dx.doi.org/10.1787/9789264268821-en

Pardo Kuklinski, H. 2010, *Geekonomía: Un radar para producir en el postdigitalismo*. Universidad de Barcelona, Barcelona.

The Economist 2018, *The Battle for Digital Supremacy*, March 15 issue

SPIRITUAL INTELLIGENCE FOR ICT PERSONNEL – THE TIME HAS COME

Eugene Kaluzniacky
Applied Computer Science, University of Winnipeg, 515 Portage Avenue, Winnipeg, Manitoba, Canada

ABSTRACT

More than 30 years ago, there appeared calls for "soft skills" among IT professionals, including personality awareness and stress management techniques. More recently, calls for Emotional Intelligence among IT personnel were noticed. The ICT professional of today and tomorrow will need to develop even stronger psychological skills parallel to technical competence. While Intellectual Intelligence is about thinking and Emotional Intelligence is about feeling, the recently-promoted Spiritual Intelligence is about being. This paper outlines the existence and functioning of a deepest inner psychological reality: the "inner core" or "being". It outlines the impact of this human psychological component on intellectual work in the vision of "I Am, therefore I think" rather than its Cartesian converse. It then elaborates on the related concept of Spiritual Intelligence and argues for timeliness of recognizing, developing and implementing this psychological capability in various ICT professions. Such development would allow for the truly human capability of *being*, a capability that will enable the necessary psychological robustness, inventive creativity and convincing communication capacity among human *beings* centrally involved in ICT work in the 21st century.

KEYWORDS

Emotional Intelligence, Spirituality, Inner Being, Psychological Awareness, Personal Wholeness, Spiritual Intelligence, ICT Professionals

1. INTRODUCTION: A DEEPER PSYCHOLOGICAL CONSCIOUSNESS

Activities that form the steps in information systems analysis, design, development and implementation are not acts of physical labour – they involve psychological aspects of a person. Today increasingly, ICT professionals are being guided to areas such as stress management, music therapy, personality type awareness, emotional intelligence and mindfulness meditation so as to develop and keep their psychological self functioning at a high level. But, is there yet another, central element of personal growth? Are we simply "rational animals" with an intellect, feelings and a body? Is there, from a conceptual as well as experiential point of view, yet another component to the human person?

This paper promotes the view that, indeed, the deepest inner self (inner core, center, being, soul) is a human psychological reality that does exist separately from the intellect, feelings and body and that this self can become consciously accessible to a well-adjusted, aware human person. In a time of rapid change and increased pressure from users, inner stability and regenerative fulfillment can be rare and valuable commodities for the ICT worker. To develop ability to recognize and to access such a "core energy" within oneself and to act from that energy can be referred to as "Spiritual Intelligence". Below is outlined introductory material on this recently advocated human consciousness along with suggestions for how such personal growth could be promoted in ICT professions.

2. DISCOVERING THE PSYCHOLOGICAL "INNER CORE"

The psychological perspective presented here is that of the "perfect inner core", which, although present in all, is regularly accessible to only a few. According to theory (as promoted, for example by the PRH International growth program), most of us are hardly in touch with our truest self (core energy) since five basic growth needs for the emergence of this energy have not been met adequately in our formative years.

To become a "whole" human person, one needs to be *seen, heard, trusted, recognized* as unique and infinitely valuable and he/she needs to be *safe to open deeply* what is important without being judged. One needs this especially in childhood and adolescence from parents, siblings, teachers, classmates, etc. To the degree that the above five needs are met, the deepest inner self emerges and becomes accessible to the intellect, feelings and body in work and personal life. To the degree that the growth needs are not met, one remains ignorant of one's deepest stabilizing strength and essential identity and emotional blockages develop in the conscious and the unconscious mind..

2.1 Programs for Personal Wholeness

Two very influential personal growth programs following this view are PRH International, (Personnalite et Relations Humaines - Personality and Human Relations) headquartered in France and the Hoffman Institute centered in the U.S. Both programs are worldwide and have a web presence. It is worth noting that such personal growth in deep inner awareness is a step in the evolution of humanity and is thus applicable to (nearly) everyone. We have blockages from our deepest self because we are humans living at this point in humanity's inner evolution and have not been recognized in all of our potentialities, and not necessarily because of an unusually traumatic past.

The end result of personal growth in being, "wholeness", is always positive - a person becomes gradually able to connect his / her mind, body and feelings to a deeper, stable, creative and vivifying self. A "whole" person becomes rooted in his core energy, from which comes his essential inner security, deepest fulfillment and creative potential. He is then able to be emotionally *interdependent* with his work (close and yet free), rather than *dependent* on it (close and not free). Such a person is self-actualized in a true, experiential way.

Becoming experientially in touch with our deep inner core reveals to us our deepest and most stable identity. When we are so connected, we experience truth, peace, personal dignity, tolerance, security, confidence and meaning. This reality is *free* within us, like oxygen, but we must remove what blocks us from experiencing this reality. We cannot think ourselves into our inner core nor can we will ourselves there; but, we can become inwardly open to this valuable, personal reality. This is an ongoing process - we disconnect, at least partially, then reconnect to this deepest inner self.

3. HUMAN *BEING* - WHOLENESS IN INTELLECTUAL WORK

Since ICT work largely involves intellectual analysis, it is essential to clarify the intended relationship between the intellect and the deepest "core energy" (inner self / being). Indeed, the intellect has all autonomy, but it is not intended to work on its own. The intellect can reject, condemn, be indifferent to, doubt, impose its agenda on, or, rather, it can co-operate with and eventually *live at the service of the deepest inner self.*

3.1 "I Think, therefore I Am?"

While much of the Western world has apparently adopted the mindset of Descartes' "I think, therefore I am", neurologist Anthony Damasio, in his book *Descartes' Error* (1994), provides compelling evidence that the work of the intellect cannot be separated from other components of the human person. The two-tier model, in fact connotes being *before* thinking as the optimal mode of living and functioning. First comes our centeredness in our deepest self (core energy, often sensed in the body in the chest and abdomen), then this energy expresses itself through our intellect and feelings.

Dr. Joan Borysenko, former cellular biologist and psychoimmunologist says "the experience of the self requires making the mind one's servant rather than letting it be our master", "when we experience ourselves as this essential center rather than as any one of our roles, we can function optimally", and such optimal functioning is certainly centrally desirable in ICT work today. Thus, a more appropriate phrase that refers to human integration would be "I AM - Therefore I think".

3.2 Becoming in Touch with "I AM"

Many people can get in touch, at least partially, with their deepest inner self when experiencing a "felt truth". For example, if we were asked, "of what are you absolutely sure about yourself?"; "where do you feel an 'unwavering certitude' ?", we may experience a deep "gut feeling" in addition to intellectual knowledge. A person may be deeply certain that he is an excellent project manager or system designer, or that she is an understanding person. In personal growth in being, we are led to eventually experience the unwavering certainty that "I am deeply good and infinitely valuable just because I AM" - all other action is then connected to the essential energy of this undeniable felt truth. We then meet our work challenges with a reservoir of actual psychic energy which is regenerative and not at all dependent on our environment. An intellect psychologically connected in such a way is working and living in the mode of "I AM – Therefore I think".

3.3 Emotional Intelligence – A Step Forward

From the beginnings of various types of occupations within Information Technology, focus, as expected, was almost exclusively on intellectual capability. The post-war knowledge and technological explosion and subsequent "Space Race" made intellectual prowess supreme. The Intelligence Quotient (IQ) became *the* only way of measuring and declaring a human being as "intelligent". Thus, there is little wonder that in an intellectually focused IT and later ICT world, the intellect was nearly "worshipped" for some time. As time went on, however, the scope of ICT involvement, enabled by more recently developed technologies, broadened. A necessity for effective human interaction (in larger project management and in user interaction) and for effective stress management among ICT professionals became more obvious and began to receive more focus.

In the same era, recognition of other forms of human intelligence (e.g., Gardner, 1994) were gaining acceptance. In 1998, psychologist Daniel Goleman gave recognition to "emotional intelligence". Such intelligence has been described as "the ability to identify and manage your own emotions and the emotions of others. It is generally said to include three skills: emotional awareness; the ability to harness emotions and apply them to tasks like thinking and problem solving; and the ability to manage emotions, which includes regulating your own emotions and cheering up or calming down other people." (Psychology Today website).

Around this time, inquisitive and daring IT professionals also took note of this recent and expanded awareness of human capability. For example, Jim McCarthy, who had been product development and program management positions at Microsoft, the Whitewater Group, and Bell Laboratories, provided the following profound and eloquent words in his book *Dynamics of Software Development (1995)*: "Leaders, good ones at any rate, are expected to provide a vision to their followers. In my opinion, a leader's empathic perception of the psychological state of his or her team is the beginning of what we call vision. If the leader

can then resonate with the team's complex emotional state—identify with it, articulate it, and give the whole constellation of feeling and thought a visible, concrete reality in his or her own personal voice or gesture—the boundaries among individual team members and the leader will collapse...Empathy will be established: the leader and the team will feel and know as one, giving voice and identity to what was an incoherent psychological community substrate."

An article in *Computerworld* called for "emotional literacy among IS professionals" in the context of personality awareness. At a national convention of the Canadian Information Processing Society (CIPS) (see Kaluzniacky, 2004), a keynote speaker proposed that "love" and not confrontation be the model for organizational communication, and received a standing ovation. A job advertisement for IT professionals within an insurance company in a prominent U.S. software center points out that this employer is

interested in contributing to the employee's professional and personal life, and advises candidates to "listen to their inner voice." Thus, towards the end of the 20th century, Emotional Intelligence had been at least "semi-legitimized" in increasingly influential ICT professions.

3.4 Spiritual Intelligence – Beyond EI

While the emotional side of a human person can contribute beyond the use of mainly the intellect, in time, downsides of EI have also been recognized. For example, emotionally intelligent persons can anticipate feelings of others, but, consequently, may have difficulty making unpopular decisions, and some may even be more adept at psychological manipulation. As well, controlling one's feelings, for example, with willpower, may eventually result in an emotional breakdown if the skill of dissipating the anger has not yet been mastered. Thus, if emotions are to assist in intellectual endeavours, what further psychological reality can assist in emotional involvement, including emotional aspects of ICT work?

Promoters of personal growth for more effective work have been admitting to and identifying a reality beyond the intellect and well-developed human emotion. Here we approach the territory of the "deepest inner self", "inner being", "soul" or "inner core", that were introduced earlier in this paper. This inner territory is being recognized as "spiritual". Thus, the term "spiritual intelligence" has begun to emerge and attract attention. Yet this term appears vague and seems to connote different realities to different people. Thus, it is useful to allude to products of high-level academic research to seek out a unifying vision for the spiritual intelligence paradigm.

A unique, thorough, and soundly academic perspective on the nature of human spirituality is evident in the work of Daniel Helminiak. In the minds of many, spirituality is necessarily connected with religion. Yet, in his authoritative book, *Spiritual Development—An Interdisciplinary Study* (1987), Helminiak argues otherwise. He points out the differences between psychology in the positivist viewpoint (which encompasses scientific aspects such as memory, cerebral functioning, cognition patterns) and psychology in the broader, philosophic viewpoint (which concerns the human need to be open, honest, and authentic). In essence, *spirituality is human authenticity.* Helminiak thus argues convincingly that the proper discipline to study spirituality is psychology within the philosophic viewpoint, rather than theology. However, a person can be spiritual and also religious by living with the intellect, emotions, and body, significantly connected to the inner core energy, and by recognizing and relating to this core energy in a revealed, transcendent context.

Helminiak also foresees development of a separate academic discipline of "human spirituality," distinct from psychology and theology. He states that: *"Just as technologies for production, travel, communication, and so on {notably information} have transformed the face of modern living around the world, so a spiritual technology might transform postmodern life."*(Helminiak, 1998) Helminiak proposes that a spirituality worthy of all humanity would actually *inform secular culture.* In such a well-grounded vision lies considerable potential, both for theoretical research and for very pragmatic applications for the work culture of ICT professionals.

4. SQ RECOGNITION, DEVELOPMENT AND RESEARCH IN ICT ORGANIZATIONS

4.1 Respectability of Spiritual Intelligence in ICT-related Areas

There is now a respectable theoretical foundation as well as increasing anecdotal evidence for considering emotional and, further, spiritual intelligence as very desirable areas of professional development, alongside intellectually-oriented skills for ICT professionals. Already a number of years ago, the inventor of Extreme Programming, Kent Beck, dismissed the idea that IT work is "some Vulcanic world of pure rationality" and declared "to be successful, we must learn and grow our *whole selves*" . It is proposed here that the "whole self" consists of the intellect, feelings and body (with its sensations) on one level, receiving and expressing the vivifying spiritual energy of the "inner core" on a second, deeper level in a psychologically integrated human being.

A first question in the minds of many persons who may be considering the proposals and challenges for spiritual development among ICT workers will likely be "is such a type of intelligence being accepted and legitimized in at least areas of human endeavour related to information technology and systems?" A truly pioneering comment regarding recognition of the deepest inner self came at the start of the popular IS text,

Introduction to Information Systems—Essentials for the Internetworked Enterprise (1998). Author James O'Brien tells his readers, *"May you love the Light within you, And in everyone you meet, And in everything you experience."* A clear reference to the deepest inner self in such a context, already twenty years ago, is indeed enlightening. It can certainly be understood as a call to a deeper inner awareness while developing an IS orientation.

Within the last ten years, other information sources on spiritual intelligence have begun to appear and attract considerable attention, e.g. the popular "Spiritual Intelligence" portal by Richard Griffiths. A number of the sources are focused on leadership (e.g., Draper, 2010, George, 2006, Wigglesworth, 2006). However, what can be considered a groundbreaking work is the paper "The Business Case for Spiritual Intelligence" by Sebastian Salicru (2010), an Australian business and performance psychologist . This author describes spiritual intelligence (coded as SQ to parallel IQ) as "the science of human energy management that clarifies and guides the structure of awareness". He aptly clarifies the position of this type of intelligence: "IQ refers to thinking, EQ to feeling and SQ to *being.*" He declares that SQ "provides individuals with the capability of sustainable performance at extraordinary levels."

As an example of the corporate world taking on SQ, Salicru outlines an initiative implemented by Pricewaterhouse Coopers who, in partnership with the Findhorn Foundation, a spiritual and educational community in Scotland, produced a program called "Developing Emotional and Spiritual Intelligence in the Workplace". One of the goals of this program is to enable employees to be *"more able to respond to the needs of the moment"*. It is not difficult to see that such an ability is centrally and increasingly necessary for many effective ICT professionals. Thus, the time has come for ICT professionals, individually and corporately to embrace the call to develop their "whole selves" and to initiate specific efforts to promote and develop spiritual intelligence in the ICT workplace.

4.2 Towards Specific Efforts within the ICT Environment

Just as alarm, by itself, will not put out a fire, vision alone will not transform and elevate the psychological robustness and creative potential of 21st century ICT professionals. Vision must lead to strategy, then tactics and operations. Psychological awareness and its most impactful form, Spiritual Intelligence, needs to be recognized, understood, and accepted as a desirable developmental aspect among ICT professionals. Can such contemporary professionals ignore such a central reality and yet serve the needs of their employers with effectiveness, creativity and impact?

For further understanding of Spiritual Intelligence, ICT management will need to investigate firstly organizational applications of emotional intelligence and relate this to various dimensions of ICT work. Then, these persons will need to become familiar of how Spiritual Intelligence (SQ - about *being*) extends Emotional Intelligence (EQ - about *feeling).* Here is potential for interdisciplinary researchers and consultants to produce theoretical and empirical research as well as applied case studies on the impact of Emotional and then Spiritual intelligence specifically on ICT work. Action research on the part of ICT workers themselves can also provide valuable contributions. A starting point in developing awareness may be the comprehensive perspective developed in the book *Managing Psychological Factors in Information Systems Work – An Orientation to Emotional Intelligence* (Kaluzniacky, 2004). The author calls for emergence of a new interdisciplinary field called *Psycho-Informatics* to address ICT-related psychological issues in a focused manner. Seminars can be given at specific locations to gather ICT workers on a specific continent to learn, discuss and share benefits and challenges of psychological development, as part of Kent Beck's challenge "to grow our whole selves". Recognized professional bodies can be involved in joint efforts with human development organizations such as Findhorn to develop specific programs for spiritual development and application among ICT employees.

There will undoubtedly be several areas of benefit for the spiritually-grounded ICT professional. Ability to handle work uncertainty and role ambiguity is significantly enhanced when the worker is not emotionally dependent on having "things go his way" for emotional security (since the security comes from a deeper source within). Work stress and inclination to burnout will undoubtedly decrease for those capable of being very emotionally close to their work but simultaneously free of addictive attachment to the work and expected results. Capacity and motivation for creativity will be significantly noticeable for one who is continually being inwardly refreshed by an unending source of psychic energy coming gratuitously from the deepest inner self. Similarly, ability for managing change and continuous learning will require a deeper and more robust sense of security that comes from the inner core. As well, the ability to interact with systems' end users with differing perspectives and oft-changing requirements so as to create a win-win communication is undoubtedly enhanced by personal psychological solidity. The stage has been set and the benefits are many for emotional and spiritual development to be strongly accepted by ICT professionals.

5. CONCLUSION

A well-timed emancipation and massive humanization of a technically-oriented profession will undoubtedly require individual courage. The words of Joseph Weizenbaum, a prominent MIT computer scientist of the 1970s can be very fitting in this context. *"If the teacher, if anyone, is to be an example of a whole person to others, he must first strive to be a whole person. Without the courage to confront one's inner as well as outer worlds, such wholeness is impossible to achieve. Instrumental reason alone cannot lead to it. And there, precisely is a crucial difference between man and machine"*

In addition to being aware of personality types, cognitive styles and creativity approaches, an awareness of and capability in emotional and furthermore spiritual intelligence will enable an ICT worker's *being* part to fortify the *human* part in a quest for effectiveness, impact and professional fulfillment .

REFERENCES

Borysenko, Joan, (1990), Guilt is the Teacher – Love Is the Lesson, *Warner, USA.*

Damasio, A., (1994), Descartes' Error, *Avon Books, New York.*

Draper, B., (2010), Spiritual Intelligence – A New Way of Being, *Lion Hudson, UK*

Gardner, H., (1983), Frames of Mind: The Theory of Multiple Intelligences, *Fontana, New York.*

George, M. (2006), Practical application of spiritual intelligence in the workplace, *Human Resource Management International Digest*, 14(5), 3-5.

Goleman, D., (1995), Emotional Intelligence, *Bloomsbury, USA.*

Helminiak, D., (1987), Spiritual Development: An Interdisciplinary Study, *Loyola, USA.*

Helminiak, D., (1998), Religion and the Human Sciences; An Approach via Spirituality, *SUNY Press, USA.*

Kaluzniacky, E., (2004), Managing Psychological Factors in Information Systems Work – An Orientation to Emotional Intelligence, *INFOSCI, USA.*

McCarthy,J., (1995), Dynamics of Software Development, *Microsoft Press, USA.*

O'Brien, J., (1998), Introduction to Information Systems – Essentials for the Internetworked Enterprise, *McGraw-Hill, USA.*

Psychology Today, website: https://www.psychologytoday.com/us/basics/emotional-intelligence

Salicru, S., (2010), The Business Case for Spiritual Intelligence, available at: www. pts. net. au

Spiritual Intelligence - portal available at website: https://sqi.co

Wigglesworth, C. (2006), Why Spiritual Intelligence Is Essential To Mature Leadership, *Integral Leadership Review.*

ACCESS TO USER'S DATA IN AN ADAPTABLE APPLICATION

Maritzol Tenemaza[1], Angélica de Antonio[3], Jaime Ramirez[3] and Sandra Gutiérrez[2]
[1]*Department of Informatics and Computer Science*
[2]*Department of Mathematics*
Escuela Politécnica Nacional, Quito, Ecuador
[3]*Laboratory E.T.S.I., Universidad Politécnica de Madrid, Campus Montegancedo, Madrid, España*

ABSTRACT

Mobile applications currently being developed require among other features to be adaptable to the user. In order to do so, it is necessary the user's permission to collect its digital footprint on the web or any other source. Additionally, it is necessary to know what characteristics of adaptability and custom recommendations would the user be fonder of. In this paper, we present an empirical investigation for which we have developed and analyzed a user's survey to gain first-hand knowledge about how confortable he would feel when an application has access to their data and what kind of data he is willing to share with the application. Results of this work have been placed in the context of an adaptable application to the user and will be useful to improve the adaptability and recommendations to the user when developing such applications.

KEYWORDS

Adaptable Applications, User's Interest Analysis, Recommendations

1. INTRODUCTION

Starting in 1996, adaptable hypertext and hypermedia were the first applications that demonstrated signs of adaptability (Brusilovsky, 1996). In the first adaptable user interfaces and models adaptation was oriented towards content and navigation. Adaptable hypermedia systems arose with the purpose of improving usability of traditional hypermedia systems (Benyon, 1993). After hypermedia, web systems arrived. According to Brusilvosky (Brusilovsky et al., 2001), the initial web applications did not satisfy the heterogeneous needs of users. The applications worked in the same way for all users, because it was difficult to interpret a user's actions and interactions with the applications.

Nowadays, web applications reflect the progress made in terms of adaptability. Applicatons developed by Google, Microsoft, Amazon and Facebook could be considered as the pioneers in adaptable web applications. Their success is based on the collection and storage of the digital footprint of each user. They stock huge amounts of data generated by users for predicting their interests and needs.

Mobile applications require being ubiquitous, pervasive, and context-sensitive. Additionally, they must be intelligent and sensitive to a user's preferences, objectives, and real-time interests in order to return useful information. However, a common issue faced by most applications is not having direct access to the users' data, because of matters of privacy. For this reason, such access is acquainted by the user and under his accordance. This study has been developed in order to measure users' attitude towards the access to their data and the adaptability of a mobile application.

The structure of this paper is as follows. Section 2 discusses the methodology, the instruments employed in this study and main findings. In Section 3 we outline some conclusions.

2. METHOD

2.1 Data Collection and Methodology

Data were collected through web survey in order to analyze users' preferences and attitudes towards the use of their own as well as others' web footprint while using an Augmented Reality (AR) mobile application called "Quito's guide". A no probabilistic sampling method was used to get a valid set of 77 observations which was used to describe relationships. The participants were 55 male and 22 female. The distribution between age groups was 27 % between 20-29; 38% between 30-39; 16% between 40-49 and the last 19% of participants were older than 50.

The survey was composed by 8 questions with a Likert-type scale (P1 to P8), measuring attitudes, considerations and recommendations for adaptability and 7 questions collecting other information about personal preferences and concerns. The scale had 5 categories labeled from 0 to 4, where 0 represents the lowest punctuation or worst perception and 5 the highest punctuation or best perception.

In this study, we have considered a twofold analysis: first we present a reflective structural model considering "users' attitudes" as exogenous variable. We chose users' attitudes as an exogenous variable since it can not be controlled by the developer. Second, we have used information collected in the other 7 questions to formulate research questions and answer them using a Chi-Square Test of hypotheses with significance level $\alpha = 0.05$. We pretend to show that that the preference for adaptability at the beginning of an application is influenced by the users' attitudes, and users' attitudes have an effect over the considerations for adaptability and the preference for using other people's information.

2.2 Hypotheses

Our hypotheses are not based in any previously tested theoretical model and are mostly empirical. They are the following:

- H1: There is a positive relationship between Users' attitudes and Adaptability at the start of application.
- H2: There is a positive relationship between Users' attitudes and the Considerations for adaptability.
- H3: There is a positive relationship between Users' attitudes and the Adaptability from information generated by other people.
- H4: There is a positive relationship between Adaptability from information generated by other people and the Considerations for adaptability.

Note that we consider a positive relationship between two variables whenever higher values of a variable result in higher values for the other.

2.3 Questions

We have classified the survey questions into the following topics: Considerations for adaptability; adaptability at the start of application; adaptability from information generated by other people; Adaptability while the application works and User's attitudes. In general for adaptability, we would like to know:

- Which factors influence positively and negatively in the perception of adaptability?
- What kind of recommendations would be better accepted?
- What is the most effective way to provide these recommendations?
- What variables can influence on recommendations?
- What elements help to distinguish some recommendations from others?
- What other aspects related to the use of this technology are there?

With the answers obtained in this survey, we can establish qualitative characteristics required for the application, and different ways to propose recommendations to the user.

The survey's questions are observed in Table 1.

Table 1. Questions to the user's (a: attitude, c: considerations, r: recommendations)

No	Questions
P1A	How important is it for you that the AR-mobile application be adaptive?
P2a	Would you allow your footprint to be analyzed (from the social web, internet activity records, or in your device) to infer your possible interests and preferences when starting your visit to Quito?
	Is there any aspect that concerns you regarding the security of your data by allowing us to analyze your footprint left on the web?
	What data would you NOT like to be analyzed? • Personal, • Education, • Search history, • Location records, • How often pages are visited, • Other pages with the same search object.
	What would you like the application to suggest when visiting Quito? • Places to visit, • Best routes to visit, • Places to eat according to your interests or tastes, • A complete plan for each day, • Others
	How would you like to receive such recommendations? Remember that we have an Augmented Reality interface: • Voice, • On-screen text on the actual image, • Icons (arrows, alerts, etc.) on screen over real image, • Photos with additional information about the site of interest that is being focused with the camera of the mobile, • Others.
P3c	Would you like us to analyze the footprint left by other people, to recommend what others liked about their visit to Quito?
	What do you think it is the most relevant to generate recommendations: • Routes followed by other people when visiting the same area • Opinions of other people about the sites of interest • Preferences about types of places of interest (churches, Italian restaurants, etc.) inferred by the app when analyzing the user's activity record in the mobile or in the cloud.
P4ar	You agree to change recommendations while the application works, adjusting to your interests and requirements in real time.
P5a	How far would you follow the recommendations faithfully?
P6a	I would always pay attention to the mobile when it wants to give me a recommendation.
P7ac	How do you prefer that your mobile recommend: proactively or reactively?
	In what way do you want to be warned when the mobile screen is off and the app proactively wants to provide you with a recommendation based on the current context (position, orientation, etc.)? • Voice. • Sound. • Vibration • None
P8arc	When information is displayed on screen about a certain object, would it seem interesting that indication about parts of the object observed through the mobile with the current image are overlapped?

2.4 Data Analysis

Using the PLS algorithm in SmartPLS 3 software (Ringle et al., 2015), the associate graphical model is presented in Figure 1.

P5a and P6a are considered for analyzing "user's attitudes". P3c is considered for analyzing "considerations for adaptability". P1A and P2a are considered for analyzing the importance of adaptability. P4ar, p7ac, P8arc are considered for analyzing the type of recommendation. The rest are open questions that offer important information when designing a program.

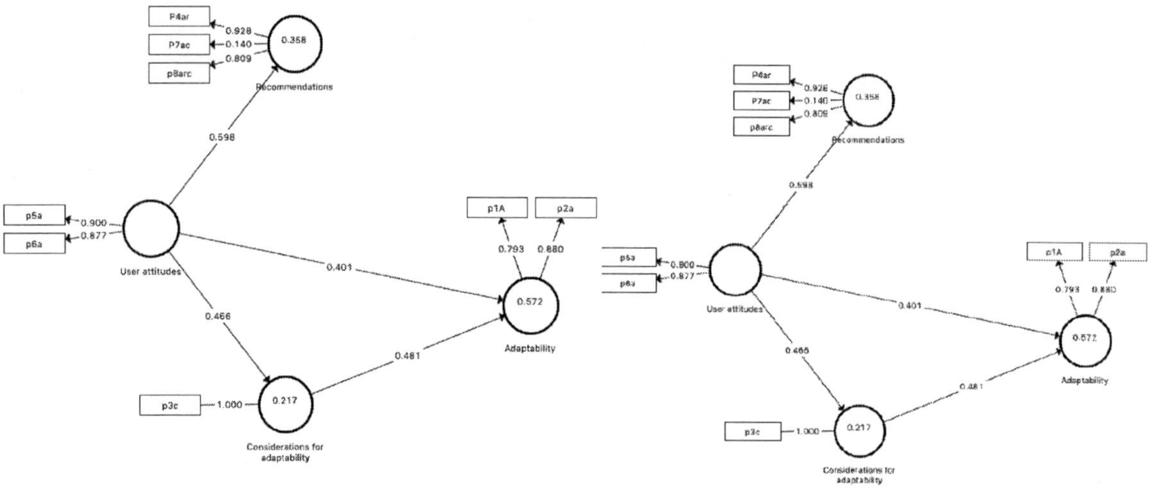

Figure 1. Proposed Reflective Structural Model

We have found that all composites (exogenous and endogenous variables) have a composite reliability of at least 0.703, above the suggested threshold of 0.7; the lowest value of Average variance extracted (AVE) is 0.511, which means that factors explain at least half the variance of their respective indicator variables. Thus we have discriminant validity.

As a result, we have that 57% of the variance of adaptability can be explained with the proposed relationships.

In Figure 1, the following information can be found:

- Outer loadings over the arrows connecting indicator variables to composites. Outer loadings represent the absolute contribution of the indicator to its latent variable.
- Path coefficients over the arrows connecting composites, the higher the value, the stronger that path.
- R^2 values inside the circles.

The structural model shows that all hypotheses were supported.

2.5 Findings

Being this study an empirical investigation and given the sample's size, results must be treated with caution, however, the heterogeneity of the sample's individuals in terms of gender and age support our results.. We can highlight the following findings:

- The most important concerns about security of data according to the sample's individuals are: loss of privacy, fear of having the footprint stored for other uses, misuse of keywords,, vulnerability of the system, misuse of information, confidentiality.
- 87.5 % of people consulted would not allow the application to access their personal data, but would be willing to provide educational data, search history, frequency of access to pages associated with the subject. In this manner, the application would have interesting user data.
- 87.6% of people prefer to be told the best places to visit but they would like additional data such as better visiting times. In addition, 79.6% are interested in places to eat, but it is low the percentage of people who want to be suggested complete plans. These are important data that define priorities when recommending.

- 75.2% of people interviewed are interested in the information left by other users who have visited these places, but they are not interested in the route followed by them, so the application must evaluate how to combine this information with the user's interests.
- The way the user wants to be alerted depends on the age. In this sample, 25% of users over 50 years old want to be alerted by voice, while no other age category wants that type of alert and prefers vibration.
- Gender influences on the considerations for adaptability. This study showed that these two variables were not independent through a Chi-Square test. (p-value=0.003)
- People who would not like their footprint to be analyzed, would not like other people's fingerprint to be analyzed either. Chi-Square test. (p value=0.013)

3. CONCLUSIONS

According to the analysis and assumptions presented in this work, for most of the users of the AR mobile application it is very important its adaptability since it might ,present real-time information according to the user's interests.

- H1: There is a positive relationship between Users' attitudes and Adaptability at the start of application. Results have shown that it's more probable that users enable access to their data at the start of the application. Also, users that prefer adaptability will follow the recommendations more faithfully. (Chi-square test, p-value=0.010)
- H2: There is a positive relationship between Users' attitudes and the Considerations for adaptability. Based on the survey's data, we found that most of the users agree with adaptability, but there are specific preferences according to age and gender.
- H3, H4: These two hypotheses are related in a sense that people who would have a positive perception of adaptability are more likely to follow recommendations, accept image overlap when using the application and are more likely to share their data.

ACKNOWLEDGEMENTS

The authors gratefully acknowledge the financial support provided by the Escuela Politécnica Nacional, for the development of project PII-16-04 "Acceptability of an A2R application".

REFERENCES

Benyon, D. 1993. Adaptive systems: a solution to usability problems. *User modeling and User-adapted Interaction,* 3, 65-87.

Brusilovsky, P. 1996. Methods and techniques of adaptive hypermedia. *User modeling and user-adapted interaction,* 6, 87-129.

Brusilovsky, P., Karagiannidis, C. & Sampson, D. The benefits of layered evaluation of adaptive applications and services. Empirical Evaluation of Adaptive Systems. Proceedings of workshop at the Eighth International Conference on User Modeling, UM2001, 2001. 1-8.

Ringle, C. M., Wende, S. & Becker, J.-M. 2015. SmartPLS 3. *Boenningstedt: SmartPLS GmbH, http://www. smartpls. com.*

MADE IN ITALY. AN INTEGRATED E-COMMERCE AND E-LEARNING INNOVATIVE PLATFORM TO PROMOTE EDUCATION FOR ITALIAN DESIGN SYSTEM (PRODUCT & PROCESS)

Dalia Gallico
Università San Raffaele Roma, Via Daverio 7 Milano – via Val Cannuta 247 Roma, Italy

ABSTRACT

DesignBest concept is born as response to the necessity to create a virtual community (a square) about italian design matters. The extraordinary achievement of communities demonstrated the enormous potentials of technologies in supporting communication and self-production of contents to create strong communities.
The program proposes to apply technologies to support a marketplace and to build a community around the education of Italian design.
Leveraging the development of ITC technologies, the program integrates innovative web technologies 2.0, in order to:
1) Create, applying the most advanced web technologies 2.0, an environment where all active users (designers, design students, etc..), vendors or buyers, can develop ideas, concepts, events, opinions, about Made in Italy Design, joining potentials of vendors, designers, students, amateurs and curious;
2) Explain Culture of Project, display products' characteristics, allowing the user/buyer to interact to capture the extra value of the product;
3) Use buyer's auto profiling and geo localization data to increase the offer and create feed-back flows which could be translated into networking;
4) Create data exchange with the logistic platform to assure the quality of logistic performance expected from the market;
5) Enable the sale of Design objects between the Italian SMEs (small, medium enterprises) and the potential buyers around the world (the business will focus mainly on the following markets: Brazil, Russia, India, China, Canada, USA, Mexico, Japan, Arab Emirates) trying to support Design SMEs in:
- Reach the "critical mass" needed to build a wide catalogue of "Made in Italy" Design learning corses
- Have access to an alternative sales channel, since they cannot bear investments in IT by themselves
- Take advantage of DesignBest international visibility to reach new markets
- Have access to e-commerce in an easy way (without problems related information technology structure, order management, invoicing, payment and shipment, returns and complaints management)
- Leverage some supporting services to international sales and marketing (information, advice, training, data bank, market analysis, etc)
- Spread the Italian Design culture through a platform of e-learning addressed to foreign companies and universities.
Critical success factors of the project are:
- A wide adhesion of Italian design schools to the marketplace, where they can offer part of their own catalogue (also ad hoc courses) exploiting a new channel (e-commerce) and entering on new markets;
- The ability to attract the Design community (lovers, designer, students)

KEYWORDS

Higher Education, Design of Societal Systems – Rethinking, Managed Learning Environments (Mles), Learning Theories and Approaches, E-Learning Models, E-Commerce

1. INTRODUCTION

Design industry plays a main role in the Italian productive system and it contributes significantly to support the trade balance (the value of design export in 2016 was of about 15 mld € between jeweller's craft and furniture accessories).

Lombardy, with its districts and SMEs, is the sector leader in Italy with an export that, only in terms of furniture accessories, is closed to 2 bln of €, of which more than 950 mln € made in Milan (around 11,3% of total Italian production).

Design is, therefore, an Excellence recognized to our country in all the world, although there are still few structures supporting internationalization of the e-commerce and e-learning of design education.

Università San Raffaele Roma decided to cooperate to start an initiative called Design Best aiming to support SMEs' commerce (above all abroad) related to Italian design sector and to spread in the world the Italian design culture.

Institutions MISE Economic Ministery and the Municipality of Milan and, recognizing to Milan the pre-eminence of design in Italy, showed interest for the initiative.

2. E-LEARNING PLATFORM TO DEVELOP CULTURE OF MADE IN ITALY DESIGN AROUND THE WORLD

A virtual research laboratory promotes design culture, distance relationships and creation of ideas shared from all design system. This pattern aims to communicate to the world, history, concepts, design art, building brand awareness and benefits for all chain companies and linked industries (including exhibitions, events, eg).

The pattern supports a network of creativity, giving visibility to companies who will benefit through talent scouting and through an open window on creative concepts worldwide.

The concept is able to involve different kind of users:

- "amateurs" who have the opportunity to widen their knowledge and test their creativity
- "aspiring professionals" who have the opportunity to start training on specific subjects, or add something new in a context of high visibility
- "professionals" who can improve their competences
- "linked", meaning all professionals , like innovation project managers, who want to learn more about design, start to be creative and know more about design industry
- companies, which can observe what is being done, look for human resources and new talents.

The project contains also a platform for e-learning with different characteristics according to the target countries. The platform covers topics "Design for Business and Business for Design" such as:

- Market and design marketing
- Networks for sharing innovation
- Research and innovation for materials and technology
- Design and Environment
- Brand Protection and legal aspects

The platform has to support creativity through technology, allowing the building of contents with a bi-directional process. Concerning to background part of training and spreading information, DesignBest provides SMEs an experimental system based on narrowcasting technologies.

The work packages to reach the objective of the stream have been :

1 - Market solutions scouting: This activity analyzes the market solutions (national and international) available about the E-learning solutions, focusing on different variables of analysis to orient the identification of the best solutions for the Design Best e-learning platform

2 - User Requirements Analysis: This activity had l carried out the users requirements through structured interview to the key users identified for each functionalities area of the E-learning platform. Afterwards the result from the interviews will be elaborated to define the necessary functional requirements of the Portal

3 - Functional specification: This activity designed the set of functionalities area of the E-learning platform. Information architecture will be define as well as interface and interaction aspects. The latter will be define in a beta version successively harmonized through the activity

4 - e-Learning platform development: This activity analyzes the technological solutions available moving from the result of the market solutions scouting national and international planned for each module to define the overall architecture of the design Best Platform, focusing on open source scenario requirements, interoperability aspects and diffusion level of technological standards

5 - Content production: aims at producing a scientific program, multimedia contents in different languages. New and scientific program will be collected, in order to define coherent and innovation program.

6 Validation: This activity defines first a validation plan to verify the consistency between the result carried out from the activity and the result carried out from the activity .During the validation will be define actions to get more consistency, where a gap will be pointed out

7 - E-Learning platform test and update on collected feedbacks: The activity objective is the definition of a coherent testing plan, able to ease the introduction of the E-learning platform in the real business contexts of the different actors involved in the testing phase. The PL aims at defining organizational, commercial and technical support processes, able to assist educational relationships based on the services delivered by the platform.

3. COMPETITIVE ADVANTAGES W.R.T THE STATE-OF-THE-ART EXPECTED IMPACTS OVER THE INDUSTRY AND OVER SPECIFIC EDUCATION COMPANIES

The project brings high improvement concerning technology and organization and allows to differentiate from competition systems of other countries.

With regard to technology, the difference is be done by web 2.0 technologies, used to build a virtual environment that will enable bottom-up creativity; support the development of sophisticated experience systems representing design as an interaction for example with interactive tables and walls or texters.

The logistic platform needs to develop advanced information management systems to create a single virtual warehouse.

Today production chain is made by many small companies usually hand-crafted. The project brings development and new elements for companies that will be able to improve their processes and take advantage of new technologies, all with the purpose of changing today's innovation processes and supply chain.

The creative process will be radically changed due to new technology allowing users to participate and develop a real bottom-up creative network. Today, creative processes involve one person or small groups, tomorrow it will be the many-to-few allowing companies to test a wide range of ideas, concepts, format, design in a very short time and obtain directly information market needed to business development.

The project plans to develop prototypes fully scalable, which may, after testing, be taken as a model to create real and complete platforms.

These models could be used also for other Made in Italy sectors, such as fashion, art and media.

The main Italy's competitors on design are France, Sweden and Germany. Compared to those competitors, the project purpose is to be the first mover and create an advantage which competitors could not reach.

The programme impacts all companies taking part of the initiative (vendors); to be more precise those of home design objects, home furnishing accessories, jewellery and trinkets:

- Knowledge development of marketing intelligence in the BRIC Brazil, Russia, India and China, NAFTA Canada, USA, Mexico, ME Middle East, above all United Arab Emirates and Japan Markets;
- Widen the reach of medium and small companies, over such markets, with positive outcomes on sales and sector employment;
- Increase of talents network through training programs and expansion of the surplus value perception supporting the spread of Made in Italy design culture;

Companies of sectors involved, are an important part of the Made in Italy export:

- About volumes, the Italian goldsmith industry is second only to the Indian, at the same level of the Chinese one, double of United States, and at least 10 times above the first European competitor, Spain. Instead, from a qualitative point of view it is absolutely the first player, and the trade balance, shows, historically, a positive credit balance around 2.8 billion of Euro, with the production value of almost 6 billion Euro. United States are the main country buyer, followed by various European countries, but China and the Middle East are in strong rising. In this sector operate about 10.000 companies of which 8.500 craftsman and 1.500 industrial, for the total of about 47.000 employed;
- Design industry, with home furnishing accessories production, is another important part of our balance of trade.Trade surplus for this sector amount to about 9 billion Euros, including crosswise sectors such as

wood carving, glass and ceramics manufactory. Companies involved are tens of thousands, with a number of employers around 120.000, 80% of these companies have hand-crafted structure Information data processed by Federorafi/ISTAT.

- Italy produces also "design culture", there are many and qualified educational programs, as IED, Istituto Europeo di Design, and others are SPD, Scuola Politecnica di Design, the Design Degree at Politecnico, ISAD, NABA or the Domus Academy, all placed in Milan; the ISIA in Florence, Faenza ed Urbino; Fabrica, related to Benetton Group, or the Interaction Design Institute situated in Ivrea. This educational offer is not addressed only to talents of Italian market but also to the international one. It may seems that Italian position as a worldwide design leader for contents production could be penalized by spreading and exporting skills. The truth is that the country is strengthened by the network, first of all the laboratory allow to share new ideas developed through cultural exchange and this means on-going improvement for Made in Italy design concept. Secondly, talents trained in Italy become real ambassadors of our design, able to spread our products' value proposition in other countries.

Chains can benefit of direct and indirect positive outcomes on their competitiveness:

- Direct outcomes:

Through an active presence on the website vendors will widen their operational horizon, being directly in touch with their customers. This is an opportunity to increase sells volumes and also the chance for companies which are not big enough to use marketing methods based on web 2.0. Companies will be able to make the most of active participation of users, bidirectional relationship, opened system, easy access, and collect information for their innovation processes and marketing;

Through services helping internationalization processes, companies will find easier to enter new booming markets and they will find instruments to reduce risks.

- Indirect outcomes:

Distant learning activities, as well as contents directed to users and generated by users through web 2.0 will increase international awareness of value offered by Italian design.

Both direct and indirect outcomes will have the consequences:

1. Increase sales volumes, first for vendors linked to the platform, secondly for all Italian producers
2. Improve of margin product due to design's culture spread
3. Increase of employment levels due to higher sales volumes

4. CONCLUSION

Exploitation of research and industrial skills and excellences.Skills development involves, among the others, the next fields:

- Advanced communication media, based on web 2.0 technologies, suitable to support B2C sale for home furnishing accessories, design objects, jewellery and trinkets;
- Analytical instruments able to produce channels analysis and customer segmentation in automatic way;
- Logistic organizational models with high performance driven to offer a worldwide service at top level, also flexible and affordable.

New technologies allows to create communities with a lot of users sharing contents. An example, among many others, is given by Wikipedia, Facebook, eBay, Del.icio.us, LinkedIn. This proves that management technologies can do more and more. The project aims to use these technologies for the marketplace development to build a community on the Italian design concept.

Related to growth of ICT technologies, the project allows the development and joining of innovative technologies web 2.0, able to:

1. Create a place where all users are connected, both vendors and buyers, could develop ideas, contents, events, opinions, all related to design Made in Italy. Advanced technologies web 2.0 will join potential of vendors, design students, interested people and keen on design;
2. Translate the characteristics of products in a way that user/buyer can exactly understand and feel the additional value of the product just through the web (experience interaction);
3. Use buyer's profile and location data to increase the offer and create return information flows which could be translated into marketing information available for vendors;

4. Create data exchange with the logistic platform able to assure the quality of logistic performance expected from the market

The final product is unique and knowledge, concepts and system supports developed can be used later in other contexts to support the Italian Industry.

REFERENCES

Gallico D, (2016). *Rapporto nazionale sul Design nelle imprese italiane (dalla A alla Z)* isbn 9788868741518 Logo Fausto Lupetti editore.

Gallico D, Giacobone T (2011). *Adi lombardia design codex 001-il paesaggio dell'innovazione*. Bologna:Logo Fausto Lupetti editore.

Gallico D (2013). *Contam-Innovation Cinquecento icone tra design, cultura e innovazione.* Bologna:Logo Fausto Lupetti editore .

Bentley, T. (1998). *Learning beyond the classroom: Education for a changing world.* London: Routledge.D'Andrea A., Ferri F., Fortunati De Luca L..

Guzzo T (2009). *"Mobile Devices to support advanced forms of e-Learning"* Handbook of Research on Multimodal Human Computer Interaction and Pervasive Services: Evolutionary Techniques for Improving Accessibility. Editor:

Patrizia Grifoni, IGI Global Dillenbourg, P., Baker, M., Blaye, A., & O'Malley, C. (1996). The evolution of research on collaborative learning. In E. Spada & P. Reiman (Eds.), *Learning in humans and machine: Towards an interdisciplinary learning science.* Oxford, UK:

Grew, P., & Pagani, E. (2005). Towards a wireless architecture for mobile ubiquitous e-learning. *Workshop on Learning Communities in the Era of Ubiquitous Computing*, Milan, Italy.

Haythornthwaite, C., Kazmer, M.M., Robins, J., & Shoemaker, S. (2000). Community development among distance learners: Temporal and technological dimensions. *Journal of Computer Mediated Communication.*

Hamdani, M. (2012)*Advance principles for visual aesthetics in designing the contents of e-learning,* Majid Hamdani, Mohammad hosseinpour, Rio Sumarni Bt. Sharifuddin Science Series Data Report, No.6, volume 4.

Horton, W, Horton K *"E-learning Tools and Technologies"* (2003) Wiley Publihing Köse, U. *Artificial Intelligence Applications in Distance Education* International Book Editorship Masterman.

E., Jameson, J., & Walker, S. (2009). *Capturing teachers' experience of learning design through case studies* Distance Education, 30(2): 223–238.

Masterman, L. , *The challenge of teachers' design practice* Chapter accepted for H. Beetham & R. Sharpe (Eds.), Rethinking Pedagogy for the Digital Age, 2nd edition. London: Routledge.

McGreal, R., & Roberts, T. (2001). *A primer on metadata for learning objects.* Retrieved from http://www.elearningmag.com/issues/Oct01/learningobjects.asp.

Pathak, A. (2005) *Medium is the Message: Web-based Lecture Presentations in Distance Education* ISSN 1347-9008 Asian J D E vol 3, no 2, pp 13 - 22.

Paz González-Brignardello, M. (2008) *E-Learning Uses of Concept Maps* Published in "Concept Mapping: Connecting Educators" Proc. of the Third International Conference on Concept Mapping A. J. Cañas, P. Reiska, M. Åhlberg & J. D. Novak, Eds. Tallinn, Estonia & Helsinki, Finland.

Ramadhanie, M. A. *Design and Implementation of Learning Object Ontology for e-Learning Personalization* Ramadhanie, M. A, Siti Aminah, A. Nizar Hidayanto, Adila A. Krisnadhi. (7-8 Dec 2009) *Proceeding International Conference on Advanced Computer Science and Information Systems (ICACSIS)*, University of Indonesia, Depok, , pp. 427-433 staf.cs.ui.ac.id 20

JUNIORTUKKIE APP: A COMMUNICATION APPROACH TO ATTRACT QUALITY STUDENTS FOR THE UNIVERSITY OF PRETORIA

Martie Kilian
Department of Enrolment and Student Administration
University of Pretoria, South Africa

ABSTRACT

Thousands of tertiary institutions are embracing the social media revolution and the opportunities that it creates for their institutions. Dwindling budgets, geographical location, time constraints and limited venue space at higher institutions make it difficult to meet the demand for, for example, contact visits with student advisors who provide study assistance and career counselling, mass events on campuses, campus tours, information sessions and career exhibitions. The JuniorTukkie mobile application (JT APP) is beneficial to the University of Pretoria's (UP) marketing strategy in various meaningful ways. It fills a gap by enabling the University to make that valuable first connection by communicating critical study-related information to prospective students in the way they prefer to be reached – on their mobile devices. By using the JT APP's push technology, which is a highly visible, real-time form of communication, we provide our target market with a mix of information and real-time services. Well-informed prospective students make wise and clear study and career decisions, which contribute to a higher retention rate for the University. According to the University of Pretoria's Strategic Plan, its projected growth in student admission over the next 15 years will be focused on enrolling an appropriate ratio of quality undergraduate students. In support of the University's strategic vision and in line with its hybrid learning model, which is a blend of face-to-face teaching enriched by virtual/online learning and communication products, the JT APP (JT APP) ticks all the boxes. In order to enhance the University of Pretoria's recruitment strategy, it is of critical importance to research the impact of the JT APP as a communication tool designed to attract more applications from high-quality prospective undergraduate students who wish to study at UP.

KEYWORDS

Mobile Application, Communication, Prospective Students, Recruitment

1. INTRODUCTION

The University of Pretoria (UP), which comprises nine faculties and the Gordon Institute of Business Science, is a diverse and dynamic university community with 47 441 registered students and 5 320 under- and postgraduate distance education students from all over the world. The website of such a large university is leveraged to contribute to an enormous range of stakeholders, including prospective and current undergraduate and postgraduate students, parents, donors, researchers, alumni, staff, sports, support services, facilities, governance, faculty-specific information, fees and funding, and much more!

The strategic vision of UP includes recruitment of the most talented students in terms of intellectual interest and abilities. The aim of the JuniorTukkie (JT) Office, within the Student Recruitment Division, is to coordinate various initiatives in an effort to recruit top-achieving undergraduate students for the University. Grade 9 to 12 learners with an academic average of 70% or higher may become members of the JT Club. The JT Office communicates with prospective students via various communication media, one of which is the JT APP. The JT member database consists of 4 698 (as on 30 November 2017) Grade 9 to 12 learners and students.

According to a research study (Potgieter, 2015) that included 1 161 (in 2013) and 1 256 (in 2014) undergraduate students, '79% of the respondents (in 2013) indicated that they owned a smartphone compared to 84% (in 2014). The respondents' most popular reasons for downloading an App is when they need information on a brand, its products or its services, as well as to be educated'.

Although the University of Pretoria's website speaks clearly to its huge target market, a need was identified for a communication tool with an information structure that is narrowed down to information that speaks specifically to the prospective undergraduate student market. 'Apps serve a single purpose – unlike websites where visitors are happy to explore, app users want to meet their goal in as few taps as possible' (Jefferson, 2015). The JT APP is one of several JuniorTukkie initiatives and was developed to improve our communication with the prospective undergraduate student market and our JT Club members. The JT APP enables the University to communicate critical information in different content formats that resonate with the young target market. 'Mobile apps provide students with a convenient and accessible platform for immediate relevant information retrieval and keeps them connected through the updates they receive in areas that interest them the most. One of the main reasons behind the growth of mobile app usage in third level is that students are accustomed to using these apps in their personal life and have an expectation that their third-level institutions will also provide similar apps to help them navigate college services and resources' (Briz-Ponce, 2016). Concise and consistent communication via the push notification function on the JT APP builds user loyalty and an informed target market that improves the throughput rates of top-achieving students at UP.

The aim of this paper is to demonstrate that the JT APP, specifically designed to provide critical study-related information to prospective undergraduate students at the time when the information is most relevant, is a source of information that can be successfully used to recruit quality students for UP. For this purpose an investigation was undertaken into the admission rate of undergraduate students to the University.

2. THE JT APP AS A COMMUNICATION TOOL

The JT APP is accessible on Android and iOS mobile phones and tablets. To download the JT APP, go to Google Play Store or Apple App Store and search for 'juniortukkie'. The JT APP complements, but in no way replaces its existing communication media, such as email, SMS, Facebook, Twitter and the JT website. The JT APP serves as a consistent source of information for our undergraduate student market. The information structure on the JT APP has a streamlined, uncluttered look with minimal scrolling, simplified forms and prominent and easily completed calls-to-action. By using the 'share' and 'enquiry' functionalities on the JT APP, users can interact directly with the JuniorTukkie Office, faculty staff and department representatives. The interactive approach creates optimal user experiences. The new and existing features allow communication of critical information in different content formats to a wider audience as there are no limitations regarding, for example, venue space, geographical location or financial constraints.

2.1 The JT APP Information Structure

The JT APP information structure is aimed at the prospective undergraduate student market. It has a three-tier menu that ensures easy navigation: (1) Main categories, (2) Subcategories and (3) Articles. Adding video's, documents, programmes, e-books, infographics, images and pamphlets to support articles or push notifications provides a much better user experience. Go to www.up.ac.za/juniortukkie > JT APP for a detailed overview of the JT APP information structure.

2.2 JT APP Functionalities

As mobile devices become more sophisticated, new features are added, and from 30 June 2017 (when the upgrade was completed), the number of JT APP downloads increased from 5 417 to 11 633 on 31 March 2018. The JT APP features include:

- *Push notifications*: Notifications are sent at the time when the specific information is most relevant to users, who are alerted by a small notification icon that appears on the screens of their mobile devices. Viewer statistics confirm that this feature is effective as the notification about online applications attracted 1 957 views within two weeks, the Health Sciences Open Day attracted 1 423 views and the Veterinary Science Open Day attracted 1 344 views.

- *Re-visit push notifications function*: Re-visit most recent push notifications by selecting the small bell icon.
- *Sliding banners:* These banners are used to promote the University's integrated campaigns and #JoinJT.
- *Upload supporting media:* Users have the option to view videos, documents, programmes, e-books, infographics, images, or pamphlets that are uploaded to any article or push notification on the JT APP.
- *Enquiry function:* Every push notification and article has an enquiry option which can be utilised by users to direct their enquiries to the JuniorTukkie Office for quick answers.
- *Share function:* Any article or notification can be shared by users with friends and family on social media.
- *Fan Cam function:* Users can take a photo of anything and share it with us or add an enquiry and, with just one click, send it to the JT Office for quick feedback and assistance.
- *Live streaming* of information sessions, open / career days, Summer / Winter Schools.
- *Content update settings:* Users may choose the categories in which they would like to receive notifications.

2.3 Target Audience

The information published on the JT APP is easy to understand, resourceful, interesting and worth sharing, and caters for UP's diverse target audience (refer to the results of the opinion poll in paragraph 3. Our target audience includes:

- *Grade 9 to 12 learners:* The JT APP contains critical information that enables them to make sound study and career choices, to apply for admission, and to finally register for studies at the University of Pretoria.
- *Life Orientation and Career Guidance teachers* find the JT APP informative and state that it enables them to provide up-to-date information about study opportunities at the University of Pretoria.
- *Parents* find the information on the JT APP very useful as it enables them to assist their children during the career and university exploration process.
- *UP Staff* are well-informed and able to effectively respond to enquiries.

2.4 User Database

Within six months after it was launched on 1 June 2016, the JT APP was already downloaded 2 535 times, as illustrated in Figure 1 below. By the end of March 2018, the JT APP user database had grown to a staggering 11 633 (refer to Figure 2 below). This is proof of the JT APP's popularity as an effective communication tool.

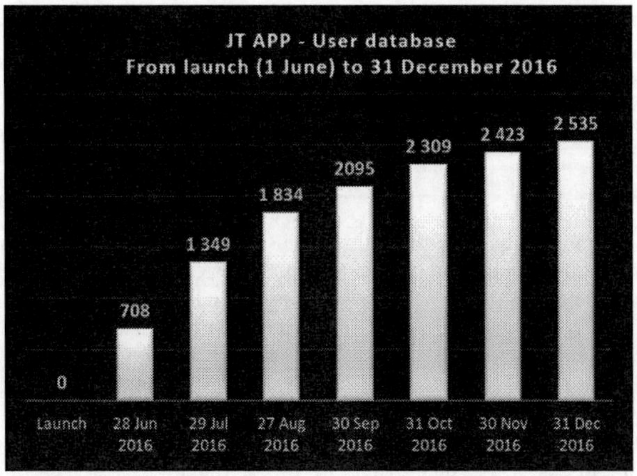

Figure 1. JT APP downloads within first six months

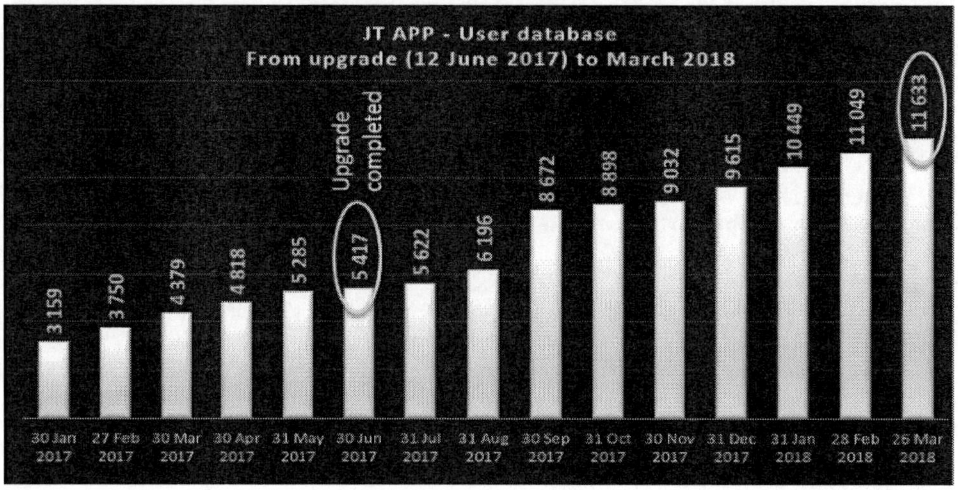

Figure 2. Total JT APP total downloads until 31 March 2018

2.5 The Approach

The decision to add a mobile app to JuniorTukkie's communication media was driven by the University's strong focus on enrolling an appropriate ratio of academically achieving undergraduate students and the fierce competition in the SA market for top-achieving students. The creation of a competitive advantage in the undergraduate student market requires continuous renewal and improvement of communication methods (refer to the goals and objectives in Table 1). The JT APP enables us to communicate important study-related information in different forms, such as text, infographics, multimedia and presentation methodologies to prospective students right where they are – and they are almost always ONLINE!

In line with the University's hybrid learning approach, the JT APP enables improved communication performance and user experiences. The new live-stream feature enables the University to communicate important information to a wider spectrum of top- achieving prospective students, such as those who would not otherwise have been reached due to geographic location, limited space or financial constraints. Live-streaming information sessions reduce travel and accommodation costs for Student Advisors and create more opportunities for recruitment initiatives. A study titled 'The role of live streaming in marketing communications and corporate branding' (Keinänen, 2017) reveals that 'the biggest benefits of live streaming include authenticity, which allows companies to differentiate and create brand awareness, and the possibility to reach a wider audience. However, in order to attract and engage audiences with live streaming, companies need to understand their target groups and create content that resonates with the audience'.

2.6 Goals and Objectives

Table 1. Goals and Outcome Objectives

Output objectives	Outcome objectives
Ensure internal stakeholder satisfaction by publishing information such as faculty programmes on the JT APP.	UP internal stakeholders are convinced that the JT APP is a sustainable and resourceful communication platform.
Improve the information structure on the JT APP to ensure easy navigation.	Organise the content in a three-tiered menu with six main categories, sub-categories and articles / push notifications. Users experience a user-friendly information structure.
Compile an animated slideshow to illustrate the new information structure to users.	Users understand the information structure, navigate easily and know exactly where to find information on the JT APP.
Establish a consistent flow of up-to-date study information on the JT APP in order to promote user growth and increase the number of informed prospective	The number of JT APP users increased enormously – refer to Figure 1 and Figure 2. Well-informed prospective students do not only apply for studies, but are admitted and finally

Output objectives	Outcome objectives
students.	register at UP because they were well-informed.
Provide concise, up-to-date and valuable information on the JT APP to ensure that users want to read more.	It is evident that the content is most valuable: - 3 379 users read 'Considering the University of Pretoria - 4 952 users read 'Become a JuniorTukkie'
Improve brand recognition.	The enormous growth in the number of JT APP users is indicative of improved brand recognition.
Build relationships with more top achieving prospective students via the JT APP by encouraging users to become members of the JuniorTukkie Club.	By 30 November 2017, the JT membership had increased to 4 698. Learners with an academic average of 70% or more qualify for JT membership.
Provide understandable, actionable and interesting content on the JT APP that is worth sharing to ensure that more prospective students benefit.	In a poll that was conducted (refer to par 3), 95.53% of the participants indicated that the information provided on the JT APP is easy to understand, actionable, interesting and worth sharing.
Establish the JT APP as a key driver of UP initiatives.	The #ChooseUP and #JoinJT campaigns are promoted.
To create a platform for users to direct enquiries to the JT Office via the 'interact' functionality on the JT APP.	Ensure quick responses to enquiries. In a poll conducted (refer to par 3), 96.09% of the participants stated that they had received satisfactory responses.
Establish a new live stream functionality to communicate with more top achieving prospective students.	Location, venue space and financial difficulties are no longer a constraint when critical information is shared.
Plan further developments for 2018.	The following features will be implemented in 2018: 1) a chat room 2) a document upload function

2.7 Awards

- **Platinum Excellence Award 2017 (MACE Excellence Awards)**: All the winners of Gold Awards of Excellence in the Mobile Apps Category competed for this prestigious award, which was won by the JT APP. The judges' feedback included the following comment: 'The entry was really "Wow!". The application is original and matches the characteristics of the target audience. We commend you on the high quality of production, the superior application of best practices, strategic alignment and excellent project management principles.'
- **Best Communication Innovation Award 2017 (SA Publication Forum)**: The SA Publication Forum rewards excellence in the field of internal and external corporate communication media. According to the judges, the term 'innovation' can be defined as something original and more effective, and therefore new, that 'breaks into' the market or society. When one considers this definition, the JT APP certainly complies with all the requirements. This App communicates with its market (young people) in a new, innovative way that speaks to their needs on a platform that will intrigue, entertain and educate.
- **Gold Excellence Award 2016 (MACE Excellence Awards)**: Excellence in communication and a constant striving to improve and find new ways of communicating effectively, won the JT APP a Gold Excellence Award in the 2016 Marketing, Advancement and Communication in Education (MACE) Competition.

3. KEY FINDINGS

The impact of the JT APP can be regarded as a factor that contributed to the 6.8% increase in the admission rate of prospective undergraduate students from the launching of the JT APP in 2016 to the 2018-intake. A survey was conducted to determine how users experience the JT APP. A total of 179 responses were received. The focus group included Grade 9–12 (75.42%); parents (10.06%); students (8.94%);a and staff (5.59%). The findings were as follows:

- 96.65% of the respondents indicated that the images on the JT APP appeal to them.
- 98.32% of the respondents found the design and information structure appealing and easy to navigate.
- 96.65% of the respondents indicated that the download instructions are clear and easy to follow.
- 95.53% of the respondents indicated that they had posted a query on the JT APP.

- 96.09% of the respondents indicated that they had received satisfactory responses to queries.
- 95.53% indicated that the content on the JT APP is easy to understand, actionable and worth sharing.
- 97.21% indicated that they found 'How to apply for studies at UP' beneficial and easy to follow the steps.
- 98.88% rated the JT APP at between 5 and 10 (1: not satisfied; 10: extremely satisfied).

4. CONCLUSION

Research done on the impact of the JT APP on the rate of admission of quality students to UP indicates that since its introduction the JT APP has made a positive contribution to the 6.8% increase in admissions of new undergraduate students. The results of the opinion poll regarding the JT APP proved that it is an effective communication platform for easy access to concise and critical study-related information, and that it enables users to make informed study and career choices. The continuous updates and steady flow of relevant information, as well as the inclusion of the topics that interest them the most, build loyalty and trust. As the logo is visible on mobile phones all the time, the mind unconsciously records the well-designed icon and seeing it several times a day is what ensures that the institution is truly noticed. The JT APP creates a direct marketing channel, with the result that well-informed and motivated prospective students apply for studies at UP. The JT APP builds the brand and therefore recognition of the University (UP). Information is always up to date and to the point to enable prospective students to make informed study and career decisions. The JT APP helps the University to stand out from the competition as it strives to be a leader in the higher education sphere and wants to be noticed for its forward-thinking approach.

REFERENCES

Briz-Ponce, L. et al, 2016. *Handbook of Research on Mobile Devices and Applications in Higher Education Settings*. Information Science Reference, Hershey PA, USA.

Jefferson, S. et al, 2015. *Valuable Content Marketing: How to Make Quality Content Your Key to Success*. Kogan Page Limited, Great Brittain and United States.

Keinänen, K., 2017. The role of live streaming in marketing communications and corporate branding. School of Business Management, Lappeenranta University of Technology.

Potgieter, A., 2015, The mobile application preferences of undergraduate university students: A longitudinal study. *South African Journal of Information Management,* 17(1), Art. #650, 6 pages.

SOCIO-TECHNICAL SELF-DEVELOPMENT USING A MICROSERVICE ARCHITECTURE

Johann Sell and Niels Pinkwart
Humboldt Universität zu Berlin, Germany

ABSTRACT

Considering *joint optimization* and *organizational choice*, socio-technical systems have to align their technical component with the ongoing change of the social system. Focusing on decentralized, loosely coupled socio-technical systems, this challenge becomes more complex, because the various parts of the social system will develop different requirements for the technical support. Focusing on a case study, we develop requirements and compare different possible architectures. We approach to use a microservice architecture to support maintenance and further development of the technical tool.

KEYWORDS

Microservices, Socio-Technical Systems, Evolutionary Organizations, Joint Optimization, Organizational Choice

1. INTRODUCTION

Our work aims to support a loosely coupled socio-technical organization in terms of self-development. We focus on ensuring compliance of *joint optimization* and *organizational choice* as most important challenges (Trist 2013; Emery 1959). *Joint optimization* implies that every change in one of both systems (technical or social) also requires changes in the opposite part. *Organizational choice* describes a needed degree of freedom, thus users are allowed to adjust their usage of the supporting system.

The case study Viva con Agua de St. Pauli e.V. (VcA, vivaconagua.org, April 2018) has developed very heterogeneous requirements regarding its technical part, because of the loosely coupled nature of its social system. Next to VcA there is an increasing set of organizations boasting the same characteristics, like foodsharing (foodsharing.de, May 09, 2018), Buurtzorg (buurtzorg.com, May 09, 2018) or FAVI (favi.com, May 09, 2018) (Laloux 2014). This paper discuses several possible architectures to address the issue of heterogeneous requirements considering a loosely coupled socio-technical system. We suggest to use microservices (Newman 2015; Zimmermann 2017; Dragoni et al. 2017) to tackle this issue.

VcA is a non-governmental organization raising awareness and collecting donations for WASH ("Water, Sanitation and Hygiene") projects. It is based on decentralized volunteering, organized in loosely coupled regional groups, called *crews* (Weick 1976). As already described by (Sell & Pinkwart 2016) its success is based on *open participation* and coordination using a web-based tool named *Pool*. Hence the tool became part of the social systems self-description and both joined into a socio-technical organization (Kunau 2006). Unfortunately, *joint optimization* and *organizational choice* became impossible to apply, due to limited men power and inflexibility of the *Pools* architecture as we will describe in this paper. Additionally, the current implementation is neither expandable nor maintainable.

We have identified three high level requirements for the technical support system: (R1) Since the *crews* are loosely coupled, new technical support may fit only requirements of a few *crews*. So, the possibility to add support for complete work processes only for a few *crews* is needed. (R2) Obviously, *crews* have to replace existing technical support of work processes. For example, the general technical support for handling finances has to be replaced for some *crews* due to alternative roles that have been evolved. Normally, a *crew* has one person responsible for finances, but there are *crews* separating between a person responsible for donations and another for housekeeping money. (R3) Since development of software requires a lot of time, and volunteers already invest a huge amount of time regarding to the goals of VcA, a distinct community of developers has to evolve. Such a community differs from Open Source communities by the motivation that is not to help themselves and others, but to acquire new skills, and investigate ideas with a huge number of users. Thus, it is quite likely that it is too much effort to take the time to learn a new technology.

In the end, a decentralized, loosely coupled social system of software developer evolves. Each one focuses small parts of the technical system and participates in terms of *open participation*. We are convinced that such a social system is obliged to consider the ongoing change of VcA and will be able to apply *joint optimization* and *organizational choice*.

2. SOLUTION CONCEPT

We approach to foster self-development of VcA. We aim to create a new loosely coupled, decentralized and volunteering community of software developers that by themselves are able to build new support functions for the social system. Thus, dynamic integration of developing teams without the need for training of specific technologies (R3) or cumbersome bureaucracy is required. This new community of software developers (Pool-volunteers) has to be separated from the existing volunteers focusing the WASH projects (WASH-volunteers), since both activities require a lot of time.

2.1 Characterization of Possible Architectures

Basically, we have identified four types of possible architectures: Monolithic systems, modular implementations, app based extension of core functionality, and microservices. There are socio-technical communities using their own special purpose tools for loosely coupled volunteering (Dittus et al. 2016). Other communities compose various technical tools (Gray et al. 2016; Ambe et al. 2016; Reuter et al. 2015).

In general, monolithic architectures exhibit *high cohesion* between implemented functions (Newman 2015). Thus, software enhancements normally have to use the same set of technologies as the origin.

In contrast, there are applications implemented very modular, like Moodle (moodle.org, February 2018) or MediaWiki (mediawiki.org/wiki/MediaWiki, February 2018). In general, these modular applications have a core and can be extended by plugins. Free choice of technology is bounded to its purpose. While developers can use one technology to extend the core, a second can be used to implement plugins.

Facebook (facebook.com, February 2018), Instagram (instagram.com, February 2018), Slack (slack.com, February 2018), and Trello (trello.com, February 2018) allow to extend the functions of its software by third-party apps. Therefore, those apps get user data and a sandbox area to deploy their functionality.

The last opportunity are microservices. First, microservices are independent processes and cohesive in bounded contexts (Dragoni et al. 2017; Namiot & Sneps-sneppe 2014; Newman 2015; Viennot et al. 2015). Second, they use message based communication via well-defined interfaces, like RESTful webservices (Dragoni et al. 2017; Rodriguez 2008). A set of connected services is called a microservice architecture (Dragoni et al. 2017; Namiot & Sneps-sneppe 2014). Such architectures force the designer to minimize communication between microservices (*loosely coupling*) and to implement related behavior inside the same scope (*high cohesion*) (Newman 2015).

2.2 Choosing an Architecture Fitting the Requirements

Volunteers of VcA use several different tools to balance their needs (Sell & Pinkwart 2016). However, VcA is an organization, thus there are important and sensible processes, like the handling of finances. Furthermore, the need for different user accounts and invisibility of used tools are serious obstacles considering *open participation*. Therefore, the usage of many independent tools breaks the organizational structures. A monolithic or module based architecture requires specific technologies. Thus, (R3) becomes impossible to fulfill, since developers would need to learn about the technologies.

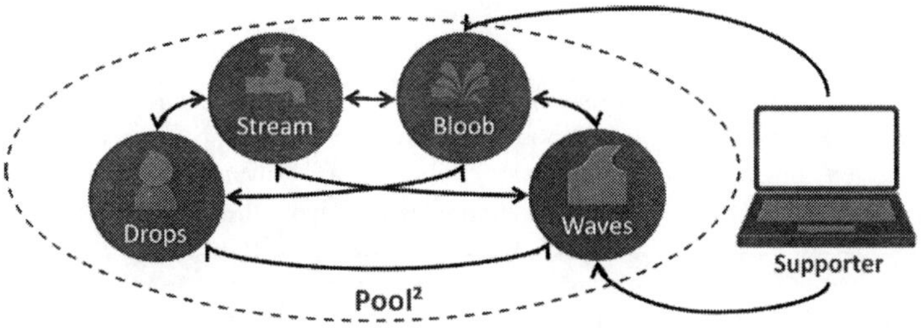

Figure 1. Microservice architecture for Pool² with arrows indicating possible communication channels. Drops *(user management),* Stream *(finances),* Bloob *(email broadcasting), and* Waves *(activities and events)*

Adopting the concept of apps for VcA requires its enhancement. First, apps are normally introduced to allow additional functions, but not to replace existing ones. Thus, the hard separation between core application and apps has to be softened due to (R1). Furthermore, it should be possible to replace all core functions with new implementations. Thus, apps become loosely coupled, as in the case of microservices.

Microservices spread responsibility for maintains and support among different teams. Thus, it supports the management of growing complexity and delegates systems modularity from a technical implementation to an organizational layer. Modularity represented in organizational structures supports independent development and deployment (Dragoni et al. 2017) (R1; R2). Furthermore, a microservice architecture allows different programming paradigms, languages, and database technologies for different parts of the system (Dragoni et al. 2017; Namiot & Sneps-sneppe 2014; Viennot et al. 2015). Thus, developers with various expertises can contribute to the new system (R3). SAP, for instance, is developing an additional mobile App.

In the end, self-development of a socio-technical organization will be possible, if the organization needs decentralized further development. Pool-volunteers can implement specific requirements of individual *crews* without changing the whole system.

2.3 Implementation

First, we implement the existing features of the *Pool* using the new architecture. We have decided to introduce four microservices addressing the functional requirements, as shown in Figure 1. *Drops* handles the representation of volunteers and the authentication. *Bloob* implements communication support, while *Stream* supports the handling of finances and *Waves* the coordination of volunteers. The new system is called *Pool²*.

The microservice architecture implies several challenges. As described in section 2.1, exchange of data between services is implemented using RESTful webservices, while the event message broker nats (nats.io, February 2018) informs about updates. OAuth 2 (oauth.net/2/, February 2018) is used to implement a shared session to prevent different accounts. Docker (docker.com, March 2018) wraps the microservices inside standalone container next to each other and allows deployment of heterogeneous technologies.

One central microservice handling the user interface would imply one team of developers has a lot of responsibility. Thus, every microservice has to implement its own user interface. Ensuring a consistent user interface across service boundaries becomes a serious obstacle. We propose to adopt the *UI Fragment Composition* (Newman 2015). The microservice *Dispenser* distributes templates containing basic HTML elements, like a global menu, header or footer. Additionally, it allows reused CSS and a corporate design.

Furthermore, all microservices have to implement the concept of widgets. A widget is a HTML element with optional CSS. It is used by other services to show content or a function that is in *high cohesion* to the widgets provider. Therefore the service is responsible for definition, delivery, and maintenance of the widget. Its delivery takes up the idea of *transclusions* (Nelson 1965; Kolbitsch & Maurer 2006). For instance, the *Drops* service should implement a widget for an auto completion field that could be used to select a user. We will implement a JavaScript function, that watches an HTML input field, communicates with *Drops*, and shows a list of users. Both concepts apply the paradigms *loose coupling* and *high cohesion* to the view layer.

3. CONCLUSION AND OUTLOOK

This paper derives requirements from the characterization of the decentralized, loosely coupled, socio-technical organization VcA. Different architectures for a technical system are compared to support socio-technical self-development. A microservice architecture has been identified assisting *joint optimization* and *organizational choice* for organizations like VcA. Additionally, a rough introduction to the implementation is given and solutions to important challenges are outlined.

Currently, we are implementing a proof of concept. Afterwards, the implications for the social system of WASH-volunteers have to be explored, like a possibly needed quality assurance or coordination support for the Pool-volunteers. Moreover, the process of further development has to be investigated: Can a new loosely coupled social system of Pool-volunteers be established and does the *Pool²* really enable loosely coupled and decentralized technical artifacts to form a coherent collaboration tool for daily work?

First, we have to figure out how much additional work is required to integrate a new technology. Effort can be measured by tracking or estimating time needed by experts. Valid estimations need complete descriptions of the non-functional requirements regarding the integration of a microservice. Such descriptions can be analyzed qualitatively, with focus on *loosely coupling* and *high cohesion*. The new architecture has to enforce these principles, since new developers could have few experiences in software architecture.

Obviously, we also have to investigate further adaption by the volunteers. Basic questions are, if they formulate new requirements and find developers to implement those requirements. At least, our approach should be validated by applying it to other organizations. Our further research will focus these questions.

REFERENCES

Ambe, A.M.H., Brereton, M.F. & Rittenbruch, M., 2016. Vendors- Perspectives of Coordination in the Information Technology Offshore Outsourcing Industry: An Exploratory Study from the Philippines. In *Proc. CSCW '16*. New York, USA: ACM Press, pp. 318–333.

Dittus, M., Quattrone, G. & Capra, L., 2016. Analysing Volunteer Engagement in Humanitarian Mapping: Building Contributor Communities at Large Scale. In *Proc. CSCW '16*. New York, USA: ACM Press, pp. 108–118.

Dragoni, N. et al., 2017. Microservices: yesterday, today, and tomorrow. Available at: https://arxiv.org/pdf/1606.04036.pdf [Accessed February 10, 2017].

Emery, F.E., 1959. *Characteristics of sociotechnical systems: A critical review of theories and facts about the effects of technological change on the internal structure of work organisations; with special reference to the effects of higher mechanisation and automation*, London.

Gray, M.L. et al., 2016. The Crowd is a Collaborative Network. In *Proc. CSCW '16*. New York, USA: ACM Press, pp. 134–147.

Kolbitsch, J. & Maurer, H., 2006. Transclusions in an HTML-Based Environment. *CIT*, 14(2), pp.161–173.

Kunau, G., 2006. *Facilitating computer supported cooperative work with socio-technical self-descriptions*. Technische Universität Dortmund. Available at: http://hdl.handle.net/2003/22226.

Laloux, F., 2014. *Reinventing Organizations* 1st ed., Brussels: Nelson Parker.

Namiot, D. & Sneps-sneppe, M., 2014. On micro-services architecture. *INJOIT*, 2(9), pp.24–27.

Nelson, T.H., 1965. Complex information processing: a file structure for the complex, the changing and the indeterminate. In *Proc. ACM '65*. New York, USA: ACM Press, pp. 84–100.

Newman, S., 2015. *Building Microservices* 1st ed. M. Loukides, B. MacDonald, & K. Brown, eds., O'Reilly Media.

Reuter, C. et al., 2015. XHELP: Design of a Cross-Platform Social-Media Application to Support Volunteer Moderators in Disasters. In *Proc. CHI '15*. New York: ACM Press, pp. 4093–4102.

Rodriguez, A., 2008. RESTful Web services: The basics. Available at: https://www.ibm.com/developerworks/library/ws-restful/ [Accessed March 17, 2017].

Sell, J. & Pinkwart, N., 2016. Rambla: Supporting collaborative group creativity for the purpose of concept generation. In T. Yuizono et al., eds. *Proc. CRIWG '16*. Kanazawa, Japan: Springer, pp. 81–97.

Trist, E.L., 2013. *Organizational Choice: Capabilities of Groups at the Coal Face under Changing Technologies - The Loss, Re-discovery & Transformation of a Work Tradition* 1st ed., London: Routledge.

Viennot, N. et al., 2015. Synapse: a microservices architecture for heterogeneous-database web applications. In *Proc. EuroSys '15*. New York, USA: ACM Press, pp. 1–16.

Weick, K.E., 1976. Educational Organizations as Loosely Coupled Systems. *ASQ*, 21(1), pp.1–19.

Zimmermann, O., 2017. Microservices Tenets: Agile Approach to Service Development and Deployment. *CSRD*, 32(3), pp.301–310.

REDUCING EGOCENTRIC BIAS BY PERSPECTIVE SHIFTING IN VIRTUAL WORLD WITH DIFFERENT CONSTRUAL-LEVEL DISTANCE BETWEEN SELF AND OTHER

Aekyoung Um, Young June Sah and Daeho Lee
Sungkyunkwan University, South Korea

ABSTRACT

Social game as a future prevalent service for collaborative work has a huge potential for transforming our work life. Collaborative work has been studied through self-awareness; it has been not yet fully to be discovered by perspective-taking strategy. Construal-level theory as a highly effective way to demonstrate self and other relationship will help to explain how we perceive team member in the two different ways. The lab experiment will be conducted by 2 (first person perspective vs. third person perspective) X 2 (the level of distance between self and other) between-subjects experiment and see how equality or equity theory will be applied. It will also show how we see ourselves in different perspective and give insight for the effective way of doing collaborative work in virtual reality.

KEYWORDS

Perspective-taking, Egocentric Bias, Construal-level Theory, Social Distance, Collaborative Work in Virtual Reality, Equity and Equality Theory

1. INTRODUCTION

There's a tiny door in empty office. It's a portal, Maxime. It takes you inside John Malkovich. You see the world through John Malkovich's eyes, then, after about fifteen minutes, you're spit out into a ditch on the side of The New Jersey Turnpike.

-Being John Malkovich

How would it be if we can experience other person's life just for fifteen minutes like a movie 'Being John Malkovich'? Out-of-body experience is an abnormal experience, which is possible in dreams, imagination and hallucination. This awkward cognitive experience usually has been used in three-dimensional role playing games. The strategy is known for making user to concentrate on shooting the target or massive multiplayer online role-playing game which needs to be seen from long distance.

Perspective-taking referred to the "ability to imagine or to represent how objects look relative to one another from another person's point of view" (Amorim, 2003) has been examined in folk psychology with respect to self and other relationship. One example is the theory of mind, which tried to describe infant's cognitive transformation by taking the ability to read confederate's mind. The schema for understanding other person's mind was developed in several field of psychology. Theories such as mind-blindness, shared-attention mechanism (SAM), Sally Anne test, theory theory and simulation theory are describing the underlying mechanism of mindreading.

The aim of this study is to examine the empirical evidence for understanding how the difference of the viewpoint cause the depth of understanding other people's mind. With the possibilities of perspective-taking in virtual reality, the mental and bodily self must be manifested in different ways than ever. In this regard, reducing egocentric bias through perspective-taking is an interesting topic which is not yet fully covered in virtual reality studies. There were some perspective-taking studies such as speech acts shaping (e.g., Schober, 1993), but the construal-level theory rarely introduced to this field.

2. BODY OF PAPER

2.1 Literature Reviews

2.1.1 Perspective-taking

Perspective-taking studies have been done mainly as a spatial and illusionary experience (Amorim, 2003). We experience the world from the point of view of our own body. At the same time, people are taking others' perspective from time to time, as explained with the mirror neuron or simulation theory (Gallese, V., & Goldman, A., 1998) to posit the attributor mimic the mental activity of the confederate. When it comes to the mental representation to the world, we can easily shift our perspective into other human's body. We call this a third-person perspective, which normally people experience when they reflect the past or imagine the future, or when they just imagine another person's thought, feelings, attitude and opinions or the reason of the action.

Visuospatial perspective-taking studies have been also actively studied in the game field. Several games provide switching between two perspective-taking options (e.g., shooting games) between first person perspective (egocentric mode) and third person perspective (exocentric, observer mode). The first person perspective is preferred in highly focusing condition for example, sniper mode or racing game whereas, the third person perspective is frequently used in dynamic action game when a person needs to see whole vision such as FIFA online or running a shop game. Both of the viewpoints provide different psychological cognitive effect to users such as immersion, identification of self, feeling of presence, feeling and even egocentric bias toward the team players. Especially with the context of collaborative work, perception for the team member might differ depending on the viewpoint. Benford et al. (1995) examined user embodiment within collaborative virtual environment. The key factor for successful collaborative virtual environment was watching one's own body in the environment because it is important to feel oneself in the simulation reality. This mechanism is based on 'What you see is what I see' paradigm, the orientation of avatar as the participants' navigation in the virtual environment.

The avatars are very flexible in the regard with visuomotor synchrony (Won et al., 2015). Using avatar body enhances a feeling of a sense of ownership and homuncular flexibility which to learn understanding our bodies through action and experience. In this study, we will use three mediums; pure imagination, image and virtual reality to conduct the perspective-taking simulation in the context of collaborative work.

2.1.2 Jerald Greenberg 'Mirror' and Reducing Egocentric Bias

Jerald Greenberg studied self-image versus impression management by using the medium 'mirror.' Seeing oneself affected to reduce egocentric bias in the collaborative work situation. His research was focused on the effect of 'self-awareness' in the collaborative work task. The relationship between self-awareness and theory of equity showed that in the self-awareness condition, which was shown the subjects' face in the mirror people, tend to consider responsibility on their performance and equity behavior was exacerbated when he subject sees oneself as responsible for relative performance differences (Jerald Greenberg, 1980).

Previous studies used self-awareness through mirror, but in this study, we will use perspective-taking strategy by using self's mental representation. Further, by measuring social distance between self and other, we expect that it will eliminate self-other discrepancy about another dimensional level. Further, we will measure whether equity or equality theory applied to see according to social distance between self and other.

2.1.3 Construal-level Theories for Self-other Study

Construal-level theory from Trope and Liberman (2003) deals with different psychological representation between spatial, temporal and social distance. This mental construal represent into one's abstraction and concreteness.

According to social distance, self-versus other and in-group and out-group comparison must be differ (Pronin, 2008). The more psychological distance, the higher level of construal will be constructed in terms of abstract thinking. On the other side, as social distance decreases, entity would be perceived in terms of concrete way also referred to as low-level construal. Self as the most proximal distance for oneself, a person's own view emphasizes the role of concrete situational factors that operate at the moment of action, whereas taking other person's view emphasizes the causal role of stable, general dispositional properties of the action.

Yet, the perspective-taking studies are not yet properly examined by construal-level theory. CLT has several benefits to explain the perception of distant others without opportunity to interact with them. Relate with the priming, knowledge activation, it is relevant to measure social distance about self and other in the visual setting environment. The representation of a visual scene from another person must be different from one's one visual scene. Not only the representation in mind but also the cognitive social distance one takes will also differ depending on the perspectives. In this study, we will measure the social distance in three mediums and see how it affects to applying equity theory to the self and others. It will be beneficial for forming self-other schema in collaborative work in virtual reality.

2.2 Method

In the case of virtual reality, it enables us to see our body through visual settings, which is not possible in the real world. The real world experience, which sometimes disturb us not to see our body often leads to cognitive misunderstanding of our situation. When we see our body through virtual world, even that visual setting is not our own body, we can perceive that avatar as ourselves which is similar phenomenon with proprioceptive drift in the rubber hand illusion (Rohde, M., Di Luca, M., & Ernst, M. O., 2011). We expect that imaging the perspective of another person, especially the unfamiliar one, produce distance to the self to perceive more in abstract way. This can be demonstrated into self-objectification. Self as the most proximal entity, it often leads us subjectivity dilemma.

2.3 Experiment Design

The aim of the study is figuring out the how do we perceive self-other relationship in different perspectives view setting. The lab experiment will be conducted using the game 'Minecraft' by building a house with confederate and we will see how participants represent the distance between self and other. To make sure that participants see the avatar as oneself in third-person perspective setting, we will give tutorial time for participants.

Hypothesis 1: The mode of presentation would affect the perspective-taking mental representation.

We are expecting the cognitive change into the shift of view-taking experience. This framework will provide insights relate to understand perceiving self and other in multiple virtual reality. During the conditions, it is important that participants would be properly primed to a new viewpoint (Amorim, 2003).

Hypothesis 2: When participants use third-person perspective in the collaborative work game, people would apply equality rule than first-person perspective.

2.3.1 Participants

Fifty gender-balanced participants will be recruited through website age over 20 less than 50 without any psychiatric illness in this study. They will be either university students or faculty members in Sungkyunkwan University. In the study, 2 (first person perspective vs. third person perspective) x 2 (construal level between self and other) conducted by using the game 'Minecraft'. Subjects will be required to perform particular action in a given context through the eye of oneself or the eyes from the observer who is outside of the situation. In the study 2, information will be announced that in the situation of collaborative task, participants should build a house in order to get a prize. The basic scenario is that students building a house for their living in the 'Minecraft'. We will a user manual describing how to control the keyboard in order to have participants fully manipulate the game situation. To make sure that each team member's social distance is same, we will set the team members as stranger who met just for this project.

This action activation enables us to see what strategy they made for the team player. If the equity rule were applied, it means that one perceived others as distant social partner. If the rule of equality was chosen, it would mean that the one perceives team members more close to oneself which means less egocentric bias applied. The ratio scale will be taken to measure how participants distribute reward to the team member.

2.3.2 Manipulation Check of the Independent Variable

We are going to adopt Ahn et al. (2016)'s measurement to verify how participants immersed to the avatar's body in the third-person perspective.

Body Transfer
We will modify Slater et al.'s Spatial Presence scale (2005) to measure perceived avatar's presence as oneself. A 15-item interval scale assessed how much the participants felt as if they had become the virtual avatar. Sample question is "When you were looking at the collaborative work situation, how much you felt as if you are actually doing that work?"

2.3.3 Dependent Variable

In the study 1, we will measure how participants represent the self and other by using Liviatan et al. (2008) model which used for level of action identification.

When participants are drawing an action in terms of how it is being done, one represents on the concrete way by executing to subordinate action construal (e.g., "Putting a stone" is subordinate to "Preparing for building a house" inasmuch as it answers how one prepares for a team project). We can give participants several questions asking about the self and others describing their actions (e.g., "Getting a prize for building" is superordinate to "Preparing for a team project, "Decorating the door" can be applied as subordinate). In this way, we can categorize an action in terms of high level and low level.

In the study 2, we are going to utilize Utz and Sassenberg (2002)'s measurement for seeing how egocentric bias have been applied.

In the free rider situation, we will see whether participants get a same reward with team member, which means equity theory, applied for justice. In the overestimated situation, we will see whether participants give more effort to compensate their overestimated reward, which means equality theory applied for justice. If both strategies were used, we can interpret that egocentric bias was reduced by the effect of perspective-taking experience.

3. CONCLUSION

3.1 Expected Results and Further

We are expecting the participants would report different self-other distance between first-person perspective and third-person perspective. When the visuospatial perspective taking offered, the significantly different effect arises with task-related activity during first-person perspective than third-person perspective according to Vogeley (2014). In the experiment 1, we are expecting to see less discrepancy in the third-person perspective compared to first-person perspective. The users thus see their own situation from the third person perspective as the most objective mental representation. The future research, it would be very useful to examine attribution-retribution assignment depending on differentiation of the teammates.

With this study, we expect that collaborative work in virtual world will grow in the fruitful way. When programmer sets the design of the app, they will refer this study and benefit users to cooperate with others harmoniously. With the VR contents also, we can easily become others even non-human being. As doing so, I think we can constitute low-level concepts of others and understand other people better than ever. The possibility of perspective taking enable us becoming being whatever we want, and dimension of our perception see this world must be broaden. The cognitive transformation into the other is crucial ability to live with other people in this life and, the perspective-taking make people to see each other more and better with technological advance. This perspective-taking framework will provide the virtual reality designer to consider the visuospatial point of view of the subject who are using the technology. As multiple embodied virtual selves will replace our work and relationship, it is crucial to understand how we perceive self and other in the different mediation. This mechanism of flexible self will expand our possible manifestation supporting successful collaborative work in virtual reality.

It is beneficial to understand others' views in communication. The awareness of the others' perspective enables us to facilitate mind reading through exposure to multiple perspectives. This allows us to enhance a psychological paradigm to improve folk-psychology model to understand others mind in the collaborative context further design space for supporting teamwork in multiple collaborative work simulation.

REFERENCES

Ahn, S. J. G., Bostick, J., Ogle, E., Nowak, K. L., McGillicuddy, K. T., & Bailenson, J. N. (2016). Experiencing nature: Embodying animals in immersive virtual environments increases inclusion of nature in self and involvement with nature. Journal of Computer- Mediated Communication, 21, 399-419.

Amorim, M. A. (2003). " What is my avatar seeing?": The coordination of" out-of-body" and" embodied" perspectives for scene recognition across views. Visual Cognition, 10(2), 157-199.

Benford, S., Bowers, J., Fahlén, L. E., Greenhalgh, C., & Snowdon, D. (1995, May). User embodiment in collaborative virtual environments. In Proceedings of the SIGCHI conference on Human factors in computing systems (pp. 242-249). ACM Press/Addison-Wesley Publishing Co..

Botvinick, M., & Cohen, J. (1998). Rubber hands "feel" touch that eyes see. Nature, 391(6669), 756–756.

Gallagher, S. (2000). Philosophical conceptions of the self: implications for cognitive science. Trends in cognitive sciences, 4(1), 14-21.

Gallese, V., & Goldman, A. (1998). Mirror neurons and the simulation theory of mind-reading. Trends in cognitive sciences, 2(12), 493-501.

Greenberg, J. (1980). Attentional focus and locus of performance causality as determinants of equity behavior. Journal of Personality and Social Psychology, 38(4), 579.

Liviatan, I., Trope, Y., & Liberman, N. (2008). Interpersonal similarity as a social distance dimension: Implications for perception of others' actions. Journal of experimental social psychology, 44(5), 1256-1269.

Nan, X. (2007). Social distance, framing, and judgment: A construal level perspective. Human Communication Research, 33(4), 489-514.

Prentice, D. A., Miller, D. T., & Lightdale, J. R. (1994). Asymmetries in attachments to groups and to their members: Distinguishing between common-identity and common-bond groups. Personality and Social Psychology Bulletin, 20(5), 484-493.

Pronin, E. (2008). How we see ourselves and how we see others. Science, 320(5880), 1177-1180.

Pronin, E., & Ross, L. (2006). Temporal differences in trait self-ascription: when the self is seen as an other. Journal of personality and social psychology, 90(2), 197.

Rohde, M., Di Luca, M., & Ernst, M. O. (2011). The rubber hand illusion: feeling of ownership and proprioceptive drift do not go hand in hand. PloS one, 6(6), e21659.

Schober, M. F. (1993). Spatial perspective-taking in conversation. Cognition, 47(1), 1-24.

Trope, Y., & Liberman, N. (2010). Construal-level theory of psychological distance. Psychological review, 117(2), 440.

Utz, S., & Sassenberg, K. (2002). Distributive justice in common-bond and common-identity groups. Group Processes & Intergroup Relations, 5(2), 151-162.

Vogeley, K., & Fink, G. R. (2003). Neural correlates of the first-person-perspective. Trends in cognitive sciences, 7(1), 38-42.

Won, A. S., Bailenson, J., Lee, J., & Lanier, J. (2015). Homuncular flexibility in virtual reality. Journal of Computer- Mediated Communication, 20(3), 241-259.

Reflection Papers

IMPACTS PROMOTED BY TELEHEALTH SOFTWARE IN SERVICES DEVELOPED BY HEALTH PROFESSIONALS IN BRAZIL

Isabele Magaldi A. de Freitas, Nícolas V. R. Veras, Custódio L. B. Guerra Neto,
Antônio Higor F. de Morais and Ricardo Alexsandro de M. Valentim
Federal University of Rio Grande do Norte, Brazil

ABSTRACT

The Telessaude Brasil Redes Program emerged in 2007, from a telemedicine pilot project in support of Brazil's primary care. Although, a country with continental territory dimensions such as Brazil which faces social, economic and political challenges, with countless inequalities, the program accomplishment would become difficult to implement. Even with all the encountered barriers, this program is a successful tool in Brazil. The objective of this work was to highlight the impacts obtained with the implementation of Telessaúde Brasil Redes Program on the state of Rio Grande do Norte. Researches were made in databases: SciELO, Biblioteca Virtual em Saúde da Atenção Primária à Saúde - BV-APS, Periódicos CAPES, Ministério da Saúde and Núcleo de Telessaúde of the state of Rio Grande do Norte's Database. An opinion research was also made with the health employees with the purpose of evaluating the impacts of the program in the state of Rio Grande do Norte. Results have shown that the service of teleconsulting provided a reduction on the amount of routing to other assistential levels. City budget were saved. Also, health department professionals were submitted to a permanent education system. Most of the health professionals of Rio Grande do Norte were satisfied with the Telessaúde Brasil Redes Program.

KEYWORDS

Telehealth, Telemedicine, Public Health, Single Health System, Primary Care

1. INTRODUCTION

The lawsuits that guided the movement that became known as the "Sanitary Movement" in 1986 motivated the principles that would later become part of the 1988 Constitution, which established Health as the right of everyone and the duty of the State. Starting from this premise, one of the greatest social achievements emerges, the Unified Health System (SUS). The principles led to the democratization of health actions and services, leaving behind a restricted and centralized system that is now being conducted in a universal and decentralized way (Brasil, 2000).

The SUS comes from a new vision regarding health. Before, health was understood as "the state of non-disease", which made all logic turn around the cure of health problems. This logic, which meant merely remedying the effects with less emphasis on causes, gave rise to a new notion, centered on the prevention of diseases and the promotion of health. To do so, health is related to the quality of life of the population. It now consists of the set of goods that encompasses food, labor, income, education, the environment, basic sanitation, sanitary and pharmacological surveillance, housing, leisure, etc. (Brasil, 2000).

However, a country of continental territorial dimensions such as Brazil, which faces a series of social, economic and political challenges, with innumerable inequalities, would be difficult to implement. The difficulty of managers to promote integration between states, municipalities and state assistance networks with national services has led to problems in access to services and the commitment to universality and integrality. Excluding large numbers of users from enjoying the universal right to health (Guerra Neto et al., 2015).

Parallel to the evolution of SUS, there is a rampant growth of information and communication technology, changing practices in the areas of health. Through the Internet, information is transmitted more efficiently between distant sites, breaking down the barriers of time and territorial boundaries (Melo, Silva, 2006).

In this perspective, in 2005, the National Council for Scientific and Technological Development (CNPq), through the "Millennium Institute" Program, placed telemedicine as one of the themes of its public announcement, generating the "Medical Digital Station". That same year , the Ministry of Health (DEGES / SGTES) organized a pilot project entitled Telematic and Telemedicine Project in support of primary health care in Brazil, which was developed in 2007, through Ministerial Order No. 35. Its implementation as Pilot Project was a national action that aimed to improve the quality of care in Basic Care in the Unified Health System. Later, with the development of this project, it came to be called the Programa Nacional Telesaúde Brasil Redes (Wen, 2008).

The Telehealth program addresses aspects of interactive Tele-Education (Veronesi et al., 2004; Wen, 2011), development of tools and programs for health promotion, supervision of distance treatments, and teleconsultations with specialists.

According to the Federal Medical Council, Telemedicine is defined as: the practice of Medicine through the use of interactive technologies, aiming to promote health care, education and research. Thus, this service allows the exchange of information that meets all levels of health and allows a continuing education of the professional. Although Telemedicine and Telehealth have similar concepts and applicability, these are different modalities. Telehealth is a broad program, encompassing medicine and other health areas such as Nursing and Dentistry, for example (FEDERAL COUNCIL OF MEDICINE, 2002; EL KHOURI, 2003; WEN, 2008).

As the world is progressing more and more in information and communication technologies, it is important to record the impact, experience and results achieved by the implementation of the Telehealth Network Networks Program in the Northeast. The objective of this work is to highlight the impacts obtained with the implementation of the Telehealth program in the state of Rio Grande do Norte.

2. METHODS

Research was carried out in the following databases: SciELO, Virtual Health Library of Primary Health Care - BV-APS, CAPES Periodicals and Ministry of Health. In addition to consultations carried out in the database of the Nucleus of Telehealth in the state of Rio Grande do Sul North.

The publications of interest for the present study were selected taking into account words in the title that denote the interest of studying the Telehealth Brazil Networks Program. From the events that preceded its creation, until the present moment of operation of the nucleus of Telehealth in the state of Rio Grande do Norte. A comparative study was carried out on the impact of the Telehealth Brazil Networks Program in the states of Amazonas, Bahia, Pernambuco, Minas Gerais and Rio Grande do Sul and was related to the data obtained in the Telessaúde nucleus in the state of Rio Grande do Norte.

In order to deepen the analysis of the impacts obtained with the program Telehealth Brazil Networks in the state of Rio Grande do Norte, a questionnaire was created based on studies already carried out in the area, to verify the opinion of the users enrolled in the nucleus of Telehealth of the RN on the Teleconsultoria service. From the database of the Telehealth Nucleus of the RN, the e-mails of 200 health workers who used the tool the most were selected. A research presentation e-mail was sent, inviting them to participate in the study, accompanied by a questionnaire containing five objective questions about:

- Level of schooling;
- Opinion about Teleconsultoria service;
- Satisfaction of the response received;
- Resolutivity of the tool, avoiding referral of patient to another level of attention;
- Recommendation of the service to other health workers.

3. RESULTS

Based on the bibliographical review and the results obtained on the National Telehealth Network Brazil Network Program in Rio Grande do Norte, there was an improvement in the quality of care and diagnoses performed by Primary Health Care. A significant reduction in costs was observed operational costs for municipalities, reduction of patient referrals to large urban centers and reduction of demand in centers of greater complexity. The program's resolution capability also provided a high level of user satisfaction and acceptance to Teleconsulting service.

The information acquired, resulting from this research, can be used by the managers of the RN nucleus, in order to monitor, plan and execute activities necessary to the reality of the municipalities of the state. In order to better understand and visualize the results, we used graphs and tables for the data contained in the research. For a more comprehensive analysis of the results obtained in the nucleus of Telehealth of the state of Rio Grande do Norte, it was sought during the discussions, to compare the data obtained in the RN with the data obtained by other nuclei of Telehealth of the Country.

4. DISCUSSION

Taking into account that user satisfaction indicators are indispensable for evaluating the quality of the Teleconsultoria service in the RN, the research reveals, that 90% of the professionals are satisfied with the Teleconsultoria service of the Nucleus of Telehealth of the state of Rio Grande do Norte. Only 10% of users were dissatisfied with the response received from the service, but there was no place to explain the reasons. A study by Marcolino et. al in 2013, about the experience of the Minas Gerais Teleassistance Network (RTMG) with the Telehealth Program, implemented since 2005 in this state, states that 95% of users answered the question and 94% were satisfied with the service. Considering the success of the Program in this state and comparing with the results obtained in the state of the NB, a state of lower proportion when compared to MG, where the implementation of the program occurred six years later, the results are compatible. In this context, we present a successful experience of the RN against the notoriety of RTMG. We can analyze which factors have impacted this dissatisfaction. There are two points considered relevant to user satisfaction. They are, time and the adequacy of the language in the response. Teleconsulting response time is an important parameter to determine the quality of service, considering that the greater part of the doubts should be resolved before the patient returns to the Basic Unit. Language Suitability is also a key indicator for assessing the quality of the response received. The professional teleconsultor should be aware of the level of training of the applicant. The language in the response should be clear and appropriate for that specific professional. This will contribute to a greater efficiency of the service, always providing the continuing education of the requesting health professionals. However, a new assessment is needed to determine the real cause in order to devise measures to increase the satisfaction of these users. For each one plays an important role in building a quality health system for all (Marcolino et al., 2013).

Regarding the Teleconsultoria service and referrals of patients to other levels of care, 57% of the health workers interviewed for the present study concluded that the use of the Teleconsultoria service has already allowed them to solve clinical cases, without it being necessary refer the patient to other levels of care. Representing a significant impact on the budget allocated to health in the NB, which can be quantified. An economic study carried out in 2012 by the Telehealth center of Minas Gerais, referring to the costs of moving patients to other places, involved 30 municipalities and showed that the average cost of each referral is around R$ 83.00. Considering this value and the result obtained from the research with RN health workers, the Telehealth Nucleus of Rio Grande do Norte with a total of 14,853 Teleconsultories answered and a percentage of 57% of the referrals that were avoided due to the use of the Teleconsultoria service, we have that about 0.57 x 14.853 x 83.00 = R$ 702.695,43 was saved with Telehealth Program in the state of Rio Grande do Norte. On the other hand, the cost of telehealth activity in the period under analysis was R$ 10.800. The study carried out by the Telehealth Center of Minas Gerais concludes that the return on investment in Telehealth is R $ 3.75 for each R $ 1.00 invested (Alkmim, Figueira et al., 2012). Another study by Castro Filho et al. (2012) demonstrated that in the experience of Telehealth Rio Grande do Sul, every two teleconsultories requested by medical professionals, a referral of patient to other levels of attention is avoided. Although the main objective of the Telehealth Program is to improve the quality of care in ABS, the study reveals that the system is not only economically self-sustaining, but is also capable of reducing health costs in municipalities. This fact is essentially relevant when the restrictive factor in the quality of healthcare is often the municipal budget.

We can conclude that 90% of the health workers interviewed recommend using the Teleconsulting service for other workers. This result demonstrates the security that the service provides in the conduct of clinical cases, proving the importance of the program Telehealth Brazil Networks in the state of Rio Grande do Norte for more resolutive care in municipalities farther away from the reference areas of the state. This result recognizes the positive contribution of the Teleconsultoria service, as an efficient tool in the diagnostic definition and conduction of the treatment, giving support to the Basic Attention of the state. Only 10% would not recommend the service. Based on the results obtained and presented, it was possible to obtain an overview of how the Teleconsultoria service has responded to the expectations and wishes of the registered health workers. It is important to emphasize the need for future research to attest to the feasibility of the program in all regions of the country.

ACKNOWLEDGEMENTS

We thank the Federal University of Rio Grande do Norte, the University Hospital Onofre Lopes and the Laboratory of Technological Innovation in Health for providing all the necessary tools and conditions for the execution of the project.

REFERENCES

Brasil. Ministério da Saúde. Secretaria Executiva. Sistema único de Saúde (SUS): princípios e conquistas/ Ministério da Saúde, Secretaria Executiva. - Brasília: Ministério Saúde, 2000. 44p.il.

Conselho Federal de Medicina. Resolução no 1.643, de 7 de agosto de 2002. Define e disciplina a prestação de serviços através da telemedicina. Diário Oficial da União, Poder executivo, Brasília, DF, 26 ago. 2002.

El Khouri, S. G. Telemedicina: análise de sua evolução no Brasil. 2003. 238f. Dissertação (Mestrado em Ciências) - Faculdade de Medicina, Universidade de São Paulo, São Paulo, 2003.

Figueira, Renato Minelli. Impactos Econômicos da Telessaúde. Belo Horizonte: UFMG, 2013. Color.

Guerra Neto, Custódio Leopoldino de Brito et al. Teleodontologia. In: VALENTIM, Ricardo Alexsandro de Medeiros et al (Org.). A Telessaúde no Brasil e a Inovação Tecnológica na Atenção Primária. Natal: Edufrn, 2015.

Marcolino, Milena Soriano et al. A Rede de Teleassistência de Minas Gerais e suas contribuições para atingir os princípios de universalidade, equidade e integralidade do SUS - relato de experiência - DOI: 10.3395/reciis.v7i2.775pt. Reciis, [s.l.], v. 7, n. 2, p.1-21, 28 jun. 2013. Fundação Oswaldo Cruz. http://dx.doi.org/10.3395/reciis.v7i2.775pt.

Melo, M. C. B.; Silva, E. M. S. Aspectos conceituais em Telessaúde. In: MELO, M.C.B.; SILVA, E. M. S. Telessaúde: um instrumento de suporte assistencial e educação permanente. 1.ed. Minas Gerais: Belo Horizonte, 2006. P.17-31.

Veronesi, M. C. et al. Cybertutor: educação mediada por tecnologia na odontologia. Rev ABENO, Londrina, v. 4, n. 1, p. 88, 2004.

Wen, C.L. Telemedicina e Telessaúde. - Um panorama no Brasil. Informática Pública, v.10, n.1, p.7-15, 2008.

Wen, C.L. Telemedicina e a Telessaúde: uma abordagem sob a visão de estratégia de saúde apoiada por tecnologia. 2011. Disponível em: <http://avancasaudebrasil.org.br/chaowen/artigos/telemedicina.aspx>. Acesso em: 28 fev. 2018.

THE ACTIVITY RECOGNITION: ADVANTAGEOUS BUT CHALLENGING!

Chaima Bouali[1], Abderrahim Tahiri[1] and Olivier Habert[2]
[1]*National School of Applied Sciences of Tetouan (ENSATe), Tetouan, Morocco*
[2]*Université de lorraine, LCOMS, Metz, France*

ABSTRACT

The Activity Recognition (AR) has caught the attention of a number of researchers in the last years. This topic has been highlighted due to its usability in different applications such as healthcare, security systems, etc. The three main categories of the approaches depicted in different papers are: Supervised, unsupervised and semi-supervised methods. However, several challenges are faced while developing a trustworthy system for the identification of activities. In this document, a presentation of different approaches related to the AR field will take place and also a discussion about the different issues that lead to an accuracy reduction of the methods developed in the past 10 years. The aim of our study has been the extraction of the lacks and the pros from each method in order to build a reliable AR system that takes into consideration the different challenges.

KEYWORDS

Activity Recognition, Smart Homes, Healthcare

1. INTRODUCTION

The purpose of a smart home for assisted living is to enhance particularly the quality of life of a resident in his daily functioning. The user might either be an old person with specific cognitive impairment or a person affected by age related changes in functional abilities.

Humans in general, elderly people in particular, trust and accept technologies that will fit with their needs and will serve them properly in the different areas of their daily living. A key aspect of assisted technologies consists in the high performance of Activity Recognition methods. The main use of the Activity Recognition is to deal with abnormal behaviors of the user (Forkan et al., 2015). It opens the way to allowing tracking different activities and predicting safe and unsafe behaviors and beside that, to supervise changes in daily activities and permit with that a medical monitoring.

The aim of this work is to state the numerous approaches related to the human Activity Recognition that have been proposed in the last 10 years and classify them according to their accuracy and also to discuss the important role of the recognition in helping individuals perform tasks more safely at a smart home.

Even if two people have as purpose preparing coffee for example, they behave differently from each other. Thus, an efficient Activity Recognition method is the one that investigate all the ways to do one activity. Evaluating the high performance is strongly influenced by many factors such as: the accuracy of the operation, the respect of privacy and last but not least the security and the safety concern.

2. THE STATE-OF-THE-ART

Smart homes are no longer needed just for comfort and luxury life but beyond that, these technologies respond nowadays to various vital human needs in various ranges. Healthcare, security and last but not least energy management are the fields that smart residences focus on (Bouchard, 2014; Wong et al., 2017). Not only these technologies are useful in different areas but also for different age ranges. In other terms, smart homes can be adapted to elderly people daily living too. Assisting them, helping them in daily tasks and tracking their activities in order to prevent injuries and harmful behaviors are the major uses of a smart home for assisted living (Zolfaghari and Keyvanpour, 2016).

Accurate recognition processes improve the efficiency of the smart home applications (Bourobou and Yoo, 2015; Gillani Fahad et al., 2015). One of the challenges resides in building a system that learn, understand and automatically establish the resident's ADL. These characteristics provide an extreme reliability of the identification system. Among the issues, we also have the identification of an action in real time. Many works have been dedicated to this aspect (Wang et al., 2012) in order to deal with ADL instantly and reveal unwanted behaviors for example (Abidine et al., 2015).

Considerable developments in the field of Activity Recognition have seen the light. The classification of the activity recognition methods into different categories will be addressed in this section with primary focus on the three major types of AR methods adopted: Supervised, unsupervised and semi-supervised approaches.

Among the researches done in this area, the paper (Chen and Nugent, 2009) focused on a method based on the use of ontological modeling. The ontology-based modeling supposes that the activity classes in the Activities Of Daily Living (ADL) ontologies are organized in a hierarchical tree with sub-classes inheriting all properties from their super-classes. The most detailed description of the activity is found on the closest point to the leaf of the class tree. The temporal dimension in this case, play an important role in the process of recognition. Along a timeline, interesting amount of information related to the activity is collected which can be useful to identify efficiently the ongoing ADL. The results generated by this method might be very promising if the objects sensors are well used and each step of the activity steps can be detected and identified.

In addition to a hierarchical scheme, (Zheng, 2015) suggests accelerometers sensor-based approach for human activity recognition in order to generate more specific data about posture, location and human interaction. In this paper, using a body-fixed accelerator kind of sensors was devoted to the measurement of the user motion. Distinguishing all of running, walking, (forward, left, right, upstairs and downstairs) jumping, sleeping and sitting activities is the highlight of the method compared to other propositions, also based on accelerometers, that detect only a few of the ADL listed above. In recent works, using accelerometers is the most widespread due to their highly accurate results. And while talking about the on-body wearable sensors, the comfort of the user should be taken into account. That said, the user have to feel comfortable wearing the body-sensors and shouldn't be annoyed. Then preferably, the sensors need to be light weighted and well placed on the body. These details influence the results of the operation drastically.

All the methodologies discussed anteriorly are initially classified as supervised approaches. Along the years, the results of this type of approaches were promising for the recognition of both "atomic" activities (sitting, standing, walking…) and more dynamic or composed activities and the supervised methods were the most widely used. However, the downsides to these approaches are that they are time-consuming, they require a large volume of labeled training data and they mostly rely on invasive sensors.

To avoid the shortcomings of the supervised methods, (Trabelsi et al., 2013) suggests an unsupervised approach using the Expectation-Maximization (EM) algorithm where no activity labels are needed. The proposed approach based on a Hidden Markov Model (Cilla et al., 2009) in a multiple regression context reveals competitive results to the ones got by the approaches that perform in a supervised framework.

The combination of the advantages of both the supervised learning and the unsupervised one is the major goal of the semi-supervised learning. (Stikic et al., 2008) addresses the semi-supervised approach as an alternative that requires far less labeled training data than traditional supervised techniques. The work attempts to show that is feasible to generate relevant results with a modest amount of labeled training data (Yao et al., 2016).

Table 1. Classification of AR Approaches

Supervised approach		
Papers	**Pros**	**Cons**
Chen and Nugent Zheng Botros and Heskes	- Promising for both atomic activities and more dynamic activities	- Large volume of labeled data - Time consuming
Unsupervised approach		
Papers	**Pros**	**Cons**
Trabelsi et al. Bouchard	- No activity labels are needed	- Insufficient ground-truth data
Semi-supervised approach		
Papers	**Pros**	
Stikic et al. Yao et al.	- Modest amount of labeled data - A combination of the advantage of both labeled and unlabeled data	

Beside the type of the approach adopted, the key elements for a significant performance of Activity identification in a smart home are the sensors. The choice of the pertinent sensors and their placement plays an important role as much as a good combination of different sensors. And ideally, avoid cameras in homes because the user is less likely to accept invasive kind of sensors for privacy reasons. In all the papers discussed above, many challenges were outlined and many solutions were proposed. Despite the implication of many researchers in the active problematic of Activity Recognition, this topic still in its infancy due to the multiple challenges involved and still require more efforts.

3. THE INTEREST OF THE TOPIC OF ACTIVITY RECOGNITION

There are enormous advantages of the Activity Recognition in a variety of applications and specifically inside a smart home for aging in place. Being able to measure the motion of the resident and to monitor his behavior and each step he takes while performing his daily activities may significantly help providing an efficient assistance and an early intervention (Wenlong Tang and Sazonov, 2014). The efficiency of the operation depends mainly on how the rate of correct predictions is, relying to our system. Interestingly, a good understanding of the user's routine, his surroundings and his needs is a relevant factor for the reliability of the system. Theoretically, the development of a similar system may seem easy and evident which is not true. Many details interfere when studying the environment of living of an unknown. For example, the issue cited above about the fact that even if many persons have the same purpose while doing a sequence of activities, they may do it differently and each person his way for accomplishing it. Moreover, one specific user may have different ways to accomplish an Activity of Daily Living (ADL) each day. Besides, distinguishing activities that are highly similar (Botros and Heskes, 2017) might be difficult. Example of the two activities sprinting and jogging that we can only differentiate by monitoring the heart rate (Sprinting is more intense). And also two other activities that we can hardly distinguish are preparing tea and preparing spaghetti, both of them begin with boiling water; the only difference is the quantity of water needed for each one. In addition to these challenges comes the one related to the issue of handling the transitions between actions. However, taking into account the different challenges is crucial in order to develop an outperforming system.

The aim of our work will tend to present a solution that gives insights into the daily activities of the user with a high accuracy and in remarkable circumstances. We will focus on the Activity Recognition in smart homes for assisted living. Our users will be mainly elderly people who will benefit from the system for the purpose of being assisted in their everyday life. By assistance we mean that the system will be able to help the user accomplishing some daily tasks and mostly to detect abnormalities in the behavior of the resident by comparing the data collected from the current behaviors with the one collected beforehand. Adequate comparisons will prevent home accidents and will present a health estate of the resident which is important in the "home assisted staying" of the aging populace.

4. CONCLUSION

Activity recognition is an important aspect in different applications. In this work, we initially reviewed the existed developments in this topic and focused on the "smart homes for assisted living" area. The Activity Recognition is a salient answer for many issues such as tracking the daily behavior of elderly people inside their homes. Elderly people who are generally affected by impacts of old age and need support in different activities of everyday routine. Despite the large amount of researches realized in this field, the problematic of the Activity Recognition still remain not solved properly and the systems developed are not accurate and complete enough. In our work, the focus will be on all the issues discussed in this paper in order to develop a system that offer numerous services to its user. We will try to explore the different angles of Activity Recognition and use every type of data that can be collected without neglecting any part of it. Satisfying our user needs in convenient circumstances will be our number one challenge.

REFERENCES

Abdur Rahim Mohammad Forkan, I. K. Z. T. S. F., 2014. A context-aware approach for long-term behavioural change detection. *Pattern Recognition,* Volume 48, pp. 628-641.

Augusto JC, M. V., 2007. *Enhanced Healthcare Provision Through Assisted Decision-Making in a Smart Home Environment,* pp. 1-6.

Botros, M., 2017. *Supervised Learning in Human Activity Recognition based on Multimodal Body Sensing*, pp. 1-28.

Bouchard, K., 2014. *Unsupervised Spatial Data Mining For Human Activity.* Atlantic City, USA.

D. Trabelsi, S. M. F. C. L. O. a. Y. A., 2013. An Unsupervised Approach for Automatic Activity Recognition based on Hidden Markov Model Regression. Volume 10, pp. 829-835.

Johnny KwokWai Wong, J. L. M. S. L. B., 2017. Technical requirements of age-friendly smart home technologies in high-rise residential buildings: A system intelligence analytical approach. *ELSEVIER,* Volume 73, pp. 12-19.

Labiba Gillani Fahad, A. M. R., 2015. Activity recognition in smart homes with self verification of assignments. *Neurocomputing,* Volume 149, pp. 1286-1298.

Larson, E. M. T. S. S. I. K., 2004. *Activity Recognition in the Home Using Simple and Ubiquitous Sensors.* Berlin, Germany, pp. 158-175.

Liang Wang, T. G. X. T. J. L., 2012. A hierarchical approach to real-time activity recognition in body. *Pervasive and Mobile Computing,* Volume 8, pp. 115-130.

Liming Chen, C. N., 2009. Ontology-based Activity Recognition in Intelligent Pervasive Environments. Volume 5, pp. 410-430.

Lina Yao, F. N. Q. Z. S. T. G. X. L. a. S. W., 2016. *Learning from less for better: semi-supervised activity recognition via shared structure discovery.* Heidelberg, Germany, pp. 13-24.

Maja Stikic, K. V. L. B. S., 2008. *Exploring Semi-Supervised and Active Learning for Activity Recognition.* Pittsburgh, USA, pp. 81-88.

M'Hamed Bilal Abidine, L. F. B. F., 2015. Improving Human Activity Recognition in Smart Homes. *International Journal of E-Health and Medical Communications (IJEHMC),* pp. 19-37.

Rodrigo Cilla, M. A. P. J. u. G. A. B. a. J. M. M., 2009. Recognizing Human Activities from Sensors Using Hidden Markov Models Constructed by Feature Selection Techniques. *Algorithms,* pp. 282-300.

Samaneh Zolfaghari, M. R. K., 2016. *Smart Activity Recognition Framework in Ambient Assisted Living.* Gdansk, poland, pp. 1435-1443.

Serge Thomas Mickala Bourobou, Y. Y., 2015. User Activity Recognition in Smart homes using Pattern. *Sensors,* Volume 15, pp. 11953-11971.

Thi V. Duong†, H. H. B. D. Q. P. S. V., 2005. Activity Recognition and Abnormality Detection with the Switching Hidden Semi-Markov Model. *Computer Vision and Pattern Recognition,* pp. 838-845.

Thomas Kleinberger, M. B. E. R. A. H. P. M., 2007. *Ambient Intelligence in Assisted Living: Enable Elderly People to Handle Future Interfaces*, pp. 103-112.

Wenlong Tang, E. S. S., 2014. Highly accurate recognition of human postures and activities through classification with rejection. Volume 18, pp. 309-315.

Yuhuang, Z., 2015. Human Activity Recognition Based on the Hierarchical Feature. *Journal of Electrical and Computer Engineering,* Volume 2015, pp. 1-9.

ALGORITHMS ARE THE NEW DRUGS? REFLECTIONS FOR A CULTURE OF IMPACT ASSESSMENT AND VIGILANCE

Hugh Harvey[1] and Federico Cabitza[2]
[1]*Kheiron Medical Technologies, 40 islington High street, London, N1 8EQ, UK*
[2]*Università degli Studi di Milano-Bicocca, IRCCS Istituto Ortopedico Galeazzi, Viale Sarca 336, 20126 Milano, Italy*

ABSTRACT

In this paper we propose that Artificially Intelligent (AI) systems in medicine are *the new drugs*. Through study and comparison of the existing corporate pharmaceutical models, we advocate a new approach in how the performance of AI solutions (both for diagnostic and prognostic aims) is measured and promoted, on how the impact of these solutions is assessed and evaluated in terms of both outcome and safety (about which the new term *technovigilance* has been introduced), and how education and a better awareness of the impact of these systems on heath care by multiple stakeholders should be promoted. Any drug must be evaluated in terms of efficacy and safety, but also in terms of real-world effectiveness, efficiency and cost-effectiveness, that is in terms of its role as a component of a wider and more complex socio-technical system where it must fit a broader process of inquiry and care, which involves patients, care givers and many other stakeholders. For this reason, we also propose, as a priority, changes in the current wide-spread attitudes that evaluate machine performance *against* humans, and instead to adopt a *collective knowledge* approach. In this light, the leading metaphor informing the public debate on the role of AI in medicine should not be that of *cognitive augmentation*, with its risks of feeding unrealistic expectations and restoring the harmful image of the medical hero, but rather of a collaboration-oriented *adjunct* in more redundant and tighter teams of competent decision makers. Not only could this lead to a more efficient and automated health care, but we argue to a better collective endeavour aimed at the well-being of patients as well as at the satisfaction of physicians.

KEYWORDS

Artificial Intelligence, Decision Support Systems, Health Technology Assessment

1. INTRODUCTION

In their book, Hofstadter and Sander (2013) wondered if analogies were like seductive and dangerous siren songs, likely to lead us astray or more like indispensable searchlight, without which we would be plunged in total darkness. In this reflection paper, we propose the creative analogy of AI algorithms as the new drugs. This analogy allows for consideration of some important points regarding the fair and safe use of AI in medical settings and to promote awareness of its potential drawbacks and "side effects".

We discuss four elements by which the analogy could be considered. First, AI must be adopted not in virtue of its potential for change or disruption, nor for its promise to make care processes more efficient or to allow for greater workloads and throughput considering any doctor shortage and increase in demand for an ageing population (as well as increase in the health-related expectations by a wider population of patients). Like drugs, AI adoption in health care must be motivated and find justification in its capability to deliver - or contribute to delivering - better clinical outcomes, including the significant reduction of medical errors (both diagnostic and prognostic ones), the improvement of treatment appropriateness (by physicians) and adherence (by patients) and the prevention of iatrogenic adverse events (Wachter, 2015). This has an impact on three main areas of concern: Promotion, in terms of Advertising and Marketing, which covers *why* AI should be adopted; Assessment and Technovigilance, which covers *how* it should be evaluated; and Education and training, which covers *who* should use it and *when*.

2. ARGUING FOR THE DRUG METAPHOR

Today, we have a heavily regulated, fiercely competitive (and frankly ridiculously profitable) global pharmaceutical industry. Last year, the top 10 Big Pharma companies combined total revenue reached over $400 billion. Now, we are at the cusp of an entirely new billion dollar industry; that of 'medical algorithmia', poised to raise medicine to even greater height. To understand where the future lies for the AI and algorithm industry (aka "Big Tech", "Big Algorithmia", "Digital Health"), one needs only to look at how 'Pharma' got quite so 'Big' in the first place, including how new drugs are developed and reach the market, how medical practice is structured around the safe delivery and monitoring of drugs, and how doctors learn to understand drug mechanisms of action and side effects.

2.1 Advertising and Marketing

In the UK, the Association of the British Pharmaceutical Industry (ABPI)[1], and in the US the Pharmaceutical Research and Manufacturers of America (PHRMA) have codes of practice that all Pharma companies must adhere to when promoting their products and interacting with health professionals. This is known in common parlance as 'Medical Affairs'. This provides a robust ethical framework, backed up by penalty fines, citing codes detailing how companies should behave when promoting their drugs to both clinicians and patients. The aim is to prevent misinformation, exaggeration and statistical slight-of-hand.

However, critics still exist. Dr Ben Goldacre, a staunch antagonist of the morally bankrupt world he calls Bad Pharma, wrote that pharma companies, despite the ethics codes, are still getting away with "...poorly designed trials, on hopelessly small numbers of weird, unrepresentative patients, and analysed using techniques that are flawed by design, in such a way that they exaggerate the benefits..." (Goldacre, 2012). Other valid criticisms include a lack of pressure on pharma companies to publish negative trial results.

Clearly, the same can be said for algorithmic developers. In our experience, most of the published deep learning papers focus only on positive results, and rarely discuss failure or algorithms that did not perform satisfactorily. In fact, due to the non-peer reviewed nature of pre-publications, some of those with 'positive' results present methodological flaws that undermine the validity or reproducibility of their results and claims. Another concern regards the dissemination pattern for which every paper on AI is swiftly followed by an exaggerated media press release stating that "Algorithm X outperforms doctors at Y!!".

It's not just the press to blame. For example, one only has to wander the stalls of the Radiological Society of North America (RSNA) congress to get an idea of how over-exaggerated some companies' claims are in the radiology space. We must all remember that the landscape and framework for robust clinical investigation and subsequent marketing of algorithms is still in its infancy, and it is this same landscape that will see massive accelerated growth as both regulators and developers find their way. Until then, we will just have to wade through the hype to find the practical truth.

2.2 Algorithm Safety & 'Technovigilance'

AI implementation at scale cannot be achieved without solid foundations of clinical research and ongoing investigational trials. The sequelae of algorithmic implementation—phase IV or post-market surveillance in the big pharma life-cycle, describe the ongoing monitoring of algorithms once they are released into the wild inside of clinical workflows.

For example, in every hospital in the world there is a pharmacist whose job it is to oversee medicines safety. Whether they focus on prescribing errors, discharge communication, aseptic preparations, or dispensing mistakes, their task is to ensure that the harms and side effects of pharmaceuticals are minimized in clinical practice. It is a hugely important part of overall patient safety, often undertaken out of the spotlight of day-to-day clinical care. The same goes for physical medical devices in a hospital – there is always someone employed to ensure electrical compliance of kit, quality assurance and maintenance.

---

[1] http://www.pmcpa.org.uk/thecode/Pages/default.aspx

Algorithmic safety will require the same level of oversight. Not only do algorithms have the potential to do unintended harm (no algorithm can ever be perfect), but they require rigorous post-market surveillance as per regulatory requirements. The international standards for medical devices (including ISO 13485) explicitly state that developers should have in place a robust system for monitoring real-world device performance. This includes regular audit of algorithmic outputs, and a feedback mechanism to ensure that errors are acted upon.

Technovigilance is our new proposed term – inspired by the equivalent in big pharma known as 'pharmacovigilance'. Rather like the yellow card system in the UK (Avery et al. 2011), and the Center for Drug Evaluation and Research system in the US for mandatory reporting of medicine safety, technovigilance is designed to ensure that companies and end users report all new or unexpected harms to a central governing body. For example, if an algorithm that detects onset of Atrial Fibrillation (AF) in a patient fails to trigger and a patient comes to harm, that event must be reported both to the developer and the relevant overseeing safety body. A more difficult, but likewise due task regards the assessment of the extent an inaccurate AI output would trigger overdiagnosis (and overtreatment), that is unnecessary exams and procedures on patients due to false positives. The unnecessary inconveniences and additional costs that could derive from this less-than-optimal algorithmic behaviour (and related human practices) should be factored in in any serious post-marketing assessment of a technology.

As a positive remark, we observer how an entire industry arising around the concept of safety and technovigilance is already emerging. Regulatory third-party companies already offer technovigilance consultancy and help in setting up regulatory compliance processes for medical devices. This industry will extend to offer services for the monitoring of algorithmic safety and, hopefully, appropriateness assessment. Not only will this industry ensure ongoing safety and quality monitoring, but it will also link in with the vitally important Phase IV clinical studies, where algorithmic performance is clinically evaluated in a live setting to assess safety performance.

2.4 Medical Education & Allied Health Specialists

Medical education will also have to adapt to the new digital future. Just as medical students today must learn about pharmacological mechanisms of action, half-lives, bio-effectiveness and chemical cascades, students of the future will need to understand statistical bias, artificial neural network function, data structure and interpretation of algorithmic outputs. Therefore, we propose medical school training will have to adapt and start including basic data science teaching, and a stronger focus on statistical understanding and cognitive ergonomics (Lawler et al., 2011). Undergraduate education must go beyond the basic Chi Squared and T-test studies that medical students these days are briefly introduced too to get a deeper comprehension of the quality level (Cabitza and Batini, 2016) and uncertainty (Cabitza et al., 2017) engrained in all medical data and their impact on the predictions of any model grounding on those data.

In addition to learning classes of drugs and their side effects, doctors will need to know about different classes of algorithms, their indications and limitations, and how to interpret their outputs in context. Just as for drug safety, *doctors need to be aware of algorithmic safety.* For instance, a drug has a known efficacy, an intended target population, recommended dosage and monitoring requirements. A doctor will know most of these facts about a drug before prescribing. Algorithms are similar, in that they have a known accuracy (evaluated in multiple ways, like error rates, F scores, AUROC), an intended target population (whose samples were extracted for their training and optimization), a recommended usage (e.g, as a second opinion only) and require monitoring. *We argue that it stands to reason that a doctor should be as equally educated about the limitations of an algorithm as they are about the side effects of drugs.* On a JAMA Viewpoint, we began speaking of these side effects in terms of unintended consequences of AI (Cabitza eta l. 2017): we began discussing those consequences that could have the deepest long-term impact and that call for focused studies and a wider debate within both the medical and the computer science communities, like the progressive deskilling of physicians and their over-reliance on the AI output, which could in turn lead to phenomena known as context desensitization (Cabitza et al. 2017b), automation complacency and automation bias (Goddard et al. 2011).

2.4.1 Algorithmists

It's not just doctors of course who will be the end users of algorithms in clinical practice. The corridors of medicine are full of non-physician specialties, one for almost each aspect of medicine. There are sonographers for ultrasound, theatre nurses for surgery, ward clerks for administration, porters for transport and pharmacists for drugs. Therefore, we predict an entire new allied health specialty will arise, focused purely on algorithmia, that we consider 'algorithmists'.

As specialists in clinical algorithmic functionality, algorithmists will check that hospitals are using the correct algorithms in the correct situations, help with procurement processes to choose good algorithms, provide advice on which technologies to use for which use cases, oversee algorithmic safety within hospitals, and manage 'technovigilance' reporting. The skill set needed to become an algorithmist will be very niche – with both a basic clinical grounding and a background in data science. Some speak of this new role as the evolution of the current professions of radiologist and the pathologist, and evoke the idea of an "information specialist" (Jha and Topol, 2016). Others may consider the need for a new separate specialty as unnecessary, and rather envision new professional roles like the AI specialist or data manager from other ancillary professions, like medical scribes and radiologic technologists. We argue that even if we educate doctors effectively, they do not have the time or know-how to become true full-time algorithmists. In fact, it would be a waste of clinical training to take doctors away from front-line medicine. Whilst some clinical academics are well equipped with the necessary skills and may lead departments of algorithmists, in the daily clinical ebb and flow of a hospital there will be huge demand for such expertise, and much of the work will require specially trained and dedicated allied health staff. It is also far more cost effective to have a specialist workforce, rather than spend money on doctors taking time away from patients.

3. BEYOND THE DRUG METAPHOR

Every metaphor is useful for its role in bridging different semantic fields together and let us better comprehend hidden relationships or aspects that in one field would be hard to unearth. However, every metaphor is also wrong to some extent, for right in virtue of its inaccuracy it can evoke new meanings and ideas. While the previous section should have shed light on the affinities between AI and drugs, especially in relation to effectiveness assessment and vigilance on potential risks (and harms), before concluding this reflection paper we have also to acknowledge at least one limitation about taking this metaphor too literally.

In fact, it is easy to consider technology as an object, even in case of technological supports that are not physically tangible, as it is the case of software applications or chatbots. However, such an objectivistic interpretation of AI would be simplistic at best and probably misleading. Rather any technology, which possibly includes physical tools, or an equipment with which people would say to interact with as we do with objects and parts of the environment, cannot be decoupled by its use, that is by the situated practices (Suchman, 1987) and organizational processes in which that technology is put to use to create value: every technology is always a *technology-in-practice* (Orlikowski, 2008) and its impact must be considered also in terms of how knowledge circulation and power relationships may change after its full and wide-spread adoption. In the same vein, drugs are not effective per se, but in virtue of a right diagnosis (which detected an illness for which a specific drug is considered effective), a timely prescription, and last but not the least, the adherence of the patient in taking them when and how prescribed so. Out of metaphor, AI – as any technology – could not be proven effective without considering the involvement of its users, their attitudes, expectations, and preferences. This is why we deem a waste of resource to compare the performance of predictive models when this is measured in simulated environments. We rather advocate the comparison of the diagnostic performance or the outcomes in terms of primary endpoints (adjusted by case mix, or at least by age and severity of initial condition) of a team of medical doctors who are supported by the AI by integrating this kind of technological aid into their routine work processes, and of a team of comparable expertise whose members have not adopted such a tool. We denote this comparative attitude as the transition from "in labo" research to the "in labore" one (i.e., from studies conducted within a controlled experimental environment to ecological studies conducted in the field of work), which requires a more socio-technical approach in designing AI-related studies a particular care on how AI should not just be considered an equipment to augment our sensorium and cognitive capabilities (Obermeyer, 2017), but rather an *adjunct* member of a close *knowledge community* (Sloman and Fernbach, 2018, p. 127) encompassing competent experts and decision makers, as well as their tools.

4. CONCLUSION

Drugs don't deliver safe and effective healthcare, people and systems do. The same goes for AI and algorithms. By exploring the Big Pharma model, we have proposed and predicted the trends that will take place in and around the AI industry in healthcare. Being prepared to adapt and adopt to these movements will be crucial for sustained growth and the appropriate use of advanced predictive technology in health care. From medical affairs and marketing guidelines, to safety and technovigilance, to education and training a specialist workforce of algorithmists, AI developers can take part in and benefit from a whole ecosystem of supporting industries. If they engage with these then, ultimately, it is patients who will benefit the most from the promise of Big Algorithmia and minimize the impact of some of the probably unavoidable unintended consequences that these technologies will bring to human work and health care as we know it.

REFERENCES

Avery, A.J., Anderson, C., Bond, C.M., Fortnum, H., Gifford, A., Hannaford, P.C., et al., 2011. Evaluation of patient reporting of adverse drug reactions to the UK 'Yellow Card Scheme': literature review, descriptive and qualitative analyses, and questionnaire surveys. Health Technology Assessment.

Cabitza, F. and Batini, C., 2016. Information quality in healthcare. In Data and Information Quality (pp. 403-419). Springer, Cham.

Cabitza, F., Alderighi, C., Rasoini, R. and Gensini, G.F., 2017. "Handle with care": about the unintended consequences of oracular artificial intelligence systems in medicine. *Recenti progressi in medicina*, 108(10), pp.397-401.

Cabitza, F., Ciucci, D., Rasoini, R. 2018. A giant with feet of clay: on the validity of the data that feed machine learning in medicin in: Cabitza, F., Magni, M., Batini, C. (eds.) *Organizing for the Digital World, Lecture Notes in Information Systems and Organisation*, vol. 28, chap. 2, pp. 113–128. Springer, ISBN: 978-3-319-90503-7

Cabitza, F., Rasoini, R. and Gensini, G.F., 2017. Unintended consequences of machine learning in medicine. *JAMA*, 318(6), pp.517-518.

Goddard, K., Roudsari, A. and Wyatt, J.C., 2011. Automation bias: a systematic review of frequency, effect mediators, and mitigators. Journal of the American Medical Informatics Association, 19(1), pp.121-127.

Goldacre, B., 2012. The drugs don't work: a modern medical scandal. *The Guardian*, 20.

Hofstadter, D. and Sander, E., 2013. *Surfaces and essences: Analogy as the fuel and fire of thinking*. Basic Books.

Jha, S. and Topol, E.J., 2016. Adapting to artificial intelligence: radiologists and pathologists as information specialists. *JAMA*, 316(22), pp.2353-2354.

Lawler, E.K., Hedge, A. and Pavlovic-Veselinovic, S., 2011. Cognitive ergonomics, socio-technical systems, and the impact of healthcare information technologies. International Journal of Industrial Ergonomics, 41(4), pp.336-344.

Obermeyer, Z. and Lee, T.H., 2017. Lost in Thought—The Limits of the Human Mind and the Future of Medicine. *New England Journal of Medicine*, 377(13), pp.1209-1211.

Orlikowski, W.J., 2008. Using technology and constituting structures: A practice lens for studying technology in organizations. In *Resources, co-evolution and artifacts* (pp. 255-305). Springer, London.

Sackett, D.L., Rosenberg, W.M., Gray, J.M., Haynes, R.B. and Richardson, W.S., 1996. Evidence based medicine: what it is and what it isn't. *BMJ*; 312:71

Sloman, S. and Fernbach, P., 2018. *The knowledge illusion: Why we never think alone*. Penguin.

Suchman, L.A., 1987. *Plans and situated actions: The problem of human-machine communication*. Cambridge University Press.

Wachter, R., 2015. The digital doctor. Hope, Hype and at the Dawn of Medicines Computer Age, p.2015.

Posters

MEDICAL RECORDS DIGITAL PERSPECTIVE COLOMBIAN: SAFETY, QUALITY, AND MANAGEMENT OF DATA

Juan David Prieto Rodríguez and Violeta Suárez Hurtado
Corporación Universitaria Minuto de Dios, Colombia

ABSTRACT

The patient's medical record is the only valid document that demonstrates the type of care a patient has received, therefore it must faithfully reflect the quality of care provided by health professionals, as well as basic personal information but vitally important. In Colombia there is a great difficulty that affects the quality of care process, this difficulty lies in the fact that each EPS generates to clinical history to its members, but does not share it with their counterparts when a user decide to migrate to another EPS. This document presents the advantages of creating a national system that contains in digital format to unique medical record per user, the existing regulations that seek to protect the privacy of personal data, as well as the practices that ensure the accuracy of the information provided in each medical and/or legal procedure.

KEYWORDS

History Clinical Digital, Data Quality, Data Management Data, eHealth, Health Information System

1. INTRODUCTION

In Colombia, the medical records contain data of different degrees of confidentiality, obtained from various sources such as: the user, when they (users) makes the affiliation, during the process of rendering health services or other processes; Other sources are added, legal and/or medical procedures, data update, among others. SIVIGILA (Public Health Surveillance System) is responsible for the flow of this information, the aforementioned surveillance system has Primary Data Generating Units (UPGD), responsible for the initial collection of data and its transfer to the Notifying Units, which act at the municipal, departmental or district level, as appropriate; Subsequently, these Notifying Units send this information to the National Notification Unit for consolidation. The Ministry of Health and Social Protection, in compliance with international agreements on public health surveillance, reports surveillance data required by international organizations, such as the World Health Organization (OMS) and the Pan American Health Organization. Health (OPS).

Personal information collected from patients, through various sources may include, but are not limited to: name; sex; religious worship; place and date of birth; gender; physical address; of e-mail; telephone numbers: cellular and landline; employer, your location and contact information; clinical information of the patient including legal background, results of examinations such as pathological and clinical, consultations, formulations, diagnoses, general and special medical and nursing care, consultations among specialists, among others; family contacts or legal representatives; education degree; occupation; information necessary to facilitate the provision of services, including family or work information, ID number, passport or TIN, nationality and country of residence, insurance or health service provider, use of products and services; RH; personal information provided to representatives of the health services provider entity. Data that are sensitive in and demand a confidential treatment of great importance, because this information can be used by criminals who seek to obtain their own benefit, for example through extortion, kidnapping and sale of data. Another point of view that demands an important challenge for health entities in Colombia, is to determine the flow of information from its obtaining to its processing, in health, the veracity of the data determines the generation of public policies and health, states of emergency and in general decisions that directly affect the communities. This article describes the proposal that is being implemented by the Colombian government regarding the obtaining, management and processing of the data stored in the digital medical records, since currently each healthcare provider manages its own digital medical record system, preventing the centralization of data related to the health of Colombians, which causes inefficiency in the operation and provision of health services and decision making of social impact.

2. HEALTH INFORMATION SYSTEMS IN COLOMBIA

SIVIGILA is based on five basic principles that guide the actions taken to improve the quality of health services as expressed in the document SIVIGILA Generality (Gobierno Digital de Colombia, 2016): Efficacy: Ability to achieve results; Unit: Functional integration of the different levels of the health sector; Efficiency: Rational use of resources to guarantee obtaining the best results in terms of public health surveillance; Forecast: Ability to identify and characterize in advance, the possible risk conditions for the health of the population; Quality: Reliability guarantee; opportunity and reliability of the information.

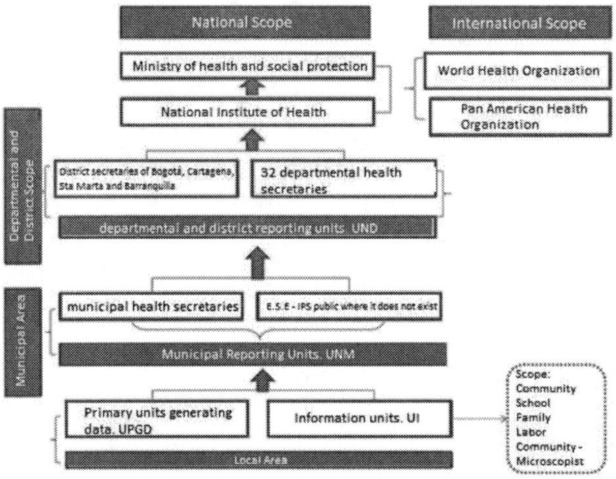

Figure 1. Data Flow Health System Colombia. Reprinted with permission, public information (Instituto Nacional de Salud, 2014)

Currently, Colombia has an inefficient information system in the health sector that does not allow obtaining updated data in an efficient and easy way, a problem that results from the lack of interoperability between existing EPS systems (health promoting entities). Although there is a legal framework, such as Decree No. 3518 of 2006 "By which the Public Health Surveillance System is created and regulated and other provisions are issued" (MINISTERIO DE LA PROTECCION SOCIAL, 2006), these information systems have not been managed to develop as expected, which has caused the information to remain segmented and there are problems related to the quality of the data, it is understood as "the quality of a set of information collected in a database or an information system which includes among its attributes the accuracy, completeness, integrity, updating, coherence, relevance, accessibility and reliability necessary for the effectiveness of operational processes. " (PowerData, 2017) (Congreso de Colombia, 2011) This is evidenced in a specific way in the management of medical records by the administrators and providers of health services, which face obstacles that hinder the process of quality care, which translates into the reduction of users' well-being, said difficulty is that each EPS creates a clinical history for its members, but does not share it with their counterparts when a user decides to migrate to another EPS, which must generate a new clinical history, which entails to losses of valuable information for people.These losses of information could cause setbacks when making emergency interventions in which patients cannot provide information about existing health conditions, allergies or other conditions, which can lead to improper practices by those responsible for the care. Exposed this scenario, the Colombian government, saw the need for the creation of a unique digital medical record that is shared by all health providers, taking into account parameters such as: accessibility, safety, integrity and confidentiality of the fact.

On January 19, 2011, the Colombian state took an important step to make the creation of digital medical records a reality through Law 1438 (Congreso de Colombia, 2011), "By means of which the General System of Social Security in Health is reformed and other provisions are issued ", Article 112. Articulation of the Information System, proposes that through the Integral Information System of Social Protection (SISPRO), the management and administration of the information of each one of the affiliates will be carried out, the of data containing such information will be articulated with the databases of the National Registry of Civil Status, the Ministry of Finance and Public Credit, the Directorate of Customs and National Taxes, the Identification System of Beneficiaries of Social Programs (SISBEN) and the Health Promoting Entities. The security of each of the user's data will be promoted through verification and consultation in the database of the National Registry of Civil Status. The digital medical records present a great challenge in terms of data security, since a large amount of such data is sensitive and private, that is why Colombian legislation has established laws that promote safety and security. Disclosure of information. The treatment of health data requires considering the availability and confidentiality of information, which are the concepts that define the main criteria of security and data protection: that the information is available when necessary, but is only available to those who are authorized for this, when authorized and for the uses that are authorized. It is also necessary to guarantee that this information has been kept intact and that whoever has accessed it cannot deny it, since this information is also part of the information.

In many countries of the world, the importance of digitizing medical records has gained great strength, and the idea of its implementation has not been limited to the borders of countries that seek to make them a reality. Due to human trafficking among the countries that make up the European Union, the need arose to create a system that allows the sharing of patient data, in order to ensure that those citizens who require medical assistance in a country that is not their own can receive the Appropriate attention, from this idea was born the EPSOS project (Smart Open Services for European Patient) (EPSOS, 2017), whose base is a system of digital medical records that allows access to those authorized regardless of where the care is provided and with independence of the place where the information was generated, thus allowing the interoperability of the health sector of the twelve participating Member States: Spain, Austria, Sweden, Czech Republic, Germany, Denmark, France, Greece, Italy, the Netherlands, Slovakia and the United Kingdom .

3. CONCLUSION

In Colombia it is essential to have a unified national system of digital medical records that allows the timely consultation of a user's information, in order to improve the quality of health service provision in both public and private entities.

Guarantee availability, identification, authentication, integrity and non-repudiation, it is necessary to have a regulatory entity in which all the data gathered from users converge in order to guarantee optimal interoperability.

The unification of digital medical records that must comply with the guidelines proposed by SIGVILA and SISPRO described in this article, for this the development of the computer system is of vital importance, which guarantees the security of the information and from the source to the delivery of the data processed for its interpretation, that is why the quality of the data (veracity of the source) are critical processes where control includes aspects of good professional behavior and good behavior of patients.

ACKNOWLEDGEMENT

Our sincere thanks to our students who belong to the SERMOV research center, especially to Ivan Rendon and David Felipe Ruiz. Your assistance in this research has resulted in an invaluable contribution that has driven our understanding of the Colombian health information management systems. We desire from the depths of our hearts many successes in their professional lives.

REFERENCES

Congreso de Colombia, 2011. *LEY 1438 DE 2011 - Por medio de la cual se reforma el Sistema General de Seguridad Social en Salud y se dictan otras disposiciones.* Bogotá - Colombia: Gobierno Nacional de Colombia.

EPSOS, 2017. *Smart Open Services for European Patients,* Stockholm Sweden: EPSOS.

Gobierno Digital de Colombia, 2016. *Datos Abiertos.* [En línea] Available at: https://www.datos.gov.co/ [Último acceso: 16 February 2018].

Instituto Nacional de Salud, 2014. *Protocolo de Vigilancia en Salud Publica,* Bogotá - Colombia: Ministerio de Salud de Colombia.

Ministerio De La Proteccion Social, 2006. *Decreto Numero 3518 De 2006.* Bogotá - Colombia: Gobierno Nacional de Colombia.

PowerData, 2017. *Calidad de Datos. Cómo impulsar tu negocio con los datos..* [En línea] Available at: https://www.powerdata.es/calidad-de-datos [Último acceso: 26 March 2018].

A PC-BASED COMMUNICATION SYSTEM FOR SCHOOL EDUCATION ABOUT SAFETY USE OF SNS

Masashi Inoue and Masayuki Higashino
Tottori University, Koyama 4-101, Tottori, Japan

ABSTRACT

In Japan, the number of children using the Internet has been increasing. With the increase, the number of children experiencing problems related to Social Networking Service (SNS) has significantly increased. Therefore, education to prevent children from such problems are strongly required. Experiencing actual behaviors is an effective way to learn correct attitudes and develop a better sense of awareness. We have developed a PC-based communication system that allows children to easily use the SNS and realize the risks and dangers of using such platforms in school education settings. A server function and a wireless LAN access point function of the system are implemented with a laptop computer. This system can be easily set up and used in places where the Internet access is not provided. We implemented a practical lesson for a special-needs school that teaches students how to safely use SNS. The results show that simulated experiences of posting content via SNS using our system caused students to have an increased interest in learning and helped them to better understand the appropriate awareness, attitudes, and behaviors for using SNS. Furthermore, according to a survey conducted approximately one month after the experiment, the students retained the knowledge on how to handle personal information and minimize the dangers of using SNS that they learned during the lesson, suggesting a long-term effect from this approach.

KEYWORDS

Computerized Educational System, Internet Literacy Education, Social Networking Service (SNS)

1. INTRODUCTION

According to a survey conducted by the Japanese government in 2015, 46% of upper elementary school students, 60% of middle school students, and 95% of high school students have either a smartphone or a mobile phone. Additionally, over 30% of middle school students and 50% of high school students use the Internet for more than two hours a day. Meanwhile, the percentage of students experiencing problems related to Internet usage (e.g., sending/receiving malicious emails, posting personal information on online forums/information boards, viewing inappropriate websites, or meeting someone of the opposite sex after communicating with them online) has significantly increased. Moreover, the incidence of detrimental behavior and slander on the Internet has been increasing the most among juveniles. Therefore, improving education about cyberbullying and appropriately using the Internet has become an urgent issue in schools. Although enabling students to acquire practical information-utilization skills and appropriately use the Internet are important goals of school education, which requires advanced social information exchange, schools have been unable to adequately address these issues (DeSmet et al. 2015).

Rachmayani (2016) emphasized the role of school education in raising awareness about the risks of the Internet. In Japan, the government has requested that schools conduct classes on how to correctly use the Internet. Various educational materials are being developed to help students learn about Internet literacy (Hasegawa and Iwayama, 2017). However, there are no teaching materials that allow children to experience actually using SNS in order to learn how to use them correctly. Although it is important to have students learn by experiencing problems on the Internet (Okada et al. 2010), it is difficult to have students actually obtain and use LINE and Twitter accounts in the classroom. Therefore, we developed a PC-based communication system that allows students to easily use an SNS in classrooms and helps teachers promote children's understanding about safely use an SNS.

2. PC-BASED COMMUNICATION SYSTEM

Experiencing actual behaviors is an effective way to learn correct attitudes and develop a better sense of awareness. Therefore, we developed a PC-based communication system that allows students to actually experience the SNS and realize the risks and dangers of using such platforms. Many schools currently provide facilities for Internet usage. However, available classrooms are often limited to specific locations such as computer rooms and laboratories. It is convenient to able to use the system in any room. In this study, a server function and a wireless LAN access point function of the system are implemented with a laptop computer (a MacBook Air). This system can be easily set up and used in places where the Internet access is not provided.

The SNS currently used the most by Japanese students is LINE. Therefore, the system were developed with the purpose of creating a pseudo-LINE experience using the web application framework Ruby on Rails. It was assumed that there would be a variety of user terminals, including tablets, laptop computers, and smartphones, so it was necessary to plan for differences in screen layouts (e.g., screen sizes and aspect ratios). This problem was easily addressed with a set of simple source code using responsive web design. More specifically, Bootstrap was used as a framework for creating responsive web design in the development of this web application. Web sockets (a communication protocol for bidirectional communication) was also used in order to transmit the messages between users in real time. Since Ruby on Rails incorporates ActionCable (a library for handling web sockets), it was relatively easy to create a real-time message transmission function. We adopted the publish–subscribe model for sending and receiving messages in ActionCable and used PostgreSQL as a database for storing messages. A number-of-times-read display and a stamp display, which were feature functions of LINE, were also implemented in the system. Additionally, a function was included that displays the message history, allowing the teachers to examine the contents of previous conversations and review them during lessons.

3. PRACTICAL LESSON

We implemented a practical lesson for special-needs schools that teaches students how to safely use SNS. All of the students used the Internet regularly, with the applications that they often used being LINE, YouTube, and games. The students used the tablets to connect to the system, selected their own individual screen names, and then attempted to post messages or stamps. After a careful explanation of how to use the system, the students quickly became comfortable with operating the system and writing messages.

Figure 1. Screen Image of Message Exchange

In the lesson, we focused contact risk, because victims of children related to contact risk has been increasing. After a while, a teacher anonymously wrote messages to the students (e.g., "What's your name?", "Where do you live?", and "Should we meet up?"). Then, the teacher was able to observe which students provided personal information such as their names in response to these questions. After this, the operation of the system was stopped, and the teacher asked the students the following questions: "What is personal information?", "Should we provide personal information to strangers?", and "Should we receive personal information in exchange?" The students noted that information such as their names, where they live, and their school name could be used to identify individuals. After this, the teacher explained the dangers of providing personal information. Students seemed able to recognize and learn from the posting behavior (Figure 1).

4. DISCUSSION AND CONCLUSION

Most of victims related to children's SNS usage are caused by contact risks, which bring users in contact with strangers. In the year 2016, more than 1,600 Japanese children were reported to have suffered from sexual problems due to this. It was revealed that many children lack awareness of the contact risks of SNS. In order to prevent problems caused by contact risks, it is important for children to understand the appropriate handling of personal information and the anonymity of the Internet. We attempted to improve students' understanding about contact risks in the school education. In the lesson, a male teacher impersonated a female student and joined the chat. This male teacher attempted to get information about the names and addresses of the children, and several students answered his questions. Afterwards, the students were surprised to realize that the male teacher had been impersonating a female student. From this experience, the students seemed to recognize the anonymity of the Internet and understand the dangers of carelessly sending personal information to strangers.

The results of the simulated experiences of posting on SNS show that our system increased the students' interest in learning and helped them understand appropriate ways to respond to people that they communicate with using SNS. In addition, according to a survey conducted approximately one month after the experiment, the students retained the knowledge about how to handle personal information and minimize the dangers of using SNS gained in this study, thus suggesting a long-term effect of utilizing this approach.

As stated earlier, we implemented a server function and a wireless LAN access point function using a laptop computer. The computer was able to respond without any time delays when 20 students accessed the system. This system is very convenient for small classes without Internet access.

5. CONCLUSION

We developed a PC-based communication system to assist a school education about safety use of SNS. We implemented a practical lesson to evaluate usefulness of the system. It was found that simulated experiences using the system helped students to better understand the appropriate awareness and behaviors for using SNS.

REFERENCES

DeSmet, Ann; Aelterman, Nathalie; Bastiaensens, Sara; et al. 2015. Secondary school educators' perceptions and practices in handling cyberbullying among adolescents: A cluster analysis. Computers & Education. 88, 192-201.

Hasegawa, Haruo; Iwayama, Naoki. 2017. Study on Information Ethics Learning Using Tablet Devices – Activities of Information Transmission among Limited People at Integrated Study in Elementary School -. Research Report of Toyama University Japan. 11(3), 107-115.

Okada, Hitoshi; Sone, Hideaki; Ogawa, Masaru. 2010. Teaching Internet Safety at Universities Using "Hikari & Tsubasa's Information Security Game". In Proceedings of the 2nd International Conference on Computer Supported Education.

Rachmayani, Dita. 2016. Internet for Children: a Review Study. In Proceedings of the 3rd International Conference on Early Childhood Educations (Bandung, INDONESIA, NOV 11-12, 2016).

Doctoral Consortium

TELEMEDICINE MANAGEMENT OF OLDER CARDIAC PATIENTS IN IRELAND - WORKFORCE TRANSFORMATION IN A PUBLIC HEALTH CARE SYSTEM: LESSONS TO BE LEARNT

Yohanca Diaz and Lucia Carragher
Dundalk Institute of Technology, Dublin Road, Dundalk, Co. Louth, Ireland

ABSTRACT

As new technologies are developed and society's expectations of healthcare change, so the delivery of healthcare itself changes. This is driven by policy commitments to provide more care outside of hospitals, closer to home and in the home, to reduce hospital stays and ensure value for money for health systems challenged by resource constraints and a growing older population. Managing chronic illness in older adults presently utilises around 70% of health services resources in Ireland. The potential of digital technology and remote monitoring to transform care and promote sustainability of the healthcare system is recognised, particularly in the management of heart failure. However, in practice little has change in how health professionals manage the care of vulnerable older adults despite long held promises that technology would transform the delivery of health care. This project explores the range of challenges associated with reshaping the Ireland's public-sector healthcare workforce to facilitate telemedicine for older adults living with chronic illnesses. It will also contribute to knowledge on how this intervention works or not to inform policy and practice.

KEYWORDS

Telemedicine, Remote Monitoring, Workforce Transformation, Older Adults, Cardiac Care, Healthcare Professionals, Barriers / Enablers, Technology Adoption

1. INTRODUCTION

As new technologies are developed and society's expectations of healthcare change, so too does the delivery of healthcare itself change. Key drivers include policy commitments to provide more care outside of hospitals, closer to home and in the home. Drivers also include pressures to reduce hospital stays and ensure value for money for health systems challenged by resource constraints and a growing older population. In Ireland, managing chronic illness in older adults utilises around 70% of health services resources (Chronic Conditions Working Group 2017) with the management of chronic cardiac conditions in particular projected to cost 660 million per annum in the years ahead (Health Policy Partnership n.d).

Clearly the potential of telemedicine, including remote self-monitoring to contribute to the sustainability of health care systems is recognised, particularly to support people living with chronic illnesses in the community (Department of Health and Children 2010; Houses of the Oireachtas 2017). Saliba et al. (2012) defined telemedicine as the provision of healthcare services utilising ICT, for example, home telemonitoring for self-management of a chronic disease. Remote self-monitoring is characterised by the patient being monitored by a number of devices set up to measure physiological data such as pulse, blood pressure, SpO2, blood glucose, and transmit this data to a monitoring service or healthcare provider. The patient's clinicians in turn access the data transmitted through a web application and assess the need for an in-clinic visit, avoiding unnecessary trips to outpatient departments and facilitating early detection of abnormal symptoms (Vinck et al. 2012). Remote self-monitoring thus offers the potential to decrease the cost of healthcare delivery by reducing the reliance on hospitals and any associated more costly care. However, in practice telemonitoring has not been widely adopted at national level and funding for pilot projects is provided largely by international organisations (e.g. the European Commission) or from regional budgets of care providers (Rojahn et al. 2016). Crucially, evidence points to a reluctance among physicians to engage in patient monitoring for fear it would be less effective than the traditional face-to-face management of chronic disease (Deloitte, 2015).

This draws attention to the challenges around new ways of working and scaling-up of pilot projects and the role of health care workers in accepting change. Acceptance of telemedicine by health care workers is a prerequisite for its diffusion on a national scale (Gagnon et al 2003). It requires care providers to understand how remote care can be integrated into existing and new care pathways (Hendy & Barlow 2014). It also means understanding cultural differences between care providers and the changes in usual practice and how acceptable [or not] these are to health care workers.

2. LITERATURE REVIEW

While some studies have explored the uptake of telemedicine and perceptions of health care professionals, further work is required to examine the attitudes and perceptions of health care professionals in Ireland regarding telemedicine in care practice. In Spain, Asua et al. (2012) found healthcare professionals were reluctant to utilise telemedicine, with subsequent negative implications for uptake by patients. In Canada, Gagnon et al. (2003) found self-awareness of technology among physicians directly influenced the level of usage in care practices. In Hong Kong, perceptions of effectiveness and efficacy were found to be critical factors for adoption in care practice (Chau & Hu 2014) and in the UK the importance of trust and a sense of security have been highlighted (Sharma et al., 2010).

To date studies in Ireland have focused on providers' experiences of their practices, telemedicine service delivery and pre-hospital emergency care (Amadi-Obi et al. 2014; MacFarlane et al. 2006). The present study will work in close collaboration with a public sector cardiac team in Ireland as they test a new model of integrated care based on remote monitoring for patients aged 65 and over living at home. Developing new integrated service models in which hospital and community care work close together is considered essential for sustainability and affordability of health care systems. In particular, evidence points to the potential benefits of telemedicine to deliver care effectively and efficiently outside of traditional in-person models (Houses of the Oireachtas 2017). Research suggests that telemedicine can allow patients to effectively self-monitor chronic illnesses at home in conjunction with health professionals (Deloitte 2015). However, in practice as alluded to above, healthcare continues to lag behind other industries in using technology with service users, in this case with older adults and their caregivers. Available evidence suggests that while health professionals appreciate the potential clinical value of telemedicine, many are not convinced about its ability to monitor patients' conditions and adherence (Deloitte 2015). Consequently, cardiac care remains very hospital centred in Ireland despite the increasing prevalence of chronic illnesses and evidence of the need for remote monitoring services for patients discharged after heart surgery (Park et al. 2011).

2.1 Home Telemedicine - Reconceptualising Cardiac Care in Ireland

Examples of telemedicine by care providers outside of Ireland show the care pathways starting before discharge and continuing into the home (United 4 Health 2016). Patients are supported in learning how to use the device(s), how to set it up in the home, how to transmit data and how to engage with clinicians remotely. In Scotland for example, cardiac patients who chose to participate in a remote monitoring service, take pulse, blood pressure and weight measurements, with the data transmitted for review to a specialist heart failure nurse working in the community setting (ibid). The nurse in turn is responsible for medication titration, changing the care plan where appropriate, and arranging a teleconsultation or a home visit if needed. All the care is contained at primary care level for as long as possible. Patients using this system report a decreased attendance to out-patient services, effective accessibility to services and higher self-confidence and autonomy because they are involved in managing their illness (Scottish Centre for Telehealth and Telecare 2018). Staff also report high compliance of care plans among patients (ibid). This study will provide much needed evidence on the attitudes of healthcare providers towards the use of telemedicine in Ireland. In addition, the pilot study will provide much needed information on barriers and facilitators as well as the impact of the technology intervention e.g. outpatient clinic visits and attendance to A & E departments etc.

Based on Summerville (2015), a description of the functional and non-functional requirements of the remote monitoring system are outlined:

Functional Requirements

• The data aggregator shall collect data from 3rd party sensors for monitoring health and wellbeing (e.g. weight, blood pressure, activity levels, sleep).
• The applicant will send data in real time to care network members, specifically the heart failure nurse, for regular review.
• The application shall be configurable at an individual level. For example, if the heart failure patient also has diabetes, a digital blood glucometer can be added with questions specific to diabetes (such as checking feet) and scheduled as well as diabetes-specific educational content.

Non-Functional Requirements

• A personalised, digital user application to help patients self-manage. It should show data from sensors, allow the patient to self-report on various aspects of well-being and contain a library of educational content to help the person self-manage.
• Health care professionals in the cardiac team can add patients with heart failure and define what sensor data to collect e.g. weight and blood pressure.
• Health professionals can set questions for patients to answer to self-assess symptoms and set the frequency with which questions are delivered e.g. a question checking for night-time breathlessness can be delivered to the patient daily.
• Health professionals can add educational content in video or text format, and schedule when this should be delivered to the patient.

2.1.1 Aims / Research Questions

The aims of this study are:

(i) To observe changing work practices of care providers engaged in home telecare monitoring of older cardiac patients

(ii) To identify factors inhibiting the implementation and integration of telecare systems for chronic disease management in the community

(iii) To support evidence-based policy-making.

Key research questions are:

(i) What attitudes do care providers hold about home monitoring for older cardiac patients?

(ii) What are the barriers/resources affecting the uptake and successful use of telemedicine for older patients?

(iii) What infrastructure is needed at a local level to support the widespread use of telemedicine for older patients?

(iv) What is the impact of telemedicine on hospital admissions (number and length of stays).

(v) What is the impact of telemedicine on the use of other health care resources such as emergency department attendances, home visits by primary care professionals, and appointments with specialists?

3. PROPOSED METHODOLOGY

Ethical approval will be sought from the Health Service Executive and Dundalk Institute of Technology prior to commencement.

This study is designed to investigate a telemedicine platform for older cardiac patients. Patients at home will be using a platform consisting of wearable sensors, a fixed measurement station, and a network infrastructure to transmit data with security mechanisms for transmission and access. The health professionals responsible for handling the various telemonitoring tasks at the operational level will be interviewed and observed by the researcher over a 26-week period to explore the processes involved in incorporating telemedicine into the care of older cardiac patients and managing this over time. Patients and relatives will also be interviewed to explore and compare experiences and outcomes of telecare with face-to-face care.

The methods of this study have been informed by Normalization Process Theory theoretical framework, widely used to evaluate how new care pathways or services are reorganised, implemented and imbedded in practice (May et al. 2009). It has been utilised in Australia to measure the factors that enable or hinder the implementation or 'normalization' of a telehealth intervention into quotidian practice (Shulver et al. 2016).

3.1 Stage One: Baseline Data

This stage is designed to gather baseline data to compare changes after the telemedicine platform has been introduced. This will involve semi-structured, in-depth interviews with clinical and managerial staff and cardiac patients / across primary, community and secondary care in Co. Louth. This will provide a qualitative, context rich picture of cardiac care for older patients and will establish baseline against which progress can be measured.

Qualitative interviews will be digitally recorded and transcribed and the data analysed using a tool called the Framework Method developed by Ritchie and Lewis (2003). It involves a matrix output: rows (cases), columns (codes) and 'cells' of summarised data which provides the structure to systematically reduce data so that it can be analysed by case (i.e. individual interviewee or predefined groups or organisations) and by code. The ability to compare with ease data across cases as well as within individual cases is a key strength of the structure and process of this tool, facilitating interpretation of data.

Qualitative data (interviews) will be matched with broad quantitative data (e.g. number of outpatient clinic visits) involving routinely collected statistical data at these three levels as well as from national databases. This data will be analysed to assess changes over time e.g. hospital admissions, length of stay, outpatient visits, home visits, and clinical outcomes.

The baseline findings will then be used to generate a few case studies as focus for examining change at primary/community/secondary care interface.

3.2 Stage Two: Telemedicine Trial

In this stage the telemedicine platform will be operationalised for older cardiac patients in Co. Louth. During this stage, the researcher will be on a clinical placement for 6 months in the hospital and community in which the trial is taking place. Previous research has shown that the reconceptualization of a service and the associated change work practices are more likely to be successful if they are planned, piloted and researched (Sligo et al. 2017). This placement will therefore not only provide an opportunity to explore the cultural and professional dimensions of change, but the technical and managerial aspects too.

The ethnographic component will allow the researcher to be immersed in the environment and see first-hand how the cardiac care is managed pre-and-post telemedicine. To ensure this, the researcher will work in collaboration with the Community Health Organisation (CHO) number 9 area staff - [in Ireland, primary care is provided in 9 CHO's]. The researcher will observe the introduction of telemedicine and attend HSE meetings to observe how practice changes over time. In this way, the complexity of the changes happening across the HSE can be better understood, and the common lessons from managing change effectively disseminated throughout the services with important implications for future policy making.

The Framework Method (2003) will capture data coming from the reconceptualization of the service. Baseline data will be revisited to compare post-pilot outcomes including quality of life of patients, working practices of staff and the impact on service usage in terms of hospital admissions, length of stay, outpatient visits, home visits, and clinical outcomes.

3.3 Stage Three: Case Studies

The third stage is case studies analysis following the pilot stage and some case studies will be based on the staff, patients and clinicians looking at the transformation and the way care is delivered. Some case studies will be based on older people using the traditional cardiac care whereas others will focus on older people with cardiac condition using telemedicine.

Case studies will inform all processes and challenges for change and at the same time - what works and what does not work. Also, case studies may be advantageous as their findings could form the basis for service replication (Noor 2008).

Normalization Theory Process has been previously used to collect the views of patients with heart failure in regard to their experiences while accessing services, treatment compliance, communication with healthcare professionals and continuity of care (Gallacher et al. 2011).

4. CONCLUSION

The rising number of chronically ill older adults, especially those with cardiac conditions and the increasing workload of health professionals have added to the growing need to transform the delivery of services. Evidence suggests that telemedicine can contribute to the sustainability of our health care system in the context of population ageing and increasing demands for care linked to chronic disease management, especially cardiac care. However, uptake of technology in care management remains low among health professionals and service users. In addition, there is a dearth of evidence on care providers' perceptions of telemedicine and its clinical value in Ireland. This study is immersed in the implementation of a telemedicine trial in Co. Louth, Ireland. As such, it will yield valuable information from health professionals at hospital and community level and how telemonitoring tasks are handled [or not] at the operational level and what this means for policymakers charged with planning for the needs of a growing older population living with chronic illnesses.

ACKNOWLEDGEMENT

This project is supported by the European Union's INTERREG VA Programme, managed by the Special EU Programmes Body (SEUPB).

REFERENCES

Amadi-Obi A. et al, 2014. Telemedicine in pre-hospital care: a review of telemedicine applications in prehospital environment. *International Journal of Emergency Medicine,* Vol 7, No. 29.

Asua J. et al, 2012. Healthcare professional acceptance of telemonitoring for chronic care patients in primary care. *BMC Medical Informatics and Decision Making*, Vol 12, No.139.

Chau P.Y.K. & Hu P.J., 2014. Examining a Model of Information Technology Acceptance by Individual Professionals: An Exploratory Study. *Journal of Management Information Systems*, Vol 18, pp. 191-229.

Chronic Conditions Working Group, 2017. *Living Well with a Chronic Condition: Framework for Self-Management Support National Framework and Implementation Plan for Self-Management Support for Chronic Conditions: COPD, Asthma, Diabetes and Cardiovascular disease* [online]. Available from: http://www.lenus.ie/hse/handle/10147/622639 [accessed 26 March 2018].

Deloitte, 2015. *Accelerating the adoption of connected health.* [online]. Available from: https://www2.deloitte.com/content/dam/Deloitte/us/Documents/life-sciences-health-care/us-dchs-connected-health.pdf [accessed 19/4/18].

Department of Health and Children, 2010. *Changing Cardiovascular Health: National Cardiovascular Health Policy 2010 – 2019.* Government Publications, Dublin.

Gagnon MP. et al, 2003. An adaptation of the theory of interpersonal behaviour to the study of telemedicine adoption by physicians. *Int J Med Inform*, Vol 71, pp.103-115.

Gallacher K et al, 2011. Understanding Patients' Experiences of Treatment Burden in Chronic Heart Failure Using Normalization Process Theory. *Annals of Family Medicine*, Vol 9, pp. 235-243.

Health Policy Partnership, "n.d.", *Heart Failure Country Barometer: Ireland* [online]. Available from: http://heartbeat-trust.ie/HFbarometer.PDF [accessed 03 February 2018].

Hendy J. & Barlow J, 2014. Adoption in practice: The relationship between managerial interpretations of evidence and the adoption of a healthcare innovation. *Health Policy and Technology*, Vol 2, pp. 216 – 221.

Houses of the Oireachtas, 2017. *Committee on the Future of Healthcare - Sláintecare Report.* [online]. Available from: https://www.oireachtas.ie/parliament/media/committees/futureofhealthcare/Oireachtas-Committee-on-the-Future-of-Healthcare-Slaintecare-Report-300517.pdf [accessed 07 November 2017].

MacFarlane A. et al, 2006. Telemedicine services in the Republic of Ireland: An evolving policy context. *Health policy,* Vol. 76, No. 3, pp 245–258.

May C et al, 2009. Development of a theory of implementation and integration: Normalization Process Theory. *Implementation Science*, Vol 4, No. 29.

Noor K.B.M., 2008. Case study: A strategic research methodology. *American journal of applied sciences*, Vol 5, pp.1602-1604.

Park D.K et al, 2011. Telecare System for Cardiac Surgery Patients: Implementation and Effectiveness. *Healthcare Information Research*, Vol 17, pp. 93-100.

Ritchie J & Lewis J, 2003. *Qualitative research practice: a guide for social science students and researchers.* Sage, London.

Rojahn K et al, 2016. Remote Monitoring of Chronic Diseases: A Landscape Assessment of Policies in Four European Countries. *PLoS ONE*, Vol 11, No. 5.

Saliba V et al, 2012. Telemedicine across borders: A systematic review of factors that hinder or support implementation. *International journal of medical informatics*, Vol 81, pp. 793–809.

Sharma U et al, 2010. Clinical users' perspective on telemonitoring of patients with long term conditions: understood through concepts of Giddens's structuration theory & consequence of modernity. *Studies in Health Technology Informatics*, Vol 160, pp. 545–549.

Shulver W et al, 2016. 'Massive potential' or 'safety risk'? Health worker views on telehealth in the care of older people and implications for successful normalization. *BMC Medical Informatics and Decision Making,* Vol 16, No. 131.

Sligo J et al, 2017. A literature review for large-scale health information system project planning, implementation and evaluation. *International Journal of Medical Informatics*, Vol 97, pp. 86-97.

Sommerville, I. 2015. *Software engineering 10th Edition*. Pearson Education Limited, England.

The Scottish Centre for Telehealth and Telecare, 2018. *Home & Mobile Health Monitoring*. [online]. Available from: https://sctt.org.uk/programmes/home-and-mobile-monitoring/ [accessed 08 April 2018].

United 4 Health, 2016. *Telehealth in Practice: Care Delivery Models from 14 Regions in Europe.* [online]. Available from: http://united4health.eu/wp-content/uploads/2016/02/U4H-Care-Process-Brochure-FINAL.pdf [accessed 06 April 2018].

Vinck I et al, 2012. Legal and organizational aspects of remote cardiac monitoring: the example of implantable cardioverter defibrillators. *Europace*, Vol 14, pp.1230–1235.

AUTHOR INDEX